Loren D. Pitt

Lecture Notes
in Control and Information Sciences 176

Editors: M. Thoma and W. Wyner

IFIP

B.L. Rozovskii, R.B. Sowers (Eds.)

Stochastic Partial Differential Equations and Their Applications

Proceedings of IFIP WG 7/1 International Conference
University of North Carolina at Charlotte, NC
June 6 - 8, 1991

Springer-Verlag
Berlin Heidelberg New York
London Paris Tokyo
Hong Kong Barcelona Budapest

Editors

Boris L. Rozovskii
Richard Sowers

University of Southern California
Center for Applied Mathematical Siences
DRB-155
Los Angeles, CA 90089-1113
USA

ISBN 3-540-55292-8 Springer-Verlag Berlin Heidelberg New York
ISBN 0-387-55292-8 Springer-Verlag New York Berlin Heidelberg

Printed in Germany

Typesetting: Camera ready by authors
Offsetprinting: Mercedes-Druck, Berlin; Bookbinding: B. Helm, Berlin
60/3020 5 4 3 2 1 0 Printed on acid-free paper

PREFACE

The International Conference on Stochastic Partial Differential Equations and their Applications was held in Charlotte, North Carolina, from June 6th to June 8th, 1991 at the University of North Carolina at Charlotte. It was preceeded by the School in SPDE's (June 3rd to June 5th, 1991). Both the Conference and the School were attended by more than a hundred scholars from all over the world.

The purpose of the conference was to bring together researchers who are working on different theoretical and applied aspects of SPDE's and to promote research in this field. The main topics of the conference were

- Modern stochastic calculuses for SPDE's
- Asymptotic problems for SPDE's
- Non-linear parabolic SPDE's
- Numerical methods for SPDE's
- Martingale measures
- Applications of SPDE's, such as in stochastic hydrodynamics and non-linear filtering theory.

Coupling the Conference with the School in SPDE's served another important goal of the organizers, namely to provide the opportunity for young researchers and scientists working in other areas to meet and have informal discussions with outstanding experts in SPDE's.

The Conference and School were held under the joint auspices of the U.S. Army Reseach Office (Grant DAAL03-91-G-0098), the National Science Foundation (Grant DMS-9015507), the Office of Naval Research (Grant N00014-91-J-1225), and the International Federation for Information Processing. We would like to thank all of them for their generous support and encouragement. We would also like to thank all of the members of the International Program Committee and the Organizing Committee for their invaluable assistance. The devoted efforts of the faculty and staff of the Department of Mathematics of the University of North Carolina and especially of Professors R. Anderson and J. Quinn contributed greatly to the success of the conference.

B. L. Rozovskii, R. Sowers

University of Southern California
Los Angeles
December, 1991

TABLE OF CONTENTS

Nonstationary Anderson Model with Lévy Potential

Ho S. Ahn, René A. Carmona * and Stanislas A. Molchanov *
Department of Mathematics
University of California at Irvine
Irvine, CA 92717, USA

September 26, 1991

1 Introduction

The purpose of the present note is to present some new results on the *nonstationary parabolic Anderson model*:

$$\frac{\partial u}{\partial t} = \kappa \Delta u + \xi_t(x) u \tag{1}$$

with the initial condition $u(t, x) \equiv 1$. Here the time variable varies through the positive real axis $[0, \infty)$ and the space variable x runs through the lattice $\mathbb{Z}^d$. The operator Δ is the discrete analog of the usual Laplacian, i.e.

$$[\Delta f](x) = \sum_{|x'-x|=1} [f(x') - f(x)],$$

and the potential function $\xi_t(x)$ is assumed to be random. In fact we assume that this potential function $\xi_t(x)$ is the time derivative of a random field $\{\zeta_t(x);\ t \geq 0,\ x \in \mathbb{Z}^d\}$ which is stationary and ergodic in x when t is fixed and which has stationary independent increments in t when x is fixed. In this way, the time derivative $\xi_t(x) = \dot{\zeta}_t(x)$ is stationary in time. There is a simple motivation for the study of such an equation. This motivation comes from the investigation of branching Brownian motion in a random environment, the rate of doubling being given by $\xi_t(x)^+$ and the rate of death by $\xi_t(x)^-$. Of course this requires

*Partially supported by ONR N00014-91-1010

the fact that the derivative $\xi_t(x) = \dot{\zeta}_t(x)$ can be decomposed into the sum of nonnegative processes $\xi_t(x)^+ = \dot{\zeta}_t(x)^+$ and $\xi_t(x)^- = \dot{\zeta}_t(x)^-$ with properties which are hopefully similar to those of the original potential field $\xi_t(x) = \dot{\zeta}_t(x)$. This decomposition property is not satisfied by fields $\zeta_t(x)$ which are Brownian motions in time (in which cases $\dot{\zeta}_t(x)$ are white noise in time). The latter were the main models covered by the extensive study of the Gaussian case done in [2] and [4]. This drawback of the white noise model was one of our motivation to study random potentials which can be written as the difference of two nondecreasing Lévy processes.

Equation (1) is rewritten as the stochastic integral equation:

$$u(t,x) = 1 + \kappa \int_0^t \Delta u(s,x)ds + \int_0^t u(s,x)d\zeta_s(x) \tag{2}$$

where the last integral can be interpreted in the sense of a Ito's stochastic integral with respect to a semi-martingale. In fact the situation is simpler in the present situation because $\zeta_\cdot(x)$ can be compensated by a constant multiple of t and the equation can be driven by a martingale of the form $\zeta_t(x) - c_1 t$. The classical Picard iteration procedure can be used to prove existence and uniqueness of a solution in a weighted Hilbert space. We also give an existence and uniqueness result in the class of nonnegative functions. We then show that the solution has moments of all orders when the Lévy measure of $\zeta_t(x)$ has moments of all orders. We then prove existence of the moment Lyapunov exponents and derive a characterization of these exponents as the spectral radii of (deterministic) multiparticle Schrödinger operators on the lattice $\mathbb{Z}^d$. As in the Gaussian case (see [2] and [4]) Ito's formula plays a crucial role in the derivation of the moment equations. But the interaction potentials appearing in the Schrödinger operators are now multi-body instead of being merely two-body potentials as in the Gaussian case treated in [2] and [4]. The investigation of the spectral radius is slightly more difficult and the results are not as complete as in [2] and [4]. We are nevertheless able to prove intermittency for the solutions of (1).

Most of the results in this note are given without proofs. A complete analysis of the parabolic Anderson model with more general Lévy potentials will appear in the forthcoming paper [1].

2 Description of the Model

In the present note we consider a family $\{\zeta_.(x);\ x \in \mathbb{Z}^d\}$ parametrized by $x \in \mathbb{Z}^d$ of independent identically distributed real valued Lévy processes $\{\zeta_t(x);\ t \geq 0\}$. We assume that the following Lévy-Kinchine representation formula holds:

$$\mathbb{E}\{e^{i\alpha\zeta_t(x)}\} = \exp\left[t \int_{\mathbb{R}} (e^{iat\lambda} - 1)\nu(d\lambda)\right]$$

for some Lévy measure ν. Recall a nonnegative measure is called a Lévy measure if:

$$\int 1 \wedge |\lambda|^2\ \nu(d\lambda) < \infty.$$

Throughout this work, we assume that this Lévy measure also satisfies the condition:

$$\int 1 \wedge |\lambda|\ \nu(d\lambda) < \infty. \tag{3}$$

This condition (3) is equivalent to the existence of independent Lévy processes $\zeta_t^+(x)$ and $\zeta_t^-(x)$ with increasing sample paths and such that:

$$\zeta_t(x) = \zeta_t^+(x) - \zeta_t^-(x).$$

The existence of this decomposition as the difference of two increasing functions is, because of the motivation coming from the interpretation in terms of branching in random environment given in the previous section, one of the main reasons for our interest in Lévy potentials. For each fixed $x \in \mathbb{Z}^d$, the sample paths of $\zeta_t(x)$ are of bounded variation in t and consequently they define a measure on $[0,\infty)$. We use the notation:

$$\zeta_{(a,b]}(x) = \zeta_b(x) - \zeta_a(x)$$

for the measure of an interval $[a,b]$.

Our investigation of the intermittency of the solution of equation (1) requires the existence of all the moments of the solution. The existence of these moments will be proved under a more restrictive condition on the Lévy measure ν. The assumption which we make is expressed in terms of the rate of decay of the tail of the Lévy measure or, equivalently, in terms of integrability properties of this Lévy measure. More precisely we shall make assumptions of the type:

$$\int |\lambda|^k \nu(d\lambda) < \infty \tag{4}$$

for various values of the integer $k \geq 1$. Notice that condition (4) is equivalent to:

$$\mathbb{E}\{|\zeta_t(x)|^k\} < \infty$$

for all $t \geq 0$ and all $x \in \mathbb{Z}^d$. Moreover, in this case one has:

$$\mu_k = \mathbb{E}\{\zeta_t(x)^k\} = \int \lambda^k \nu(d\lambda). \tag{5}$$

3 Existence and Uniqueness of a Solution

We first investigate the properties of the integral equation (2) as a stochastic integral equation in an infinite dimensional space. The most natural state space is obviously $\ell^2(\mathbb{Z}^d)$. Unfortunately, equation (2) does not make sense in $\ell^2(\mathbb{Z}^d)$. We enlarge the space and we use instead the weighted Hilbert space $\ell^2_{(w)} = \ell^2(\mathbb{Z}^d, w)$ associated to the (weight) function w. The norm of this space is defined by:

$$||f(\cdot)||^2_{(w)} = \sum_{x \in \mathbb{Z}^d} w(x)|f(x)|^2.$$

Exponential weight functions are particularly convenient. In this work we use the weight function:

$$w(x) = \alpha^{|x|},$$

for some fixed number α satisfying $0 < \alpha < 1$. In this case we use the special notation $\ell^2_\alpha(\mathbb{Z}^d)$.

Theorem 3.1 *If the first two moments of the Lévy measure ν are finite, i.e. if condition (4) is satisfied with $k = 1$ and $k = 2$, then for every $\kappa > 0$ the stochastic integral equation:*

$$u(t,x) = 1 + \int_0^t \kappa\Delta u(s,x)ds + \int_0^t u(s,x)\, d\zeta_s(x)$$

has a unique solution in $\ell^2_\alpha(\mathbb{Z}^d)$.

The various steps of the proof are standard: they rely on the classical proof of the existence and uniqueness of solutions of stochastic integral equations based on the Picard iteration procedure. It is very easy to control the effects of the Hilbert space norm. A modicum of care is needed when one uses the martingale inequalities.

The solution can be written in terms of averages over the Brownian paths $\{X_t;\ t \geq 0\}$ in $\mathbb{Z}^d$, i.e. the paths of the continuous time symmetric random walk on $\mathbb{Z}^d$ with generator $\kappa\Delta$. This representation can be regarded as a Feynman-Kac formula. In the present situation it reads:

$$u(t,x) = \mathbb{E}_x\left\{\prod_{0\leq s\leq t}(1 + J_s\zeta_s(X_s))\right\}. \tag{6}$$

Here we used the notation:

$$J_s\zeta_s(x) = \zeta_s(x) - \zeta_{s-}(x)$$

for the jump at time s of the path of the Lévy process $\{\zeta_t(x);\ t \geq 0\}$. Recall that, here and in the following, we always assume that we are working with versions of the Lévy processes which are right continuous and have left limits. We shall denote by $T_1, T_2, \cdots\cdots$ the instants at which the Brownian motion jumps (from one site to one of its neighbors) and by $N(t)$ the number of jumps prior to time t. With these notations, the Feynman-Kac formula can be rewritten in the form:

$$u(t,x) = \mathbb{E}_x\left\{\prod_{i=0}^{N(t)-1}\left(\prod_{s\in[T_i,T_{i+1})}(1 + J_s\zeta_s(X_i))\right)\right\}.$$

When we consider the representation of the solution at time t, we are only interested in the Brownian path up to time t. We shall use the convention $T_{N(t)+1} = t$ for convenience. We also use the notation $X_0 = x, X_1, \cdots, X_n, \cdots$ for the sites visited by the right-continuous sample path of this symmetrical random walk.

In fact, the uniqueness and the Feynman-Kac representation formula hold not only in the weighted Hilbert space $\ell^2_\alpha(\mathbb{Z}^d)$, but in the class of all nonnegative functions, without any apriori restriction on the growth of $u(t,x)$ as long as $u(t,x) \geq 0$. We shall not dwell further on this problem.

4 Existence of the Moments and Moment Equations

Theorem 4.1 *If the Lévy measure has moments of all orders, i.e. if condition (4) is satisfied for all the integers $k \geq 1$, then the solution $u(t,x)$ has moments of all orders in the sense that:*

$$< |u(t,x)|^p > < \infty$$

for all $t \geq 0$, $x \in \mathbb{Z}^d$ and for all integer $p \geq 1$. For each integer $p \geq 1$, each $t \geq 0$ and each $\mathbf{x} = (x_1, \cdots, x_p) \in \mathbb{Z}^{pd}$ we set:

$$m_p(t,\mathbf{x}) = \langle u(t,x_1) \cdots u(t,x_p) \rangle. \tag{7}$$

for the p-correlations of the solution $u(t,x)$. Then these moments satisfy the following parabolic equation:

$$\frac{\partial m_p}{\partial t} = \kappa(\Delta_{x_1} + \cdots + \Delta_{x_p})m_p + \left(p\mu_1 + \sum_{k=2}^{p} \mu_k \sum_{1 \leq i_1, \cdots, i_k \leq p} \delta(x_{i_1}, x_{i_2}, \cdots, x_{i_k}) \right) m_p \tag{8}$$

with the initial condition $m_p(0, \,\cdot\,) \equiv 1$. Here the constants μ_k are the moments of the Lévy measure defined in (5) and:

$$\delta(x_{i_1}, x_{i_2}, \cdots, x_{i_k}) = \begin{cases} 1, & \text{if} \quad x_{i_1} = \cdots = x_{i_k} \\ 0, & \text{otherwise} \end{cases}$$

It is convenient to write $\Delta_{\mathbf{x}} = \Delta_{x_1} + \cdots + \Delta_{x_p}$ and:

$$V_p(\mathbf{x}) = p\mu_1 + \sum_{k=2}^{p} \mu_k \sum_{1 \leq i_1, \cdots, i_k \leq p} \delta(x_{i_1}, x_{i_2}, \cdots, x_{i_k})$$

for $p \geq 1$. With these notations, equation (8) becomes:

$$\frac{\partial m_p}{\partial t} = H_p(\kappa) m_p$$

where the operator $H_p(\kappa) = \kappa\Delta_{\mathbf{x}} + V(\mathbf{x})$ is a p-particle Schrödinger operator on the lattice $\mathbb{Z}^{pd}$. In the particular case $p = 1$ one has the classical diffusion equation:

$$\frac{\partial m_1(t,x)}{\partial t} = \kappa \Delta m_1(t,x)$$

whose solution is (for the given initial condition) $m_1(t,x) \equiv 1$.

Remark 1.

The situation is slightly simpler in the Gaussian case treated in [2] and [4]. Indeed, the multiparticle potential V contains only pairwise interaction potentials while all the types of interactions are present in the present situation.

Remark 2.

In the same way the potential V was a sum of δ-functions in [2], the interaction potentials are given by δ- functions because we assume (as in [2]) that the processes $\{\zeta_t(x);\ t \geq 0\}$ are independent for different values of $x \in \mathbb{Z}^d$. More general random fields are considered in [1]. The simplest one are of the form:

$$\zeta_t(x) = \sum_{y \in \mathbb{Z}^d} \varphi(x-y)\tilde{\zeta}_t(y)$$

where the processes $\{\tilde{\zeta}_t(y);\ t \geq 0\}$ are assumed to be independent for different values of $y \in \mathbb{Z}^d$.

The solutions of the moment equations (8) have an exponential behavior in the asymptotic regime $t \to \infty$. The main purpose of this work is to investigate this behavior. The first step is given by the following result:

Theorem 4.2 *For each integer $p \geq 1$, the limit:*

$$\lim_{t \to \infty} \frac{1}{t} \log m_p(t, \mathbf{x})$$

exists and is independent of $\mathbf{x} = (x_1, \cdots, x_p)$. This limit is called the p-th (moment) Lyapunov exponent of the solution $u(t,x)$ and it is denoted by $\gamma_p(\kappa)$. It is equal to the supremum of the spectrum of the multiparticle Schrödinger operator $H_p(\kappa)$ appearing in the moment equation (8).

The proof of this result is standard. It is given in full detail at the beginning of Chapter III of [4].

5 Properties of the Lyapunov Exponents and Intermittency

This section is devoted to the study of the dependence of the Lyapunov exponent $\gamma_p(\kappa)$ upon the diffusion parameter κ. Because of its characterization as the supremum of a spectrum, the variational principal gives:

$$\gamma_p(\kappa) = \kappa \sup_{f\in\ell^2(\mathbb{Z}^{pd}),\ \|f\|=1} - \sum_{x,x'\in\mathbb{Z}^{pd},\ |x-x'|=1} |f(x)-f(x')|^2 + \sum_{x\in\mathbb{Z}^{pd}} V_p(x)|f(x)|^2$$

which shows that $\gamma_p(\kappa)$ is a nonincreasing continuous nonnegative convex function of the diffusion coefficient κ. Moreover, the case $\kappa = 0$ can be solved explicitly:

$$\gamma_p(0) = \sum_{k=1}^{p} \binom{p}{k} \mu_k = \int [(1+\lambda)^p - 1]\nu(d\lambda).$$

The main properties of the moment Lyapunov exponents are summarized in the following theorem.

Theorem 5.1 *For $d = 1, 2$ and $p \geq 2$ one has:*

$$\gamma_p(\kappa) > 0 \qquad \textit{for all} \quad \kappa > 0$$

and:

$$\lim_{\kappa\to\infty} \gamma_p(\kappa) = 0.$$

For $d \geq 3$ and $p \geq 2$ one has:

$$\gamma_p(\kappa) > 0 \qquad \textit{for } \kappa \textit{ small} \qquad \textit{and} \qquad \gamma_p(\kappa) \equiv 0 \qquad \textit{for } \kappa \geq \kappa_{p,cr}$$

for some $\kappa_{p,cr} > 0$.

In the terminology of [4] this means that the solution field $u(t,x)$ exhibits full intermittency in dimensions $d = 1$ and $d = 2$ for all the values of the diffusion coefficient κ while intermittency holds only for small κ in dimensions $d \geq 3$. As a byproduct of the proof of Theorem 5.1 one gets that:

$$\kappa_{p,cr} = \frac{1}{(2\pi)^{d(p-1)}} \int_{S^{d(p-1)}} \frac{d\varphi}{-\psi(\varphi)}$$

where the exponent function $\psi(\varphi)$ is defined by:

$$\psi(\varphi) = 2\{\sum_{i=1}^{d}[\cos(\varphi_{i1} + \cdots + \varphi_{i(p-1)}) - 1] + \sum_{i=1}^{d}\sum_{j=1}^{p-1}[\cos(\varphi_{ij}) - 1]\}$$

where $\varphi = (\varphi_1, \ldots, \varphi_{p-1})$ is a generic element of the $d(p-1)$-dimensional torus $S^{d(p-1)}$.

The proof of Theorem 5.1 relies on upper and lower bounds on the Lyapunov exponents $\gamma_p(\kappa)$ which are of interest on their own.

The Lower Estimation

As in the Gaussian case (see [4]), a simple application of the Feynman-Kac formula gives the lower bound:

$$\frac{\gamma_p(\kappa)}{p} \geq \left(\frac{\gamma_p(0)}{p} - 2d\kappa\right)_+$$

Moreover, one has the useful comparison result:

$$\frac{\gamma_p(\kappa)}{p} \geq \frac{\gamma_2(\kappa)}{2}.$$

Indeed, the case $p = 2$ is easier to control because $\gamma_2(\kappa)$ can be characterized as the positive root of the equation:

$$1 = \frac{1}{(2\pi)^d} \int_{S^d} \frac{d\varphi}{2\kappa \sum_{i=1}^{d}(1 - \cos \varphi_i) - \gamma} \tag{9}$$

and $\gamma_2(\kappa) = 0$ when this equation has no solution. It follows from there that $\gamma_2(\kappa) > 0$ for all $\kappa > 0$ in dimensions $d = 1$ and $d = 2$. Consequently the comparison result which we give above implies part of the proof of theorem 5.1.

The Upper Estimation

It is proven in [4] that an upper bound on the Lyapunov exponent $\gamma_p(\kappa)$ can be given in terms of the Lyapunov exponent $\gamma_2(\kappa)$. This is due to the fact that the multibody potential $V(x_1, \cdots, x_p)$ can be expressed as the sum of two-body potentials. This is no longer the case in the present situation. The comparison potential which we use in the Lévy case is $\delta(x_1, \cdots, x_p)$. We consider the self-adjoint operator:

$$H_a^{(pd)}(\kappa) = \kappa(\Delta_{x_1} + \cdots + \Delta_{x_p}) + a\delta(x_1, \cdots, x_p)$$

on the Hilbert space $\ell^2(\mathbb{Z}^{pd})$ and we denote by $\lambda_a^{(pd)}(\kappa)$ the supremum of its spectrum. A simple space translation gives:

$$\lambda_a^{(pd)}(\kappa) = \lambda_a^{((p-1)d)}(\kappa)$$

and a simple scaling argument gives:

$$\lambda_a^{(pd)}(\kappa) = a\lambda_1^{(pd)}(\frac{\kappa}{a}).$$

Consequently, the problem reduces to the control of the quantity $\lambda_1^{((p-1)d)}(\kappa)$. It is possible to prove that it is the unique solution of the equation:

$$1 = \frac{1}{(2\pi)^{(p-1)d}} \int_{S^{(p-1)d}} \frac{d\varphi}{2\kappa\Psi(\varphi) - \lambda} \tag{10}$$

where:

$$\Psi(\varphi) = \sum_{i=1}^{d} [1 - \cos(\varphi_{i,1} + \cdots + \varphi_{i,p-1})] + \sum_{i=1}^{d} \sum_{j=1}^{p-1} [1 - \cos \varphi_{i,j}].$$

In any case, a careful analysis of this equation leads to the upper bound:

$$\frac{\gamma_p(\kappa)}{p} \le \mu_1 + \frac{1}{p} \sum_{k=2}^{p} \binom{p}{k} \mu_k \lambda_1^{((p-1)d)}(\frac{\kappa}{\mu_k M_k})$$

which holds for any sequence of numbers M_k satisfying $1 \le M_k \le \infty$ and:

$$\sum_{k=2}^{p} \binom{p}{k} \frac{1}{M_k} = 1,$$

and to the remainder of the proof of Theorem 5.1.

Remark 3.

As a final remark, we want to emphasize that if the diffusion constant is small (and fixed for the purpose of the present discussion), then the function $p \hookrightarrow \gamma_p(\kappa)/p$ has, for small values of p, a different behavior in the Lévy case than in the Gaussian case. It is proven in [2] and [4] that:

$$\frac{\gamma_p(\kappa)}{p} = O(p)$$

in the Gaussian case, for which the Lévy measure ν is 0. If we suppose for example that $\nu(d\lambda) = \delta_1(d\lambda)$, in which case the $\zeta_t(x)$ are independent Poisson processes, then it is easy to prove that:

$$\frac{\gamma_p(\kappa)}{p} = O(\frac{2^p}{p}).$$

References

[1] H. S. Ahn, R. Carmona and S. A. Molchanov: Nonstationary Parabolic Anderson Problem with General Lévy Potential. (in preparation)

[2] R. Carmona and S. A. Molchanov: Intermittency and Phase Transitions for some Particle Systems in Random Media. (to appear in Proceedings of the Katata Conf. on Stoch. Proc. August 1990).

[3] R. Carmona and S. A. Molchanov: Large Time Asymptotics for the Solutions of Stochastic Parabolic Equations on the Lattice. (preprint)

[4] B. Fristedt (1974): Sample Functions of Stochastic Processes with Stationary Independent Increments. *Adv. in Probability* 3, ed. P. Ney and S. Port, pp. 241-396.

STOCHASTIC PARTIAL DIFFERENTIAL EQUATIONS IN CONTROL OF STRUCTURES†

A. V. Balakrishnan
Electrical Engineering Department, UCLA
Los Angeles, CA 90024, U.S.A.

1. Introduction

In this paper we are concerned with stochastic partial differential equations arising in the control of structures. Rather than present the most general version we examine in some detail a canonical example which amply illustrates the main features of the theory. Extension to the more complex models is fairly transparent. In the model we consider, the sensors and controllers are collocated at a finite number of discrete points on the structure so that the "forcing" or "input" noise is finite-dimensional and hence either the White Noise [1, 2] or Wiener process set-up can be used. The induced process in the continuum of the structure is infinite-dimensional, and we shall show that the white noise version has many advantages. The partial differential equation is treated in an appropriate Hilbert space setting — the choice of inner product turns out to be crucial and the existence and uniqueness are handled by semigroup theory as initiated in [2]. The relationship of this abstract version to the concrete is examined at some length.

The primary interest in applications is in the steady state properties. Here we have a further difficulty in that the semigroup is not exponentially stable and the notion of strong stability has to be used, the theorem of Benchimol [3] on the generation of strongly stable semigroups playing an essential role. Of main interest among our results is that the steady state covariance is not *nuclear;* but this presents no problem in the white noise framework. For the example studied we are able to calculate explicitly both the steady state covariance as well as the spectral density. The question of convergence of time-averages for sample paths will be treated elsewhere. For a generalization of the stochastic control problem in this setting see [5].

2. Main Results

We begin with the example, simple and yet canonical in many respects. Thus we consider a beam-bending model — a Bernoulli beam clamped at one end and with a lumped mass and collocated controller-sensor at the other end. The dynamics can then be represented by:

---

† Research supported in part by NASA Grant NCC 2-374.

$$\left.\begin{aligned} \rho a \frac{\partial^2 f(t, s)}{\partial t^2} + EI_\phi \frac{\partial^4 f(t, s)}{\partial s^4} &= 0 , \\ 0 < s < L, \ 0 < t & \\ f(t, 0) = f'(t, 0) = f''(t, L) &= 0 \end{aligned}\right\} \tag{2.1}$$

$$m\ddot{f}(t, L) + u(t) + N_a(t) = EI_\phi f'''(t, L) \tag{2.2}$$

$$v(t) = \dot{f}(t, L) + N_r(t) \tag{2.3}$$

where t is the time variable, s the spatial variable, superdot denotes differentiation with respect to time t, and prime denotes differentiation with respect to s. For each $f(t, s)$, $0 < s < L$ denotes the displacement of the beam at s, the coefficients in (2.1) being constants. In (2.2) the control input is $u(\cdot)$, with $N_a(\cdot)$ representing noise — Gaussian white noise with spectral density d_a. (2.1), (2.2) is our basic stochastic partial differential equation.

The sensor data is the "rate" described by (2.3) where $N_r(\cdot)$ represents sensor noise which we model as white Gaussian with spectral density d_r, independent of $N_a(\cdot)$. The function $u(\cdot)$ in (2.2) is the control input and we shall consider primarily the "direct rate feedback" case, where it is defined by

$$u(t) = \gamma v(t) \tag{2.4}$$

where γ is a fixed ("gain") constant, so that (2.2) becomes

$$m\ddot{f}(t, L) + \gamma\dot{f}(t, L) + \gamma N_r(t) + N_a(t) - EI_\phi f'''(t, L) = 0 . \tag{2.2a}$$

We are concerned primarily with the steady state properties of the stochastic partial differential equation (2.1) - (2.2a).

We shall need to formulate the problem more precisely before we can discuss the nature of the induced random process $f(t, s)$, $0 < t$, $0 < s < L$.

First we define the differential operator A_0 by

$$A_0 f = EI_\phi \frac{d^4 f}{ds^4}$$

with domain in $L_2[0, L]$ defined by

$$\mathcal{D}(A_0) = [\, f \in L_2[0, L] \text{ such that } f,\ f',\ f'' \text{ and } f''' \text{ are absolutely continuous with } f'''' \text{ in } L_2[0, L], \text{ satisfying in addition to the boundary conditions } f(0) = f'(0) = 0 = f''(L)\,] .$$

On $\mathcal{D}(A_0)$ introduce the inner product

$$[f, g] = \int_0^L f(s)\, \overline{g(s)}\, ds + f(L)\, \overline{g(L)} .$$

The space completed under this inner product is the Hilbert space $\mathcal{H} = L_2[0, L] \times C^1$. Let D denote elements x in $\mathcal{H}$ of the form

$$x = \begin{vmatrix} f \\ b \end{vmatrix}, \qquad f \in \mathcal{D}(A_0), \; b = f(L).$$

Then D is a dense subspace of $\mathcal{H}$. Define the operator A on D with range in $\mathcal{H}$ by

$$Ax = \begin{vmatrix} A_0 f \\ -EI_\phi f'''(L) \end{vmatrix}.$$

Then A is closed, self-adjoint and nonnegative definite. In particular

$$[Ax, x] = EI_\phi \int_0^L f''''(s) \overline{f(s)}\, ds - EI_\phi f'''(L) \overline{f(L)}$$

$$= EI_\phi \int_0^L |f''(s)|^2 \, ds .$$

The resolvent of A is compact and zero is in the resolvent set of A. Moreover we can express (2.1) - (2.3) in the form:

$$M\ddot{x}(t) + Ax(t) + Bu(t) + BN_a(t) = 0 \tag{2.5}$$

$$\mathrm{v}(t) = B^*\dot{x}(t) + N_r(t) \tag{2.6}$$

where

$$Mx = \begin{vmatrix} \rho a f \\ mb \end{vmatrix}, \qquad x = \begin{vmatrix} f \\ b \end{vmatrix}$$

$$Bu = \begin{vmatrix} 0 \\ u \end{vmatrix}$$

so that B maps C^1 into $\mathcal{H}$ and is a finite-dimensional linear operator, and B^* is given by:

$$B^*x = b .$$

A property relating B and A that is crucial to us is the following: suppose ϕ is an eigenvector of A:

$$A\phi = \omega^2 M\phi , \qquad \phi \neq 0, \; \omega \text{ nonzero real} .$$

Then

$$B^*\phi \neq 0 .$$

Let us prove this. Let

$$\phi = \begin{vmatrix} f \\ f(L) \end{vmatrix}.$$

Then

$$EI_\phi f''''(s) = \omega^2 \rho a f(s), \qquad 0 < s < L \tag{2.7}$$

$$-EI_\phi f'''(L) = \omega^2 m f(L).$$

Suppose

$$B^*\phi = f(L) = 0.$$

Then we have:

$$f(0) = f'(0) = 0; \qquad f(L) = f''(L) = 0$$

and it is immediate that any solution of (2.7) satisfying these conditions must be zero.

Since $-A$ is the infinitesimal generator of a strongly continuous semigroup over $\mathcal{H}$, another way of expressing this property is that $(-A, B)$ is controllable.

To proceed further we need to "convert" (2.5) into a first-order equation. Thus we write

$$Y(t) = \begin{vmatrix} x(t) \\ \dot{x}(t) \end{vmatrix}, \qquad Y(\cdot) \in \mathcal{H} \times \mathcal{H}$$

and define $\mathcal{A}$ by

$$\mathcal{A} = \begin{vmatrix} 0 & I \\ -M^{-1}A & 0 \end{vmatrix} \tag{2.8}$$

where

$$\mathcal{D}(\mathcal{A}) = \mathcal{D}(A) \times \mathcal{H}$$

and define $\mathcal{B}$ by

$$\mathcal{B}u = \begin{vmatrix} 0 \\ -M^{-1}Bu \end{vmatrix}$$

mapping C^1 into $\mathcal{H}$. Then (2.5) goes over into

$$\dot{Y}(t) = \mathcal{A}Y(t) + \mathcal{B}u(t) + \mathcal{B}N_a(t). \tag{2.9}$$

Unfortunately $\mathcal{A}$ does *not* generate a semigroup in this space, viz. $\mathcal{H}\times\mathcal{H}$. We need to change the inner product. Thus on $\mathcal{D}(\sqrt{A}) \times \mathcal{H}$ we introduce the "energy" inner product:

$$[Y, Z]_E = [\sqrt{A}\, x_1, \sqrt{A}\, x_3] + [Mx_2, x_4]$$

where

$$Y = \begin{vmatrix} x_1 \\ x_2 \end{vmatrix}, \qquad Z = \begin{vmatrix} x_3 \\ x_4 \end{vmatrix}.$$

Because zero is not in the resolvent set of A, it is easy to see that $\mathcal{D}(\sqrt{A}) \times \mathcal{H}$ is closed in this inner product, and we define the new space by $\mathcal{H}_E$. Then with $\mathcal{A}$ defined as before we see that in this inner product

$$\mathcal{A}^* = -\mathcal{A}$$

and further for Y in the $\mathcal{D}(\mathcal{A})$ we have

$$[Y, \mathcal{A}Y]_E + [\mathcal{A}Y, Y]_E = 0 .$$

Hence $\mathcal{A}$ generates a strongly continuous semigroup $S(t)$, $t \geq 0$, and

$$\|S(t)\| = \|S^*(t)\| = I .$$

Hence (2.9) has the weak sense solution (see [2]):

$$Y(t) = S(t)Y(0) + \int_0^t S(t-\sigma)\mathcal{B}u(\sigma)\, d\sigma + \int_0^t S(t-\sigma)\mathcal{B}N_a(\sigma)\, d\sigma \tag{2.10}$$

where for each $t > 0$:

$$u(\cdot) \in L_2(0, t)$$

$$N_a(\cdot) \in L_2(0, t) .$$

At this point we pause to examine the relationship of (2.1), (2.2) to the abstract version (2.9). First of all (2.9) is to be interpreted in the weak sense, cf. [2]: for each ψ in the domain of $\mathcal{A}^*$,

$$\frac{d}{dt}[Y(t), \psi]_E = [Y(t), \mathcal{A}^*\psi]_E + [(u(t) + N_a(t)), B^*\psi] , \qquad t > 0$$

and

$$\|Y(t) - Y(0)\| \to 0 \qquad \text{as } t \to 0+ . \tag{2.11}$$

Let

$$\psi = \begin{vmatrix} \phi_1 \\ \phi_2 \end{vmatrix} , \qquad \phi_1, \phi_2 \in D(A)$$

$$Y(t) = \begin{vmatrix} y_1(t) \\ y_2(t) \end{vmatrix} .$$

Then we have

$$\frac{d}{dt}[y_1(t), A\phi_1] + \frac{d}{dt}[y_2(t), M\phi_2]$$
$$= [y_1(t), A\phi_2] + [y_2(t), M\phi_1] - [(u(t) + N_a(t)), B^*\phi_2] .$$

Taking $\phi_2 = 0$, we have

$$\frac{d}{dt}[y_1(t), A\phi_1] = [y_2(t), A\phi_1] .$$

Since zero is in the resolvent set of A, the range of A is the whole space and hence for every ϕ in $\mathcal{H}$

$$\frac{d}{dt}[y_1(t), \phi] = [y_2(t), \phi] . \tag{2.12}$$

Hence $y_2(t)$ is the weak derivative of $y_1(t)$. We cannot of course claim it is the strong derivative, but in the "modal" expansion of $y_1(t)$:

$$y_1(t) = \sum_1^\infty a_k(t)\phi_k$$

where the M-normalized eigenvectors of A defined by

$$A\phi_k = \omega_k^2 M\phi_k \,, \qquad [M\phi_k, \phi_k] = 1 \,,$$

we do have that

$$y_2(t) = \sum_1^\infty \dot{a}_k(t)\phi_k \,.$$

Again

$$\frac{d}{dt}[y_2(t), M\phi_2] = \frac{d^2}{dt^2}[y_1(t), M\phi_2] = -[y_1(t), A\phi_2] - [(u(t) + N_a(t)), B^*\phi_2] \,.$$

In other words letting

$$x(t) = y_1(t)$$

we obtain (2.5) interpreted in the weak sense:

$$\frac{d^2}{dt^2}[x(t), M\phi] + [x(t), A\phi] + [B(u(t) + N_a(t)), \phi] = 0 \,, \qquad \phi \in \mathfrak{D}(A) \tag{2.13}$$

or,

$$\ddot{a}_k(t) + \omega_k^2 a_k(t) + [(u(t) + N_a(t)), b_k] = 0 \,, \tag{2.14}$$

where

$$b_k = B^*\phi_k \,.$$

Also:

$$\mathfrak{B}^*Y(t) = -B^*y_2(t)$$

and from (2.12), taking $\phi = Bu$, we have

$$[B^*y_2(t), u] = [y_2(t), Bu] = \frac{d}{dt}[y_1(t), Bu] = \frac{d}{dt}[B^*y_1(t), u] \,, \qquad u \in E^1$$

Hence it follows that $B^*y_1(t)$ is absolutely continuous in t, with derivative equal to $B^*y_2(t)$. However, setting

$$y_1(t) = \begin{vmatrix} f(t, \cdot) \\ B^*y_1(t) \end{vmatrix} \,, \qquad f(y, \cdot) \in L_2[0, L]$$

we *cannot* assert that

$$f(t, L) = B^*y_1(t)$$

unless $y_1(t) \in \mathfrak{D}(A)$. In particular we cannot prove that (2.2) holds.

A sufficient condition for $y_1(t) \in \mathfrak{D}(A)$ is (see [2]) that

(i) $y_1(0) \in \mathcal{D}(A)$

(ii) For each t, $u(\cdot)$ and $N_a(\cdot)$ are $C^1[0, t]$. For the direct feedback case, we only need to require this of $N_a(\cdot)$.

In that case (2.9) holds in the strong topology as well, and so does (2.10); and $Y(t) \in \mathcal{D}(\mathcal{A})$ for each $t \geq 0$. Hence (2.2) and (2.2a) hold. Moreover (2.1) also holds for each s, $0 \leq s \leq L$ and $0 < t$. But the sample paths of the noise $N_a(\cdot)$ have to be restricted if the basic dynamic equations (2.1), (2.2) are to hold. Of course in the Wiener process setup these paths would have zero probability, and hence may be defined in any way we want. We already can see the advantage of white noise theory, where the sample paths $N_a(\cdot)$, $N_r(\cdot)$ are in $L_2[0, t]$ for every t, and for (2.2a) to hold we require that they be $C^1[0, t]$, in addition to the initial condition restriction.

To discuss the steady state properties we return to the abstract formulation and now set

$$u(t) = \gamma v(t)$$

so that we have:

$$\dot{Y}(t) = \mathcal{A}Y(t) - \mathcal{B}\gamma\mathcal{B}^*Y(t) + \mathcal{B}(\gamma N_r(t) + N_a(t)) . \tag{2.15}$$

To treat (2.15) we need to invoke results from semigroup theory.

Lemma 1

$$(\mathcal{A} - \gamma\mathcal{B}\mathcal{B}^*)$$

for each $\gamma > 0$, is the infinitesimal generator of a strongly continuous semigroup. Denoting it $S_\gamma(\cdot)$ we have

(i) $\|S_\gamma(t)Y\|_E \to 0$ as $t \to \infty$ for every Y in $\mathcal{H}_E$ (strong stability)

(ii) $\displaystyle\int_0^\infty \|\mathcal{B}^*S_\gamma(t)^*Y\|^2 \, dt = \frac{\|Y\|_E^2}{2\gamma}$.

Proof

We invoke Benchimol's theorem [3]. Since $\mathcal{A}$ generates a contraction semigroup, has a compact resolvent and $(\mathcal{A}, \mathcal{B})$ controllability holds (as a consequence of (A, B) controllability), the result follows. See [4] for details.

Interpreting (2.15) in the weak sense, it has the unique solution:

$$Y(t) = S_\gamma(t)Y(0) + \int_0^t S_\gamma(t-\sigma)\mathcal{B}(\gamma N_r(\sigma) + N_a(\sigma)) \, d\sigma . \tag{2.16}$$

defining Gaussian process for $0 \leq t \leq \infty$.

From (2.16) it follows that the covariance operator:

$$\mathcal{R}(t, t) = E[Y(t)\ Y(t)^*]$$

is given by

$$\mathcal{R}(t, t)Y = S_\gamma(t)\Lambda S_\gamma(t)^*Y + \int_0^t S_\gamma(\sigma)\mathcal{B}(\gamma^2 d_r + d_a)\mathcal{B}^* S_\gamma(\sigma)^* Y\ d\sigma$$

where Λ is the covariance of $Y(0)$. If Λ is nuclear, so is $\mathcal{R}(t, t)$ for every $t \geq 0$. On the other hand, by virtue of properties (i) and (ii) in Lemma 1

$$\lim_{t\to\infty} [\mathcal{R}(t, t)Y,\ Y] = (\gamma^2 d_r + d_a)\int_0^\infty \|\mathcal{B}^* S_\gamma(\sigma)^* Y\|\ d\sigma = \left(\frac{\gamma^2 d_r + d_a}{2\gamma}\right) \|Y\|_E^2 .$$

Hence

$$\mathcal{R}_\infty Y = \lim_{t\to\infty} \mathcal{R}(t, t)Y = \left(\frac{\gamma^2 d_r + d_a}{2\gamma}\right) I$$

and is not nuclear. Now for $t \geq s$, the steady state covariance:

$$\lim_{L\to\infty} E[Y(s + L)\ \ Y(t + L)^*]Y = \mathcal{R}_\infty S_\gamma^*(t - s)Y$$

and is *not* nuclear. Hence in the steady state we have Gaussian stationary process with non-nuclear covariance. Let $R(t)$ denote the steady state covariance function:

$$\begin{aligned} R(t)Y &= \mathcal{R}_\infty S_\gamma^*(t)Y & t > 0 \\ &= R(-t)^*Y & t < 0 . \end{aligned} \tag{2.17}$$

To calculate the spectral density of the steady state process we cannot take the Fourier transform of (2.17) since the semigroup $S_\gamma(t)$ is not exponentially stable. But we can proceed in the following alternate way exploiting property (ii). Thus we can write

$$\mathcal{R}_\infty S_\gamma^*(t)Y = \int_0^\infty S_\gamma(\sigma)\mathcal{B}\mathcal{B}^*\ S_\gamma(\sigma)^*\ S_\gamma^*(t)Y\ d\sigma , \qquad t > 0$$

and hence

$$[\mathcal{R}_\infty S_\gamma^*(t)Y,\ Y] = \int_0^\infty [\mathcal{B}^* S_\gamma(\sigma)^*\ S_\gamma^*(t)Y,\ \mathcal{B}^* S_\gamma(\sigma)^* Y]\ d\sigma \tag{2.18}$$

which we can express as:

$$\int_{-\infty}^\infty [F(\sigma + t),\ F(\sigma)]\ d\sigma , \qquad t > 0$$

where

$$\begin{aligned} F(\sigma) &= \mathcal{B}^* S_\gamma(\sigma)^* Y & \sigma \geq 0 \\ &= 0 & \sigma < 0 . \end{aligned}$$

Thus defined, $F(\cdot)$ is in $L_2[-\infty, \infty]$, and by standard Fourier transform theory, (2.18) can be expressed in terms of Fourier transforms as:

$$\int_{-\infty}^{\infty} e^{2\pi i f t}[\mathcal{B}^* R(2\pi i f, \mathcal{A})^* Y, \ \mathcal{B}^* R(2\pi i f, \mathcal{A})^* Y] \ dt \, , \qquad t > 0$$

since

$$\int_{-\infty}^{\infty} e^{2\pi i f t} F(\sigma) \ d\sigma \ = \ \mathcal{B}^* \mathcal{R}(2\pi i f, \mathcal{A})^* Y \, .$$

Hence it follows that the spectral density of the process is given by

$$P(f) \ = \ \mathcal{R}((2\pi i f, \mathcal{A})\mathcal{B}\mathcal{B}^* \ \mathcal{R}(2\pi i f, \mathcal{A})^* \, .$$

For the example this reduces further to:

$$P(f) \begin{vmatrix} z_1 \\ z_2 \end{vmatrix} \ = \ \begin{vmatrix} ([z_1, h_1] + [z_2, h_2])\phi \\ p([z_1, h_1] + [z_2, h_2])\phi \end{vmatrix}$$

where

$$h_1 \ = \ \frac{1}{[\psi(p)|^2} \begin{vmatrix} -p^2 \rho a \phi(\cdot) \\ T(p) \end{vmatrix} \qquad\qquad h_2 \ = \ \frac{1}{[\psi(p)|^2} \begin{vmatrix} p \rho a \phi(\cdot) \\ mp \end{vmatrix}$$

$$\phi(s) \ = \ \frac{(\sin \lambda s \ - \ \sinh \lambda s)(\cos \lambda L \ + \ \cosh \lambda L) \ - \ (\sin \lambda L \ + \ \sinh \lambda L)(\cos \lambda s \ - \ \cosh \lambda s)}{(\sin \lambda L \ - \ \sinh \lambda L)(\cos \lambda L \ + \ \cosh \lambda L) \ - \ (\sin \lambda L \ + \ \sinh \lambda L)(\cos \lambda L \ - \ \cosh \lambda L)}$$

$$T(p) \ = \ EI_\phi \lambda^3 \, \frac{1 \ + \ \cos \lambda L \cosh \lambda L}{\cosh \lambda L \ \sin \lambda L \ - \ \sinh \lambda L \ \cos \lambda L}$$

$$\psi(p) \ = \ mp^2 \ + \ \gamma p \ + \ T(p)$$

$$\lambda \ = \ \left(\frac{\rho a}{EI_\phi}\right)^{1/4} \sqrt{ip} \qquad\qquad (\ \sqrt{p} > 0 \ \text{ when } \ p > 0 \ )$$

$$p \ = \ 2\rho i f$$

$$\phi \ = \ \begin{vmatrix} \phi(\cdot) \\ 1 \end{vmatrix} .$$

The steady state properties of the boundary process

$$B^* y_1(t)$$

"corresponding to" $f(t, L)$, are readily calculated, the variance in particular being

$$= \ \frac{\gamma^2 d_r \ + \ d_a}{2\gamma} B^* A^{-1} B \ = \ \left(\frac{\gamma^2 d_r \ + \ d_a}{2\gamma}\right) \frac{3EI_\phi}{L^3} \, .$$

Similarly, the (rate) process

$$\frac{d}{dt} B^* y_1(t) = B^* y_2(t)$$

is asymptotically stationary with covariance

$$\lim_{L\to\infty} E[(B^* y_2(t)) \ (B^* y_2(t))] = \left(\frac{\gamma^2 d_r + d_a}{2\gamma}\right) \text{Tr. } B^* M^{-1} B = \left(\frac{1}{m}\right)\left(\frac{\gamma^2 d_r + d_a}{2\gamma}\right).$$

References

1. G. Kallinapur and R. L. Karandikar. *White Noise Theory of Prediction, Filtering and Smoothing*. New York: Gordon & Breach, 1988.
2. A. V. Balakrishnan. *Applied Functional Analysis,* 2nd edition. New York: Springer-Verlag, 1980.
3. C. E. Benchimol. "A Note on the Weak Stabilizability of Contraction Semigroups," *SIAM Journal on Control and Optimization,* Vol. 16 (1978).
4. A. V. Balakrishnan. "Compensator Design for Stability Enhancement with Collocated Controllers," *IEEE Transactions on Automatic Control,* Vol. 36, No. 9 (1991).
5. A. V. Balakrishnan. "Stochastic Regulator Theory for a Class of Abstract Wave Equations," *SIAM Journal on Control and Optimization,* Vol. 29 (1991).

SPLITTING UP METHOD IN THE CONTEXT OF STOCHASTIC PDE

A. Bensoussan
University of Paris Dauphine and
INRIA, BP 105, F-78153 Le Chesnay Cedex

1. INTRODUCTION

In this paper, we survey some methods and results, concerning stochastic partial differential equations. The main idea is to use the splitting up method, which permits to separate the deterministic and stochastic part of the equation.

We use in this article the papers of A. BENSOUSSAN, R. GLOWINSKI, A. RASCANU [2], [3], A. BENSOUSSAN [1]. For related work see F. LEGLAND [5].

2. THE DETERMINISTIC SPLITTING UP METHOD.

We survey here the main idea of the splitting up method. We refer to PEACEMAN, RACHFORD [9] and DOUGLAS-RACHFORD [4] for details. See also P.L. LIONS, B. MERCIER [7].

Consider the problem

$$\begin{aligned} &\frac{dy}{dt} + A_1(y) + A_2(y) = 0 \\ &y(0) = y_0 \end{aligned} \tag{2.1}$$

to be solved on $(0, T)$.

We split the interval $(0, T)$ into intervals of length k, setting

$$(N+1)k = T.$$

Considering the sequence

$$0, k \dots rk, \ (r+1)k \dots (N+1)k = T.$$

Define a sequence y_k^r with $y_k^0 = y_0$ and $y_{1k}(t)$, $y_{2k}(t)$ by the following 4 possible schemes.

I -
$$\begin{aligned} &\frac{dy_{1k}}{dt} + A_1(y_{1k}) = 0 && \frac{dy_{2k}}{dt} + A_2(y_{2k}) = 0 \\ &y_{1k}(rk) = y_k^r && y_{2k}(rk) = y_k^{r+1/2} = y_{1k}((r+1)k - 0) \\ &y_k^{r+1} = y_{2k}((r+1)k - 0) \end{aligned}$$

II -
$$\frac{dy_{2k}}{dt} + A_2(y_{2k}) = 0 \qquad \frac{dy_{1k}}{dt} + A_1(y_{1k}) = 0$$

$$y_{2k}(rk) = y_k^r \qquad\qquad y_{1k}(rk) = y_k^{r+1/2} = y_{2k}((r+1)k-0)$$

III - $\quad \dfrac{dy_{1k}}{dt} + A_1(y_{1k}) = 0 \qquad \dfrac{dy_{2k}}{dt} + A_1(y_{1k}) + A_2(y_{2k}) = 0$

$$y_{1k}(rk) = y_k^r \qquad\qquad y_{2k}(rk) = y_k^r$$

$$y_k^{r+1} = y_{2k}((r+1)k-0)$$

IV - $\quad \dfrac{dy_{2k}}{dt} + A_2(y_{2k}) = 0 \qquad \dfrac{dy_{1k}}{dt} + A_1(y_{1k}) + A_2(y_{2k}) = 0$

$$y_{2k}(rk) = y_k^r \qquad\qquad y_{1k}(rk) = y_k^r$$

$$y_k^{r+1} = y_{1k}((r+1)k-0)$$

The above schemes can be rewritten in sequential form, as follows :

I–

$$\frac{y_k^{r+1/2} - y_k^r}{k} + A_1(y_k^{r+1/2}) = 0$$

$$\frac{y_k^{r+1} - y_k^{r+1/2}}{k} + A_2(y_k^{r+1}) = 0$$

(which is the Peaceman-Rachford iteration)

II–

$$\frac{y_k^{r+1/2} - y_k^r}{k} + A_2(y_k^{r+1/2}) = 0$$

$$\frac{y_k^{r+1} - y_k^{r+1/2}}{k} + A_1(y_k^{r+1}) = 0.$$

Note the following variant :

$$2\,\frac{y_k^{r+1/2} - y_k^r}{k} + A_1(y_k^{r+1/2}) + A_2(y_k^r) = 0$$

$$2\,\frac{y_k^{r+1} - y_k^{r+1/2}}{k} + A_1(y_k^{r+1/2}) + A_2(y_k^{r+1}) = 0$$

III–

$$\frac{\hat{y}_k^{r+1} - y_k^r}{k} + A_1(\hat{y}_k^{r+1}) + A_2(y_k^r) = 0$$

$$\frac{y_k^{r+1} - y_k^r}{k} + A_1(\hat{y}_k^{r+1}) + A_2(y_k^{r+1}) = 0.$$

This is the Douglas-Rachford scheme.

IV–

$$\frac{\hat{y}_k^{r+1} - y_k^r}{k} + A_1(\hat{y}_k^r) + A_2(\hat{y}_k^{r+1}) = 0$$

$$\frac{y_k^{r+1} - y_k^r}{k} + A_1(\hat{y}_k^{r+1}) + A_2(\hat{y}_k^{r+1}) = 0.$$

Remark : for fixed k, $r \to \infty$, $y_k^r \to y$ solution of

$$A_1(y) + A_2(y) = 0$$

(see P.L. LIONS [7]).

III. SPLITTING UP FOR ZAKAI EQUATION

We consider a linear stochastic infinite dimensional equation, given by

$$\begin{aligned} &dy + Ay\,dt = By \cdot dw \\ &y(0) = y_0 \end{aligned}$$

with the following standard notation of J.L. LIONS [6]

$$\begin{aligned} &V \subset H \subset V' \\ &A \in L(V;V') \quad < A\varphi, \varphi \geq \alpha \|\varphi\|^2 \\ &\alpha > 0, \quad \forall \varphi \in V. \end{aligned}$$

As far the stochastic part is concerned, we assume

$$\begin{aligned} &B \in L(H; H^m) \\ &By \cdot dw = \sum_{j=1}^{m} B_j\, y\, dw_j \end{aligned}$$

Remark :

$$\begin{aligned} &A(t) \in L^\infty(0,T; L(V;V')) \\ &B(t) \in L^\infty(0,T; L(H;H^m)) \end{aligned}$$

i.e. time dependent operators can be considered. To simplify, we assume time independence. The equation (3.1) as a unique solution such

$$y \in L^2_F(0,T;V) \cap L^2(\Omega, \mathcal{A}, \mathcal{P}; C(0,T;H))$$

where $F^t = \sigma(w(s), s \leq t)$, and $L^2_F(0,T;V)$ represents the set of L^2 integrable processes adapted to the filtration F^t.

One has also

$$y \in L^\infty(0,T; L^4(\Omega, \mathcal{A}, \mathcal{P}; H)).$$

In addition the following Ito's calculus rules are available,

$$d|y(t)|^2 + 2 < Ay(t), y(t) > dt = 2\sum_j (y, B_j y) dw_j + \sum_j |B_j y|^2 dt$$

$$\begin{aligned} d|y(t)|^4 = 2|y(t)|^2[-2 < Ay, y > dt + \sum_j |B_j y|^2 dt \\ + 2\sum_j (y, B_j y) dw_j] + 4\sum_j (y, B_j y)^2 dt. \end{aligned}$$

We apply the method of splitting up as follows. Consider the sequence y_{1k}, y_{2k} defined by

$$\begin{aligned}
&\frac{dy_{1k}}{dt} + Ay_{1k} = 0\\
&y_{1k}(r_k) = y_k^r\\
&y_k^{r+1/2} = y_{1k}((r+1)k - 0)\\
&dy_{2k} = By_{2k} \cdot dw\\
&y_{2k}(rk) = y_k^{r+1/2}\\
&y_k^{r+1} = y_{2k}((r+1)k - 0)\\
&y_k^0 = y_0
\end{aligned}$$

Remark : One can rewrite y_{1k} and y_{2k} as follows

$$y_{1k}(t) + \int_0^t Ay_{1k}\,ds = \int_0^{k[t/k]} By_{2k} \cdot dw + y_0$$

$$y_{2k}(t) + \int_0^{k([t/k]+1)} Ay_{1k}\,ds = \int_0^t By_{2k} \cdot dw + y_0$$

which shows more clearly the link with the initial equation.

Theorem 3.1. *One has the following result*

$$\begin{aligned}
&y_{1k}, y_{2k} \to y \quad \mathit{in} \quad L_F^2(0,T;V)\\
&y_{1k}(t), y_{2k}(t) \to y(t) \quad \mathit{in} \quad L^2(\Omega, \mathcal{A}, P; H), \quad \forall t \in [0,T[\\
&y_{1k}(T-0), y_{2k}(T-0) \to y(T) \quad \mathit{in} \quad L^2(\Omega, \mathcal{A}, P; H)
\end{aligned}$$

see F. LEGLAND [5] for application to non linear filtering.

For the proof of Theorem 3.1, we refer to A. BENSOUSSAN, R. GLOWINSKI, A. RASCANU [2].

4. NON LINEAR EQUATIONS :

4.1. The continuous problem.

Consider the case of a monotone operator A, as follows

$$\begin{aligned} &dy + A(t, y(t))dt = B(t, y(t))dw \\ &y(0) = y_0 \end{aligned} \tag{4.1}$$

where the notation is the following

$$V \subset H \subset V'$$

Now V is a reflexive Banach space, and

$$\begin{aligned} &A(t,\cdot) \text{ is a family of operators from } V \text{ to } V' \\ &B(t,\cdot) \text{ is a family of operators from } H \text{ to } H^m \end{aligned}$$

whith the assumptions

$$\begin{aligned} &2 < A(t,\varphi), \varphi > +\lambda|\varphi|^2 + \nu \geq \alpha\|\varphi\|^p \\ &2 < A(t,\varphi) - A(t,\psi), \varphi - \psi > +\lambda|\varphi - \psi|^2 \geq 0 \\ &\|A(t,\varphi)\|_{V'}^{p'} \leq \beta(1 + \|\varphi\|^p) \quad \frac{1}{p} + \frac{1}{p'} = 1 \\ &\theta \to < A(t, \varphi + \theta\psi), \chi >: R \to R \text{ is continuous } \forall \varphi, \psi, \chi \in V \\ &t \to < A(t,\varphi), \psi >:]0, T[\to R \text{ is measurable} \\ &|B(t,\varphi) - B(t,\psi)|^2 \leq K|\varphi - \psi|^2 \\ &|B(t,\varphi)|^2 \leq \gamma(1 + |\varphi|^2) \\ &t \to (B(t,\varphi), \psi) :]0, T[\to R \text{ is measurable.} \end{aligned}$$

We can state the following result due to E. PARDOUX [8].

Theorem 4.1. *Let $y_0 \in H$, then there exists one and only one solution*

$$y \text{ in } L_F^p(0,T;V) \cap L^2(\Omega, \mathcal{A}, P; C(0,T;H)),$$

of the equation

$$\begin{aligned} &dy + A(t,y)dt = B(t,y)dw \\ &y(0) = y_0 \end{aligned}$$

4.2. Splitting up

We define the following schemes, inspired from the linear case

$$
\begin{aligned}
&\frac{dy_{1k}}{dt} + A(t, y_{1k}) = 0 \\
&y_{1k}(rk) = y_k^r \\
&y_k^{r+1/2} = y_{1k}((r+1)k - 0) \\
&dy_{2k} = B(t, y_{2k}) \cdot dw \\
&y_{2k}(rk) = y_k^{r+1/2} \\
&y_k^{r+1} = y_{2k}((r+1)k - 0) \\
&y_k^0 = y_0
\end{aligned}
$$

or written in a way close to the continuous equation

$$
\begin{aligned}
y_{1k}(t) + \int_0^t A(s, y_{1k}) ds &= y_0 + \int_0^{k[t/k]} B(s, y_{2k}) \cdot dw \\
y_{2k}(t) + \int_0^{k([t/k]+1)} A(s, y_{1k}) ds &= y_0 + \int_0^t B(s, y_{2k}) \cdot dw
\end{aligned}
$$

We can state the following result

Theorem 4.2.

$$
\begin{aligned}
&y_{1k}, y_{2k} \to y \textit{ in } L_F^p(0, T; V) \textit{ weakly} \\
&y_{1k}(t), y_{2k}(t) \to y(t) \textit{ in } L^2(\Omega, \mathcal{A}, P; H), \quad \forall t \in [0, T[\\
&y_{1k}(T-0), y_{2k}(T-0) \to y(T) \quad \textit{in} \quad L^2(\Omega, \mathcal{A}, P; H)
\end{aligned}
$$

For the proof we refer to A. BENSOUSSAN, R. GLOWINSKI, A. RASCANU [3].

5. NONLINEAR EQUATIONS WITH COMPACTNESS ARGUMENT

5.1. Notation and assumptions

We consider

$$V \subset H \subset V'$$

where V is a reflexive Banach space ; we assume that the injection of V into H is *compact*, and the following conditions

$$
\begin{aligned}
&< A(t;\varphi), \varphi > \geq \alpha \|\varphi\|^p \\
&< A(t;\varphi) - A(t;\psi), \varphi - \psi > \geq 0 \\
&\|A(t,\varphi)\|_{V'}^{p'} \leq \beta(1 + \|\varphi\|^p)
\end{aligned}
$$

$\theta \to < A(t, \varphi + \theta\psi), \chi >: R \to R$ is continuous, $\forall \varphi, \psi, \chi \in V$.

$t \to < A(t, \varphi), \psi >:]0, T[\to R$ is measurable $g(t;\varphi), B(t;\varphi) : (0, T) \times H \to H, \ H^m$ measurable continuous from H to H, H^m for a.e.t.

$$
\begin{aligned}
&|g(t,\varphi)|^2 + |B(t;\varphi)|^2 \leq \gamma(1 + |\varphi|^2) \\
&|B(t,\varphi) - B(s,\varphi)|^2 \leq 0(t-s)(1 + |\varphi|^2)
\end{aligned}
$$

with $0(h)$ monotone increasing, $0(h) \to 0$ as $h \to 0$.

We consider the equation

$$(5.1)\qquad \begin{aligned} &dy + A(t, y(t))dt = g(t, y(t))dt + B(t, y(t))dw \\ &y(0) = y_0 \in H \end{aligned}$$

One has to be careful in defining the concept of solution. We state

Definition : A solution of (5.1) on $[0, T]$ is a system $(\Omega, \mathcal{A}, P, F^t, w(t), y(t))$ such that

$$\begin{aligned} &\Omega, \mathcal{A}, P \text{ is a probability space, } F^t \text{ is a filtration} \\ &w(t) \text{ a } F^t \text{ standard Wiener process with values in } R^n \\ &y(t) \in L^p_F(O, T; V) \cap L^2(\Omega, \mathcal{A}, P; C(0, T; H)) \end{aligned}$$

and the following relation holds

$$y(t) + \int_0^t A(s, y(s))ds = y_0 + \int_0^t g(s, y(s))ds + \int_0^t B(s, y(s))dw \quad \forall t \in [0, T].$$

□

In the present case, the splitting up method will be used to prove the existence of a solution, by a constructive algorithm.

5.2. Existence result

Theorem 5.1. *There exists a solution of (5.1).*□

The splitting up approximation is defined as follows. Set $k = T/N + 1$, and define a map $\Psi_{1k} : C(0, T; R^m) \to L^2(0, T; H)$ as follows

$$\Psi_{1k}(b(\cdot)) = y_{1k}(\cdot)$$

where

$$\begin{aligned} &\frac{dy_{1k}}{dt} + A(t; y_{1k}) = 0 \\ &y_{1k}(rk) = y_k^r \\ &y_k^{r+1/2} = y_{1k}((r+1)k - 0) \qquad \overline{y_k^r} = \frac{1}{k}\int_{rk}^{(r+1)k} y_{1k}(s)ds \\ &y_{2k}(t) = y_k^{r+1/2} + \int_{rk}^t g(s, y_{1k}(s))ds + B(rk; \overline{y_k^r})(b(t) - b(rk)) \\ &y_k^{r+1} = y_{2k}((r+1)k - 0) \\ &y_k^0 = y_0 \end{aligned}$$

Remark : Ψ_k is a continuous functional and takes values in $L^p(0, T; V)$. We set also $\Psi_{2k}(b(\cdot)) = y_{2k}(\cdot) : C(0, T; R^m) \to L^2(0, T; H)$. Let us next define

$$\begin{aligned} &\tilde{\Omega} = C(0, T; R^m), \qquad \tilde{\mathcal{A}} = \text{ Borel } \sigma - \text{algebra} \\ &\tilde{P} = \text{ Wiener measure} \end{aligned}$$

and set

$$\begin{aligned} &b(t; \tilde{\omega}) = \tilde{\omega}(t) \\ &y_{1k}(t; \tilde{\omega}) = \Psi_{1k}(b(\cdot; \tilde{\omega})) \\ &y_{2k}(t; \tilde{\omega}) = \Psi_{2k}(b(\cdot; \tilde{\omega})) \end{aligned}$$

We begin with several Lemmas :

Lemma 5.1.

$$\tilde{E}\int_0^T \|y_{1k}(t)\|^p dt \leq C$$
$$\tilde{E}|y_{1k}(t)|^2, \quad \tilde{E}|y_{1k}(t)|^4 \leq C$$
$$\tilde{E}|y_{2k}(t)|^2, \quad \tilde{E}|y_{2k}(t)|^4 \leq C$$

Lemma 5.2 :

$$\tilde{E} \sup_{t\in[0,T]} |y_{1k}(t)|^2 \leq C$$
$$\tilde{E} \sup_{t\in[0,T]} |y_{2k}(t)|^2 \leq C$$

Lemma 5.3. :

$$\tilde{E} \sup_{|\theta|\leq\delta} \int_0^T \|y_{1k}(t+\theta) - y_{1k}(t)\|_{V'}^{p'} dt \leq C\delta^{\frac{1}{p-1}}$$

where y_{1k} is extended by 0, outside $(0,T)$. □

We then proceed with a tightness argument. Set

$$S = C(0,T;R^m) \times L^2(0,T;H),$$

and let π_k be the probability on S image of $\tilde{P}$ by the map

$$\tilde{\omega} \rightarrow (b(\cdot,\tilde{\omega}), \psi_{1k}(b(\cdot,\tilde{\omega})))$$

We have the

Lemma 5.4. *The family π_k is uniformly tight.*

By Prokhorov's theorem, the family π_k is relatively compact in the set $\mathcal{P}(S)$ of probability measures on S, equipped with the weak convergence topology

Pick $\pi_{k_j} \rightarrow \pi$, a weakly converging subsequence

By Skorokhod's theorem, there exists a probability space $\Omega, \mathcal{A}, P$, and random variables

$$w_{k_j}(\cdot,\omega), \quad y_{k_j}(\cdot,\omega) \text{ on } (\Omega,\mathcal{A},P) \text{ with values in } S$$
$$w(\cdot,\omega), \quad y(\cdot,\omega)$$

such that the probability law of $w_{k_j}(\cdot), y_{k_j}(\cdot)$ is π_{k_j}

$$w_{k_j}(\cdot,\omega), y_{k_j}(\cdot,\omega) \rightarrow w(\cdot,\omega), y(\cdot,\omega) \text{ a.s., as } j \rightarrow \infty,$$

and the probability law of $w(\cdot), y(\cdot)$ is π.

Let

$$F^t = \sigma(\mathbb{1}_t w(\cdot,\omega), \mathbb{1}_t y(\cdot,\omega)),$$

then $w(t;\omega)$ is a F^t Wiener process, and the system $\Omega, \mathcal{A}, P, F^t, w(t), y(t)$ is a solution.

5.3. Navier-Stokes equations (dim 2).

We use the following notation

$$V \subset H \subset V' \text{ where } V \text{ is a Hilbert space,}$$
$$\text{and } A \in L(V;V'), \quad < Av, v > \geq \alpha \|v\|^2$$
$$B : V \to V', \text{ with the assumptions}$$
$$\|B(v_1) - B(v_2)\|_{V'} \leq C(|v_1|^{1/2}\|v_1\|^{1/2} + |v_2|^{1/2}\|v_2\|^{1/2} + 1)$$
$$|v_1 - v_2|^{1/2}\|v_1 - v_2\|^{1/2}$$
$$|< B(v), v >| \leq \beta\|v\|^2 + C\beta(|v|^2 + 1)$$

Remark : B continuous from V weakly to V'. Consider next functionals,

$$g : H \to H \text{ continuous } |g(v)| \leq \bar{g}(1 + |v|)$$
$$G : H \to H^m \text{ continuousJ } |G(v)|_{H^m} \leq \bar{G}(1 + |v|)$$

We consider the following equation

$$\text{(5.2)} \qquad dy + (Ay + B(y) - g(y))dt = G(y)dw$$
$$y(0) = y_0$$

A solution of (5.2) on $[0,T]$ is a system $\Omega, \mathcal{A}, P, F^t, w, y$ where $(\Omega, \mathcal{A}, P)$ is a probability space, F^t a filtration, $w(t)$ a standard F^t Wiener process, $y(t)$ an element of $L^2_F(0,T;V) \cap L^2(\Omega, \mathcal{A}, P; C(0,T;H))$, and a.s.

$$y(t) + \int_0^t (Ay + B(y) - g(y))ds = y_0 + \int_0^t G(y)dw \quad \forall t \in [0,T].$$

We can state the

Theorem 5.2. *There exists a solution of (5.2).*

Details of the proof can be found in A. BENSOUSSAN [1].

We just mention the splitting up approximation, which provides a constructive proof

$$\Psi_{1k} : C(0,T;R^m) \to L^2(0,T;H)$$
$$\Psi_{2k} : C(0,T;R^m) \to L^2(0,T;H)$$
$$y_{1k}(\cdot) = \Psi_{1k}(b(\cdot))$$
$$y_{2k}(\cdot) = \Psi_{2k}(b(\cdot))$$
$$\frac{dy_{1k}}{dt} + Ay_{1k} + B(y_{1k}) = 0$$
$$y_{1k}(rk) = y_k^r$$
$$y_k^{r+1/2} = y_{1k}((r+1)k - 0)$$
$$y_{2k}(t) = y_k^{r+1/2} + \int_{rk}^t g(y_{1k}(s))ds + G(y_k^r)(b(t) - b(rk))$$
$$y_k^{r+1} = y_{2k}((r+1)k - 0)$$
$$y_k^0 = y_0.$$

References

[1] A. BENSOUSSAN, A model of stochastic differential equation in Hilbert spaces applicable to Navier-Stokes equations in dimension 2, Volume in honor of Prof. Zakai, Academic Press, 1991.

[2] A. BENSOUSSAN, R. GLOWINSKI, A. RASCANU, Approximation of Zakai equation by the splitting up method, to be published in SIAM J. of Control and Optimization.

[3] A. BENSOUSSAN, R. GLOWINSKI, A. RASCANU, Approximation of some stochastic differential equations by the splitting up method, to be published in AMO.

[4] J. DOUGLAS, H.H. RACHFORD, On the numerical solution of the heat conduction problem in 2 and 3 space variables, Trans. Amer. Math. Soc. 82 (1956) pp. 421-439.

[5] F. LE GLAND, High order time discretization of nonlinear filtering equations, 28th IEEE CDC (Tampa 1989), 2601-2606.

[6] J.L. LIONS, Contrôle optimal des systèmes gouvernés par des équations aux dérivées partielles, Dunod, Paris 1968.

[7] P.L. LIONS, B. MERCIER, Splitting algorithms for the sum of two nonlinear operators, SIAM J. Numer. Anal., vol. 16, No. 6, Dec. 1979.

[8] E. PARDOUX, Stochastic partial differential equations and filtering of diffusion processes, Stochastics, 3, 1979, pp. 127-168.

[9] D.H. PEACEMAN, H.H. RACHFORD, The numerical solution of parabolic elliptic differential equations, J. Soc. Ind. App. Math. 3 (1955), pp. 28-41.

Generalized Stochastic Differential Equations on $(\mathcal{D}^*)$

David Betounes

Mathematics Department
University of Southern Mississippi
Hattiesburg, MS 39406-5045

1 Introduction

Recent work on anticipating, or non-adapted, SDE's has proceeded along various avenues and from different points of view (compare e.g., [OP 89a,b], [KP 90], [Oc 90], [PP 90], and [Kuo 91] to cite just a few papers on this subject). In this paper we briefly describe a general theory for treating such SDE's based on the Wiener algebra $(\mathcal{D}^*)$ and the natural analysis associated with this space of generalized Wiener functionals. Within this setting the weak derivative: $\dot{x} = dx/dt$ and weak Pettis integral: $\int x(t)dt$, of functions $x : R \to (\mathcal{D}^*)$, allow one to not only generalize the theory of stochastic integrals (cf. [BR 91a,b]), but also obtain (as we argue here) a straight-forward formulation and generalization of SDE theory: one which is expressed without the need of stochastic integrals or of adaptedness assumptions.

Thus an initial value problem on the Wiener algebra is naturally formulated by

$$\dot{x} = F(t,x) \tag{1}$$

$$x(0) = c \tag{2}$$

The notation and general assumptions here are: $(\mathcal{D}^*)^m = (\mathcal{D}^*) \times \cdots \times (\mathcal{D}^*)$ denotes the Cartesian product of $(\mathcal{D}^*)$ with itself m times, and we assume that F is a mapping:

$$F : J \times U \subseteq R \times (\mathcal{D}^*)^m \to (\mathcal{D}^*)^m$$

with $J \subseteq R$ an interval and $U \subseteq (\mathcal{D}^*)^m$. The initial value is an element $c \in U$, and by a solution of (1) we mean a weakly differentiable function:

$$x : J_0 \subseteq J \to U \subseteq (\mathcal{D}^*)^m,$$

such that $\dot{x}(t) = F(t, x(t))$ for all $t \in J_0$. Here $\dot{x}(t) = (\dot{x}^1(t), ..., \dot{x}^m(t))$.

While the theory for such generalized SDE's is still in a primitive stage, we describe here its relation to the SDE theory and the existence and uniqueness problem (Section 3). We also present a number of results on renormalization (Section 4) and its application to the explicit solution of a broad class of GSDE's (Section 5).

2 Background

Here we give just the bare essentials of the white noise analysis on the Wiener Algebra $(\mathcal{D}^*)$. A detailed discussion is presented in [BR 91a].

The basic probability space is $\Omega = \mathcal{S}^*(R, R^M)$, the dual of the space of rapidly decreasing vector valued functions $\xi : R \to R^M$. Endowing Ω with the standard Minlos measure μ, one obtains the Fock space (Wiener chaos) decomposition $(L)^2 \equiv L^2(\Omega, \mu) = \bigoplus_{n=0}^{\infty} K_n$, where $K_n = I_n(L_n^2)$ is the space of n-fold multiple Wiener integrals of functions $f_n \in L_n^2 \equiv L^2(R^n, (R^M)^{\otimes n})$. One obtains M independent Brownian motions $B^1(t), \cdots, B^M(t)$ on Ω via:

$$B^j(t) \equiv I_1(1_{[0,t)}\varepsilon^j)$$

where $\{\varepsilon^j\}_{j=1}^M$ is the standard basis for R^M (this dictates the choice of M in the theory).

Let $\mathcal{D}_n = C_c^\infty(R^n, (R^M)^{\otimes n})$, then the coproduct: $(\mathcal{D}) \equiv \coprod_{n=0}^\infty(\mathcal{D}_n)$ of the spaces: $(\mathcal{D}_n) \equiv I_n(\mathcal{D}_n)$ is the space of Wiener test functionals. With the inductive topology on $(\mathcal{D})$, the dual space $(\mathcal{D})^* = (\mathcal{D}^*) = \prod_{n=0}^\infty(\mathcal{D}_n^*)$ is the space of Wiener distributions.[1] We denote the elements in the Cartesian product (and coproduct) by formal sums (series) rather than by sequences. Thus a Wiener distribution $x \in (\mathcal{D}^*)$ is a formal sum: $x = \sum_{n=0}^\infty x_n$ with each $x_n = I_n(T_n)$ being a generalized multiple Wiener integral of a distribution $T_n \in \mathcal{D}_n^*$. The symmetric tensor product $\hat{\otimes}$ of distributions carries over to give a commutative product $\hat{\otimes}$ on $(\mathcal{D}^*)$: Namely, with x as above and with $y = \sum_{n=0}^\infty I_n(U_n)$ one defines:

$$x\hat{\otimes}y = \sum_{n=0}^{\infty}\sum_{s=0}^{n} I_n(T_s\hat{\otimes}U_{n-s}).$$

This makes $(\mathcal{D}^*)$ into a commutative associative algebra.

The analysis for functions $x : J \subseteq R \to (\mathcal{D}^*)$ is quite simple. Thus x is differentiable at $t \in J$ (respectively integrable over J) if (1) $t \to \langle x(t), \Psi\rangle$ is differentiable at t (resp. integrable over J) for each Ψ and (2) there is an element $\dot{x}(t)$ (resp. $\int_J x(t)dt$) in $(\mathcal{D}^*)$ such that $\langle \dot{x}(t), \Psi\rangle = \frac{d}{ds}\langle x(s), \Psi\rangle|_{s=t}$ (resp. $\langle \int_J x(t)dt, \Psi\rangle = \int_J \langle x(t), \Psi\rangle dt$) for each $\Psi \in (\mathcal{D})$.

3 Existence and Uniqueness

The usual adapted SDE theory can be modelled in our framework as follows. Suppose $H, G_1, ..., G_M$: $J \times R^m \to R^m$ (for brevity we do not consider more general situations) and let $G(t,r) = \{G_j^i(t,r)\}_{j=1,...,M}^{i=1,...,m}$ be the corresponding $m \times M$ matrix. Assume that H and G satisfy the Lipschitz conditions: $|H(t,r) - H(t,\bar{r})| \leq K|r - \bar{r}|$ and $\|G(t,r) - G(t,\bar{r})\| \leq K|r - \bar{r}|$ (where $\|G\| = [\mathrm{tr}(GG^t)]^{1/2}$ is the usual matrix norm). For a measurable function $x : \Omega \to R^m$, let $H^\#(t,x)$ denote the measurable function defined by $H^\#(t,x)(\omega) = H(t,x(\omega))$. Then, because of the Lipschitz conditions, one finds that if $x \in (L^2)^m$, then $H^\#(t,x), G_j^\#(t,x) \in (L^2)^m$. Consequently, we get a map $F : J \times (L^2)^m \to (\mathcal{D}^*)^m$ defined by

$$F(t,x) \equiv H^\#(t,x) + G_j^\#(t,x)\hat{\otimes}\dot{B}^j(t) \tag{3}$$

where there is implied summation on $j = 1, ..., M$.[2] Then the generalized SDE:

$$\dot{x} = H^\#(t,x) + G_j^\#(t,x)\hat{\otimes}\dot{B}^j(t) \tag{4}$$

is (modulo notation) an ordinary SDE. This is so because stochastic integrals of the form: $\int_0^t Y(s)\hat{\otimes}\dot{B}^j(s)\,ds$ are Skorohod integrals and reduce to Ito integrals when Y is adapted. Under the assumptions of linearity on H, G_j, namely: $H(t,r) = A(t)r + a(t)$ and $G_j(t,r) = Q_j(t)r + q_j(t)$, the SDE (4) is linear, and may be written in the following form:

$$\begin{aligned} \dot{x} &= A(t)x + a(t) + [Q_j(t)x + q_j(t)]\hat{\otimes}\dot{B}^j(t) \\ &= [A(t) + Q_j(t)\dot{B}^j(t)]\hat{\otimes}x + [a(t) + q_j(t)\dot{B}^j(t)] \end{aligned} \tag{5}$$

This then is a special case of the more general form of a linear generalized SDE in our setting:

$$\dot{x} = \mathcal{A}(t)\hat{\otimes}x + \beta(t) \tag{6}$$

where $\mathcal{A}(t)$ is an $m \times m$ matrix with entries from $(\mathcal{D}^*)$ and $\beta(t) \in (\mathcal{D}^*)^m$.

With this model of the usual SDE theory within our framework (via eq.(4)), one could ask in general for an analog of the Lipschitz condition to ensure existence and uniqueness of solutions to $\dot{x} = F(t,x), x(0) = c$. One natural analog is to require that for each $\Psi \in (\mathcal{D})^m$, there exist a constant C_Ψ such that:

$$|\langle F(t,x) - F(t,y), \Psi\rangle| \leq C_\Psi|\langle x - y, \Psi\rangle| \tag{7}$$

for all $x, y \in U$ and $t \in J$. Using (7), together a few additional assumptions on F, one can apply the Picard iteration method to prove existence and uniqueness in pretty much the standard way. This is (essentially) the approach of Kuo & Potthoff [KP 90] (slightly generalized). However, the difficulty with this is that it

[1] Note that $(\mathcal{D}_0) = (\mathcal{D}_0^*) = R$, by definition.

[2] Also for $y = (y^1, .., y^m) \in (\mathcal{D}^*)^m$ and $z \in (\mathcal{D}^*)$, $y\hat{\otimes}z \equiv (y^1\hat{\otimes}z, ..., y^m\hat{\otimes}z)$.

is not general enough to include the usual SDE case. Namely with H and G as above, satisfying Lipschitz conditions, and $F(t,x)$ as in (3), one can not in general prove the Lipschitz condition (7) for F. Of course both cases: (a) with a general F satisfying (7), and (b) with F as in (3) H, G satisfying Lipschitz conditions, lead to a Lipschitz condition:

$$||T(x)(t) - T(y)(t)||_{\alpha}^{p} \leq C_{\alpha} \int_{0}^{t} ||x(s) - y(s)||_{\alpha}^{p}\, ds, \tag{8}$$

on the fixed point map: $T(x)(t) = c + \int_0^t F(s, x(s))\, ds$. In each case the domain of T contains a function space: $\mathcal{F}(J,U)$ (in (a): $\mathcal{F} = C(J,(\mathcal{D}^*))$, and in (b): $\mathcal{F} = L^2(J,(L^2))_{adapt}$), which is invariant under T, and condition (8) is relative to a family of seminorms: $\{||\cdot||_{\alpha}\}_{\alpha\in\Gamma}$ defining the topology on U. Thus the existence and uniqueness question is still somewhat problematic (short of just assuming enough data to ensure that the fixed point scheme goes through. The following example falls within the setup of (8), and exhibits another aspect of this problem.

Example: (m = 1) Consider the GSDE: $\dot{x} = x\hat{\otimes}\dot{B}(t)$, with initial condition: $x(0) = c \in K_{\infty}^{*} \equiv \prod_{n=0}^{\infty} K_n$. Letting $\mathcal{F} = \prod_{n=0}^{\infty} L^2(J, K_n)$ (i.e the set of $x : J \to K_{\infty}^{*} \ni x_n \in L^2(J, K_n), \forall n$), one can show that $T(\mathcal{F}) \subseteq \mathcal{F}$, and $||T(x)(t) - T(y)(t)||_{(n)}^2 \leq n \int_0^t ||x(s) - y(s)||_{(n)}^2 ds$, $\forall n$ and $x, y \in \mathcal{F}$. Here $||x||_{(n)} \equiv \max\{\, ||x_j||_{K_j} \,|\, 0 \leq j \leq n\}$. Now in this example one can use renormalization (next section) to directly exhibit the solution: $x(t) = c\hat{\otimes}(\exp)(B(t))$, and prove uniqueness by looking at the Fock space structure of the DE: $\dot{x}_0 = 0$, $\dot{x}_1 = x_0\dot{B}(t)$, $\dot{x}_2 = x_1\hat{\otimes}\dot{B}(t)$, etc. We use this approach below (Theorem (3) and Corollary (1)) for a broad class of GSDE's for which condition (8) would be difficult to verify.

4 Renormalization

A useful tool in the analysis of generalized SDE's on $(\mathcal{D}^*)$ is the renormalization technique. This technique associates to each analytic function $g : R \to R$, a function $(g) : (\mathcal{D}^*) \to (\mathcal{D}^*)$ such that $(g)(r) = g(r)$, for $r \in R = (\mathcal{D}_0^*)$. This association is, as we shall see, functorial in that it preserves compositions and other relevant structure. It turns out to be useful, and economical, to enlarge from the outset the class of functions g under consideration. Thus we only require that g be analytic on an interval, say $(-\rho, \rho)$ and we also allow g to have values in $(\mathcal{D}^*)$. So for $g : (-\rho,\rho) \to (\mathcal{D}^*)$ analytic, its renormalization will again be a map $(g) : (\mathcal{D}^*) \to (\mathcal{D}^*)$. The simplest example of this is the case when $g(r) = \sum_{k=0}^{p} a_k r^k$ is a polynomial with coefficients $a_k \in (\mathcal{D}^*)$. Then the renormalized version of g is $(g)(x) = \sum_{k=0}^{p} a_k \hat{\otimes} x^{\hat{\otimes}k}$. The general definition relies on some elementary facts about power series in $(\mathcal{D}^*)$ and Theorem (1) below. Namely, for a sequence $\{a_k\}_{k=0}^{\infty}$ in $(\mathcal{D}^*)$, let a_k have Fock space decomposition $a_k = \sum_{n=0}^{\infty} a_{kn}$. Then for $r \in R$, it's easy to see that: (a) the power series

$$g(r) = \sum_{k=0}^{\infty} a_k r^k \tag{9}$$

convergences in $(\mathcal{D}^*)$ if and only if: (b) for every n the power series

$$g_n(r) = \sum_{k=0}^{\infty} a_{kn} r^k \tag{10}$$

converges in $(\mathcal{D}_n^*)$, if and only if: (c) for every n, for every $\Psi_n \in (\mathcal{D}_n)$, the power series

$$\langle g_n(r), \Psi_n \rangle = \sum_{k=0}^{\infty} \langle a_{kn}, \Psi_n \rangle r^k \tag{11}$$

converges in R. Letting $\rho(\Psi_n)$ be the radius of convergence of series (11), then (by definition) $\rho_n \equiv \inf\{\rho(\Psi_n) | \Psi_n \in \mathcal{D}_n\}$ is the radius of convergence of series (10), and $\rho \equiv \inf\{\rho_n | n = 0, 1, 2, ...\}$ is the radius of convergence of series (9). Consequently (9) gives a function $g : (-\rho, \rho) \to (\mathcal{D}^*)$ which is infinitely differentiable (weakly) with $g^{(k)}(0) = k! a_k$. The Fock space components g_n of g are given by (10), and $g_n^{(k)}(0) = k! a_{kn}$. A special case of these considerations is when $a_{kn} = 0$ for all k and all $n \neq n_0$. Then $g = g_{n_0} : (-\rho, \rho) \to (\mathcal{D}_{n_0}^*)$. In particular for $n_0 = 0$, one has $g = g_0 : (-\rho, \rho) \to R$ is an ordinary analytic function. In general, the following theorem justifies the natural definition of the renormalization of such a g.

Theorem 1 *Suppose $g : (-\rho, \rho) \to (\mathcal{D}^*)$ is a power series:*

$$g(r) = \sum_{k=0}^{\infty} a_k r^k$$

in $(\mathcal{D}^*)$ with radius of convergence ρ. Let $(\mathcal{D}^*)_\rho = \{x \in (\mathcal{D}^*) \mid |x_0| < \rho\}$ Then for each $x \in (\mathcal{D}^*)_\rho$ the series:

$$(g)(x) = \sum_{k=0}^{\infty} a_k \hat{\otimes} x^{\hat{\otimes} k}$$

converges in $(\mathcal{D}^*)$. Indeed the element $(g)(x)$ to which the series converges has Fock space decomposition:

$$(g)(x) = g_0(x_0) + \sum_{n=1}^{\infty} [g_n(x_0) + \sum_{\ell=1}^{n} \sum_{p=1}^{\ell} g_{n-\ell}^{(p)}(x_0) \hat{\otimes} \mathcal{P}_p^\ell(x)] \tag{12}$$

Here: $\mathcal{P}_p^\ell(x) = \sum (m_1! \cdots m_s!)^{-1} x_{i_1}^{\hat{\otimes} t m_1} \hat{\otimes} \cdots \hat{\otimes} x_{i_s}^{\hat{\otimes} m_s}$, where the sum is over all partitions: $\ell = m_1 i_1 + \cdots + m_s i_s$ of ℓ into p part s $(0 < i_1 \cdots < i_s$ and $m_1 + \cdots + m_s = p)$.

Remarks: It is easy to see how the above renormalization techniques extend to functions of several variables. Thus if, for example, $g : R^p \to R^s$ is analytic, then its renormalization is a map $(g) : (\mathcal{D}^*)^p \to (\mathcal{D}^*)^s$. (Similarly for $g : R^p \to (\mathcal{D}^*)^s$). Two of the main functorial properties of the renormalization functor $(\cdot)$ are given in the following theorem:

Theorem 2 *Suppose $g : R^p \to R^s$ and $f : R^s \to R^q$ are real analytic. Then renormalization is functorial with respect to composition of functions:*

$$(f \circ g) = (f) \circ (g)$$

Furthermore if $x : R \to (\mathcal{D}^*)^p$ is differentiable, then $(g) \circ x : R \to (\mathcal{D}^*)^s$ is differentiable and the **renormalized Ito formula** *holds:*

$$\frac{d}{dt}[(g)(x(t))] = \sum_{j=1}^{p} (\frac{\partial g}{\partial u_j})(x(t)) \hat{\otimes} \dot{x}^j(t).$$

5 Exactly solvable generalized SDE's

We discuss here a number of examples of generalized SDE's which can be solved explicity. In each case the initial condition is $x(0) = c \in (\mathcal{D}^*)^m$. Some of the examples present quite general results.
Example 1: A general scalar linear DE has the form:

$$\dot{x} = a(t) \hat{\otimes} x + b(t) \tag{13}$$

With the appropriate integrability conditions on the coefficients $a(t), b(t) \in (\mathcal{D}^*)$, the solution of (13) is

$$x(t) = (\exp)(\alpha(t)) \hat{\otimes} c + \int_0^t (\exp)(\alpha(t) - \alpha(s)) \hat{\otimes} b(s)\, ds \tag{14}$$

where $\alpha(t) = \int_0^t a(s) ds$
Example 2: As particular instances of Example (1), consider the following DE's and their solutions. (The first two are discussed in the Kuo-Potthoff paper, except that here I have included a general coefficient $a \in (\mathcal{D}^*)$.

(2a) The equation: $\dot{x} = a \hat{\otimes} x \hat{\otimes} \dot{B}(t)$
has solution:

$$x(t) = c \hat{\otimes} (\exp)(a \hat{\otimes} B(t)) = \sum_{k=0}^{\infty} \frac{1}{k!} c \hat{\otimes} a^{\hat{\otimes} k} \hat{\otimes} B(t)^{\hat{\otimes} k}$$

(2b) The equation: $\dot{x} = 2a \hat{\otimes} x \hat{\otimes} B(t) \hat{\otimes} \varrho t B(t)$
has solution: $x(t) = c \hat{\otimes} (\exp)(a \hat{\otimes} B(t)^{\hat{\otimes} 2})$

(2c) The equation: $\dot{x} = a\hat{\otimes}x\hat{\otimes}\ddot{B}(t)$
has solution: $x(t) = c\hat{\otimes}(\exp)(a\hat{\otimes}[\dot{B}(t) - \dot{B}(0)])$

Example 3: (linear system) The general linear system of DE's in our setting has the form

$$\dot{x} = \mathcal{A}(t)\hat{\otimes}x + \beta(t)$$

with $\mathcal{A}(t)$ an $m \times m$ matrix with coefficients in $(\mathcal{D}^*)$ and $\beta(t) \in (\mathcal{D}^*)^m$. The program then is to use the fundamental matrix: $\dot{\mathcal{E}}(t) = \mathcal{A}(t)\hat{\otimes}\mathcal{E}(t)$ to construct the solution as in the ODE case. The details of this program involve techniques too lengthy to include here (and a few technicalities which are not yet resolved), so for the sake of illustration we discuss only the following well-known special case: $\mathcal{A}(t) = A(t)$, an $m \times m$ matrix with real entries, $\beta(t) = G(t)\dot{B}(t)$ with $G(t)$ an $m \times M$ real matrix, and $\dot{B}(t) = (\dot{B}^1(t), ..., \dot{B}^M(t))$. If we let $E(t) = \{E_{ij}(t)\}$ be the fundamental matrix for the ODE: $\dot{r} = A(t)r$, then the corresponding SDE is:

$$\dot{x} = A(t)x + G(t)\dot{B}(t),$$

or written out in terms of components (using implied summation on repeated indices) is

$$\dot{x}^i = A_{ij}(t)x^j + G_{ij}(t)\dot{B}^j(t),$$

and has for solution:

$$x^i(t) = E_{ij}(t)c^j + I_1(1_{[0,t)}E_{ik}(t)E_{kp}^{-1}G_{pj}\varepsilon^j),$$

with $c = (c^1, ..., c^m) \in (\mathcal{D}^*)$ the initial value. Here $I_1 : L_1^2 \to K_1$ is the Wiener integral and it is assumed that $E_{kp}^{-1}G_{pj}\varepsilon^j \in L_1^2$.

The rest of the examples involve non-linear GSDE's. The following theorem shows how to solve a large class of such DE's.

Theorem 3 *Suppose $g : R \times R^m \times R \to R^m$ is analytic, and $Y : R \to (\mathcal{D}^*)$ is differentiable with $Y(0) = 0$. Then for a fixed parameter $a \in (\mathcal{D}^*)$, the generalized SDE:*

$$\dot{x} = (g)(Y(t), x, a)\hat{\otimes}\dot{Y}(t),$$

with initial condition $x(0) = c$, has solution:

$$x(t) = (\phi)(Y(t), c, a).$$

Here $\phi(t, r, p)$ is the flow generated by the ODE: $\dot{r} = g(t, r, p)$.

Proof: The flow ϕ for $\dot{r} = g(t, r, p)$ satisfies:

$$\frac{\partial\phi}{\partial t}(t, r, p) = g(t, \phi(t, r, p), p),$$

with $\phi(0, r, p) = r$. Thus $\partial\phi/\partial t = g \circ H$, where $H(t, r, p) = (t, \phi(t, r, p), p)$. Hence using the properties of renormalization from Theorem (2), we have:

$$\begin{aligned}
\dot{x}(t) &= \frac{d}{dt}(\phi)(Y(t), c, a) \\
&= (\frac{\partial\phi}{\partial t})(Y(t), c, a)\hat{\otimes}\dot{Y}(t) \\
&= (g \circ H)(Y(t), c, a)\hat{\otimes}\dot{Y}(t) \\
&= [(g) \circ (H)(Y(t), c, a)]\hat{\otimes}\dot{Y}(t) \\
&= (g)(Y(t), x(t), a)\hat{\otimes}\dot{Y}(t).
\end{aligned}$$

Remark: In the special case when $Y(t) = t$, the theorem says that the flow for a renormalized DE is equal to the renormalization of the flow for the DE.

Example 4: By the theorem (with $Y(t) = t$), the solution of

$$\dot{x} = 2t\, a\hat{\otimes}x\hat{\otimes}x$$

is the renormalization of the flow: $\phi(t,r,p) = r(1-t^2pr)^{-1}$ for the ODE: $\dot{r} = 2tpr^2$. Thus:

$$x(t) = \sum_{k=0}^{\infty} a^{\hat{\otimes} k} \hat{\otimes} c^{\hat{\otimes} k+1} t^{2k}$$

Example 5: By the theorem (with $Y(t) = B(t)$) the solution of

$$\dot{x} = 2a\hat{\otimes}x\hat{\otimes}x\hat{\otimes}\dot{B}(t)$$

is

$$x(t) = \sum_{k=0}^{\infty} a^{\hat{\otimes} k} \hat{\otimes} c^{\hat{\otimes} k+1} \hat{\otimes} B(t)^{\hat{\otimes} 2k}.$$

Corollary 1 *Suppose $g : R \to R$ is analytic. Then the generalized SDE:*

$$\dot{x} = (g)(x)\hat{\otimes}\dot{B}(t),$$

with $x(0) = c$, has: $x(t) = (\phi)(B(t), c)$ as the **unique** *solution.*

Proof: To prove uniqueness just look at the Fock space structure of the DE: $\dot{x}_0 = 0$, and (for $n \geq 0$): $\dot{x}_{n+1} = \sum_{p=1}^{n} g^{(p)}(x_0)\mathcal{P}_p^n(x)\hat{\otimes}\dot{B}(t)$. Thus uniqueness follows from induction: x_{n+1} is uniquely determined by $x_0, ..., x_n$.

References

[BR 91a] Betounes, D. and Redfern, M.: Wiener distributions and white noise analysis, *Appl. Math. Optim.* (to appear).

[BR 91b] Betounes, D. and Redfern, M.: Stochastic integrals for non-previsible, multiparameter processes, (preprint).

[Kuo 91] Kuo, H. -H., Lectures on white noise analysis, *Proc. Preseminar Int. Conf. Gaussian random fields* (1991) T. Hida and K. Saito (eds.) 1-65.

[KP 90] Kuo, H. -H., and Potthoff, J., Anticipating stochastic integrals and stochastic differential equations, *White Noise Analysis - Mathematics and Applications*, T. Hida, H. -H. Kuo, J. Potthoff, L. Streit (eds.) (1990) 256-273, World Scientific.

[Oc 90] Ocone, D., Anticipating stochastic calculus and applications. *White Noise Analysis - Mathematics and Applications* T. Hida, H. -H. Kuo, J. Potthoff, L. Streit (eds.) (1990) 298-314, World Scientific.

[OP 89a] Ocone, D. and Pardoux E., Linear stochastic differential equations with boundary conditions, *Probab. Th. Rel. Fields* **82** (1989) 489-526.

[OP 89b] Ocone, D., and Pardoux E., A generalized Ito - Ventzell formula. Application to a class of anticipating stochastic differential equations, *Ann. Inst. H. Poincare Probab. Statist.* **25** (1989) 39-71.

[PP 90] Pardoux, E. and Protter P., Stochastic Volterra equations with anticipating coefficients, *Ann. Probab.* **18** (1990) 1635-1655.

On invariant measure for semilinear equations with dissipative nonlinearities

Giuseppe Da Prato
Scuola Normale Superiore di Pisa,
Classe di Scienze
Pisa Italy

Jerzy Zabczyk
Institute of mathematics,
Polish Academy of Sciences,
Warsaw, Poland

1 Setting of the problem

We are concerned with the following semi-linear equation:

$$dX(t) = [AX(t) + F(X(t))]dt + dW(t), \quad X(0) = \xi, \tag{1}$$

where $A : D(A) \subset H \to H$ is a linear operator and $F : D(F) \subset H \to H$ is a non linear mapping in the separable Hilbert space H. We shall denote by $|\cdot|$ the norm and by $<\cdot,\cdot>$ the scalar product in H. Moreover $W(\cdot)$ is an $H-$ valued cylindrical Wiener process, that is such that the covariance operator of $W(t)$ is equal to tI, in a probability space $(\Omega, \mathcal{F}, \mathbf{P})$, adapted to a given normal filtration $\{\mathcal{F}_t\}$ in $\mathcal{F}$ and that ξ is an $\mathcal{F}_0-$ measurable random variable. We assume:

$$\begin{cases} (i) \ A \ \text{generates a } C_0 \text{ class semigroup } S(\cdot) \text{ in } H. \\ (ii) \ \int_0^T \operatorname{Tr}[S(s)S^*(s)]ds < +\infty, \forall\, T > 0, \end{cases} \tag{2}$$

and we write equation (1) in the integral, or mild, form

$$X(t) = S(t)\xi + \int_0^t S(t-s)F(X(s))ds + W_A(t) \tag{3}$$

where

$$W_A(t) = \int_0^t S(t-s)dW(s). \tag{4}$$

It is well known that, under hypothesis (2) the *stochastic convolution* (4) is a well defined process in H.

We assume in addition

$$\begin{cases} \text{There exists a Banach space } E \text{ continuously and densely embedded} \\ \text{in } H \text{ such that } W_A(\cdot) \text{ is a } E-\text{continuous process,} \end{cases} \tag{5}$$

and moreover

$$\begin{cases} (i) \ E \text{ is an invariant subspace for } S(t), \forall\, t \geq 0,\ S(\cdot)x \text{ is measurable in} \\ E, \forall\, x \in E \text{ and there exists } M > 0, \omega \in \mathbf{R} \text{ such that} \\ \qquad \|S(t)\|_E \leq Me^{\omega t}, t \geq 0. \\ (ii) \ E \subset D(F),\ F \text{ is uniformly continuous and bounded on bounded} \\ \quad \text{sets of } E \text{ and it is dissipative on } E. \end{cases} \tag{6}$$

We shall denote by $\|\cdot\|$ the norm in E. We recall that a mapping $F : E \to E$ is dissipative on E if $\|x-y\| \leq \|x - y - \alpha(F(x) - F(y))\|, \forall\, \alpha > 0, \forall\, x, y \in E$.

We remark that hypothesis (5) is fulfilled in several situations, see for instance [4] and [3]. Morever in [3] the following existence and uniqueness result is proved.

Theorem 1.1 *Assume (2), (5) and (6). Then, if $\xi \in E$, $\mathbf{P}-a.s.$, equation (3) has a unique $E-$continuous solution, $\mathbf{P}\,a.s.$*

By Theorem 1.1 one can construct the Markov transition semigroup on E, and show the existence of a unique invariant maesure in E, see [3], under the additional hypothesis that $M = 1, \omega < 0$, that is when the mapping $A+F$ is strictly dissipative. However, in some application, see Example 3.2 below, $A+F$ is only dissipative in E but it is strictly dissipative in H. In this paper we generalize Theorem 1.1 by extending the Markov semigroup in H. By using this extension we are able to prove he existence of a unique invariant measure in H, under the hypothesis that $A+F$ is is strictly dissipative in H.

2 An existence and uniqueness result

If $x \in H$ and conditions of Theorem 1.1 are fulfilled then, we are not able to establish the existence of a mild solution of equation (1). However, under additional conditions, we can show, see Theorem 2.1 below that there exists a generalized solution in the following sense. A process X is a *generalized solution* of (1) if for arbitrary sequence $\{x_n\} \subset E$ such that $\lim_{n\to\infty} |x - x_n|_H = 0$, the corresponding sequence of solutions $\{X_n\}$ converges to X in $C([0,T];H)$, $\mathbf{P}-$a.s. A similar result was proved recently in [1] using different method.

Theorem 2.1 *Assume that hypotheses of Theorem 1.1 are satisfied and moreover that F is dissipative in H, that is $< F(x)-F(y), x-y > \leq 0, \forall\, x,y \in D(F)$, and there exists $\eta \in \mathbf{R}$ such that $|S(t)| \leq e^{-\eta t}$, $\forall\, t > 0$. Then, for arbitrary $x \in H$, there exists a generalized solution to equation* (1)

Proof Let $\{x_k\}$ be a sequence in E that converges to x in H. By Theorem 1.1, for any positive integer k there exists a unique solution X_k to the equation

$$X_k(t) = S(t)x_k + \int_0^t S(t-s)F(X_k(s))ds + W_A(t),\ t \geq 0.$$

Set $Z_{j,k} = X_j - X_k$, then $Z_{j,k}$ is the mild solution of the problem:

$$\begin{cases} \frac{d}{dt} Z_{j,k}(t) = AZ_{j,k}(t) + F(X_j(t)) - F(X_k(t)), \\ Z_{j,k}(0) = x_j - x_k. \end{cases} \tag{7}$$

For any $n > \omega$ denote by $Z_{j,k,n}$ the solution of the approximating problem:

$$\begin{cases} \frac{d}{dt} Z_{j,k,n}(t) = A_n Z_{j,k,n}(t) + F(X_j(t)) - F(X_k(t)), \\ Z_{j,k,n}(0) = x_j - x_k, \end{cases} \tag{8}$$

where $A_n = nA(n-A)^{-1}$ are the Yosida approximations of A. We have:

$$\begin{aligned} \tfrac{1}{2}\tfrac{d}{dt}|Z_{j,k,n}(t)|^2 &= < AZ_{j,k,n}(t), Z_{j,k,n}(t) > \\ &+ < F(X_j(t)) - F(X_k(t)), Z_{j,k,n}(t) > \\ &\leq -\eta |Z_{j,k,n}(t)|^2, t \geq 0. \end{aligned}$$

Consequently, letting n tend to infinity,

$$|Z_{j,k}(t)|^2 \leq e^{-\eta t}|x_j - x_k|^2, t \geq 0.$$

This way the existence of a generalized solution has been shown. Uniqueness follows from a standard argument based on dissipativity. ∎

3 Existence and uniqueness of an invariant measure

We assume here that hypotheses of Theorem 2.1 are satisfied. We denote by P_t the Markov semigroup in $C_b(H)$, ($C_b(H)$ is the set of all funtions from H into $\mathbf{R}$ which are uniformly continuous and bounded), corresponding to equation (1) :

$$P_t\varphi(x) = \mathbf{E}\varphi(X(t,x)), \varphi \in C_b(H),\ x \in H, \tag{9}$$

where $X(\cdot, x)$ denotes the generalized solution of (1), with $\eta \equiv x$. Let $M(H)$ be the space of all bounded measures on $(H, \mathcal{B}(H))$, and $M_1^+(H)$ the subset consisting of all probability measures. For any $\varphi \in B_b(H)$, ($B_b(H)$ is the space of all bounded and Borel mappings from H into $\mathbf{R}$), and any $\mu \in M(H)$, we set

$$< \varphi, \mu >= \int_H \varphi(x)\mu(dx).$$

and introduce the dual semigroup P^* acting on $M(H)$. We have obviously that

$$< \varphi, P_t^*\mu >=< P_t\varphi, \mu >, \forall\, \varphi \in B_b(H), \mu \in M(H).$$

A measure μ in $M_1^+(H)$ is said to be *invariant measure* for (1) if

$$P_t^*\mu = \mu, \forall\, t > 0.$$

The main result of this section is the following.

Theorem 3.1 *Assume that hypotheses of Theorem* 2.1 *hold with* $\eta > 0$ *and moreover that*

$$\int_0^{+\infty} \text{Tr}\ [S(s)S^*(s)]ds < +\infty. \tag{10}$$

and

$$|F(x)| \le C(1 + |x|^m) \tag{11}$$

for some $m > 0, C > 0$. *Then there exists a unique invariant measure* μ *for system* (1). *Moreover, for arbitrary* $\nu \in M_1^+(H)$, *we have*

$$P_t^*\nu \to \nu \text{ weakly as } t \uparrow +\infty. \tag{12}$$

It will be convenient to consider equation (1) in all real line. Therefore we define process $W(t)$ for $t < 0$ by choosing independent Wiener process $\widetilde{W}(\cdot)$ and $\widetilde{W}_1(\cdot)$ with the same law as $W(\cdot)$ and $W(\cdot)$ and setting

$$W(t) = \widetilde{W}(-t),\ \ t \le 0.$$

Now, for any $\lambda > 0$, denote by $X_\lambda(t,x), t \ge -\lambda$ the unique mild solution of the equation

$$\begin{cases} dX = (AX + F(X))dt + dW, \\ X(-\lambda) = x \in H. \end{cases} \tag{13}$$

It is easily seen that

$$\mathcal{L}(X_\lambda(0,x)) = \mathcal{L}(X(\lambda, x)),\ \lambda \ge 0. \tag{14}$$

It is therefore enough, to prove the theorem, to show that

$$\lim_{\lambda\to\infty} \mathcal{L}(X(\lambda, x)) = \mu, \text{ weakly for some } \mu \in M_1^+H \text{ and all } x \in H.$$

In fact, we will prove that there exists a random variable η such that

$$\lim_{\lambda\to\infty} \mathbf{E}|X_\lambda(0,x) - \eta|^2 = 0,\ \forall\, x \in H. \tag{15}$$

We prove that (15) is true for $x = 0$, (the proof of a general x is completely similar) and, to simplify notation, we put $X_\lambda(t) = X(\lambda, 0)$.

Therefore

$$X_\lambda(t) = \int_{-\lambda}^{t} S(t-s)F(X_\lambda(s))ds + W_{A,\lambda}(t),$$

where

$$W_{A,\lambda}(t) = \int_{-\lambda}^{t} S(t-s)dW(s), \; t \geq -\lambda.$$

We show now that there exists $c_1 > 0$ such that

$$\mathbf{E}(|X_\lambda(t)|^2) \leq c_1, \; \forall \lambda > 0, \; \forall t > -\lambda. \tag{16}$$

We first remark that $Z_\lambda(t) = X(\lambda, t) - W_{A,\lambda}(t), t \geq -\lambda$ is the mild solution of the problem

$$\begin{cases} \frac{d}{dt}Z = AZ + F(Z + W_{A,\lambda}) \\ Z(-\lambda) = 0, \; ,t \geq -\lambda \end{cases}$$

It follows, by computing $\frac{d^-}{dt}\|Z_\lambda(t)\|$, and recalling the hypotheses that

$$|Z_\lambda(t)| \leq c \int_{-\lambda}^{t} e^{-\omega(t-s)}(1 + |W_{A,\lambda}(s))|^m)ds, \; t > -\lambda. \tag{17}$$

Note that

$$\sup_{s \geq -\lambda \geq 0} \mathbf{E}\|W_{A,\lambda}(s))\|^m = \sup_{s \geq 0} \mathbf{E}\|W_A(s))\|^m.$$

Since the process $W_A(\cdot))$ is Gaussian, we arrive easily at (16). In a similar way we show that for a constant $c_2 > 0$ and all $\mu > \lambda > 0, t > -\lambda$

$$\mathbf{E}\|X_\lambda(t) - X_\mu(t)\| \leq c_2 e^{-\omega(t+\lambda)}.$$

Therefore there exists an E–valued random variable η such that

$$\lim_{\lambda \to +\infty} \mathbf{E}\|X_\lambda(0) - X_\lambda(0, x)\| = 0, \forall x \in E.$$

The law $\mathcal{L}(\eta)$ is the required invariant measure. ∎

Example 3.2

Consider the problem

$$\begin{cases} du(t,\xi) = [u_{\xi,\xi}(t,\xi) - |u(t,\xi)|^p u(t,\xi)]dt + dW(t,\xi), t \geq 0, \xi \in [0,\pi] \\ u(t,0) = u(t,\pi) = 0, \; t \geq 0 \\ u(0,\xi) = x(\xi), \xi \in [0,\pi] \end{cases} \tag{18}$$

where $p \geq 1$ and $W(\cdot)$ is a cylindrical Wiener process in $H = L^2(0,\pi)$. Let A be the linear operator

$$\begin{cases} D(A) = H^2(0,\pi) \cap H_0^1(0,\pi) \\ Au = u_{\xi,\xi} \end{cases} \tag{19}$$

Then A generates a C_0 semigroup $S(\cdot)$ in H and

$$|S(t)| \leq e^{-t}, \; t \geq 0,$$

So Theorem 3.1 applies with $E = C([0,\pi])$. Notice that the restriction of A to E is dissipative but not strictly dissipative, so that Theorem 2.2 in [2] does not apply. ∎

References

[1] BUCKDAHN R. & PARDOUX E. (1991) *Monotonicity methods for white noise driven quasi linear SPDE's* in Diffusion processes and related problems in Analysis, Vol.I, M.Pinsky, Progress in Probability, **22**, BIRKHÄUSER, 219-233.

[2] DA PRATO G. & ZABCZYK J. (1988) *A note on semilinear stochastic equations*, Differential and Integral Equations, 1 , 143-155.

[3] DA PRATO G. & ZABCZYK J. (to appear) *Non explosion, boundedness and ergodicity for stochastic semilinear equations*, J. Differential Equations.

[4] WALSH J.B. (1984) *An introduction to stochastic partial differential equations* École d'eté de Probabilité de Saint Flour XIV-1984, ed. P.L. Hennequin, LNiM 1180, 265-439.

Giuseppe Da Prato
Scuola Normale Superiore di Pisa, 56126 Pisa, Italy and
Jerzy Zabczyk
Institute of mathematics, Polish Academy of Sciences, Warsaw, Poland.

RANDOM CONSERVATION LAWS AND GLOBAL SOLUTIONS OF NONLINEAR SPDE APPLICATION TO THE HJB SPDE OF ANTICIPATIVE CONTROL*

M.H.A.Davis,G.Burstein
Department of Electrical Engineering,Imperial College
London SW7

Abstract. The stochastic characteristics method provides a unique local solution (up to a stopping time) for first order Stratonovich nonlinear SPDE.We obtain conditions for the existence of a unique global solution based on random conservation laws for the stochastic characteristics system of SDE. This ensures that the first stochastic characteristic generates a global flow of diffeomorphism and not a local one like in the general case.We apply this to represent globally the solution of our Hamilton Jacobi Bellman SPDE of anticipative control and to represent the cost of information on the future.An illustration of the Poisson bracket condition obtained is given for the case of anticipative LQG.

1. Stochastic characteristics method for nonlinear SPDE : local solutions

Consider the first order nonlinear SPDE in Stratonovich form with initial condition

$$dv(t,x)=F(t,x,v_x)dt+G(t,x,v_x)\circ dw(t,\omega) \qquad (1)$$
$$v(0,x)=c(x)$$

where $w(t,\omega)$ is a n-dimensional Brownian motion,G is a vector function of components G_j,v_x denotes the gradient of v and $x\in\mathbb{R}^n$,$t\in[0,T]$. For (1) Kunita [9,Ch.6], developed a stochastic characteristics method for representing solutions.We introduce next this method that will be used throughout the paper. A random field $v(t,x,\omega)$ defined for all $t\in[0,T]$,$x\in\mathbb{R}^n$ is called global $C^{m,\alpha}$-process if for almost all $\omega\in\Omega$,$v(t,\cdot,\omega)$ is $C^{m,\alpha}$ in x for all $t\in[0,T]$(i.e. m-times continously differentiable in x with continous in (t,x) partial derivatives $\frac{\partial^k v(t,x,\omega)}{\partial x^k}$,$|k|\leq m$ where $\frac{\partial^k}{\partial x^k}=(\partial/\partial x_1)^{k_1}\cdots(\partial/\partial x_n)^{k_n}$,$|k|=k_1+\cdots+k_n$;the m-th partial derivatives being α-Holder continous in x for all $t\in[0,T]$).A global $C^{m,\alpha}$-process is a global $C^{m,\alpha}$-semimartingale if $\frac{\partial^k v(\cdot,x,\omega)}{\partial x^k}$,$|k|\leq m$ are semimartingales for each $x\in\mathbb{R}^n$.A random field $v(t,x,\omega)$ is a global $C^{m,\alpha}$- solution of (1) if it is a global $C^{m,\alpha}$-semimartingale and it satisfies for all $(t,x)\in[0,T]\times\mathbb{R}^n$

$$v(t,x)=c(x)+\int_0^t F(\tau,x,v_x(\tau,x))d\tau+\int_0^t G(\tau,x,v_x(\tau,x))\circ dw_\tau \quad \text{a.s.}$$

Assume now that $F(t,x,p)$ is continous in (t,x,p) and $C^{m+1,\alpha}$ in (x,p) ;$G_j(t,x,p)$,j=1,...,n are continous in (t,x,p) and $C^{m+2,\alpha}$ in (x,p) for $m\geq 3$; c(x) is $C^{m+1,\alpha}$.F_x,F_p will denote the gradients

* The research reported in this paper has been sponsored in part by U.S.Army E.R.O. contract DAJA 45-90-C-0024

with respect to the coressponding set of variables.Then in general (1) does not admit a unique global solution but only a unique local $C^{m,\beta}$ solution (for some $\beta>0$) v(t,x) defined for each x up to a stoping time $T(x,\omega)$ (local solutions are defined as before for $t\leq T(x,\omega)\leq T$ [9,Ch.6]) by the stochastic characteristics formula

$$v(t,x)=r_t\circ q_t^{-1}(x),\quad t\leq T(x,\omega)\leq T \tag{2}$$

with $q_t(x)=q_t(x,c_x(x))$, $r_t(x)=r_t(x,c(x),c_x(x))$, $s_t(x)=s_t(x,c_x(x))$ the flows of stochastic characteristics (we omit x from $q_t(x)$ for simplicity and we use summation convention)

$$dq_t=-F_p(t,q_t,s_t)dt-G_{j,p}(t,q_t,s_t)\circ dw^j,\ q_0=x \tag{3}$$

$$dr_t=(F(t,q_t,s_t)-F_p(t,q_t,s_t)s_t)dt+(G_j(t,q_t,s_t)-G_{j,p}(t,q_t,s_t)s_t)\circ dw^j,\ r_0=c(x) \tag{4}$$

$$ds_t=F_x(t,q_t,s_t)dt+G_{j,x}(t,q_t,s_t)\circ dw^j,\ s_0=c_x(x) \tag{5}$$

The reason for the solution (2) being only locally defined is that $q_t(x)$ is only a local flow of diffeomorphisms a.s. thus $q_t^{-1}(x)$,the inverse flow,exists only up to a stopping time .This is because of the coupling between (3) and (5).Let us explain this in detail.Denote $\partial q:=\partial q_t(x)/\partial x$, $\partial s:=\partial s_t(x)/\partial x$.Then

$$d\ \partial q=-(F_{px}\partial q+F_{pp}\partial s)dt-(G_{j,px}\partial q+G_{j,pp}\partial s)\circ dw^j,\ \partial q_0=I_n$$

$$d\ \partial s=(F_{xx}\partial q+F_{xp}\partial s)dt+(G_{j,xx}\partial q+G_{j,xp}\partial s)\circ dw^j,\ \partial s_0=c_{xx}(x)$$

Using the matrix adjoint system

$$dK_t=(K_tF_{px}+L_tF_{pp})dt+(K_tG_{j,px}+L_tG_{j,pp})\circ dw^j,\ K_0=I_n$$

$$dL_t=-(K_tF_{xx}+L_tF_{xp})dt-(K_tG_{j,xx}+L_tG_{j,xp})\circ dw^j,\ L_0=O_n$$

(I_n,O_n are the identity and zero nxn matrices) it can be seen like in [9,p.296] by applying Ito rule that

$$(K_t,L_t)\times(\partial q,\partial s)^T=I_n$$

which yields rank$(\partial q,\partial s)=n$ for all t a.s. but nothing can be said about the rank of ∂q.If however the stochastic characteristic system is decoupled (the equation for $q_t(x)$ does not depend on $s_t(x)$) then $q_t(x)$ is a global flow of diffeomorphisms a.s. and (2) is a global solution as $q_t^{-1}(x)$ exists for all $t\in[0,T]$ a.s.

2. Global solutions under Poisson bracket conditions for random conservation laws

The existence of a random conservation law $s_t(x)=b(t,q_t(x),\omega)$ for all $(t,x)\in[0,T]\times\mathbf{R}^n$ a.s. will decouple (3) from (5) yielding

$$dq_t=-F_p(t,q_t,b(t,q_t,\omega))dt-G_{j,p}(t,q_t,b(t,q_t,\omega))\circ dw^j \tag{6}$$

(6) is an SDE with random coefficients which has a nonexploding solution $q_t(x)$ for $q_0=x$ generating a global flow of $C^{m,\alpha}$ diffeomorphisms a.s. provided (see Kunita [9,p.108,173]) the following condition

(C) holds almost surely

$$\sup_{y\in\mathbf{R}^n}(|F_p(t,y,b(t,y,\omega))|/(1+|y|))+\sum_{1\le|k|\le m}\sup_{y\in\mathbf{R}^n}|\partial^k F_p(t,y,b(t,y,\omega))/\partial y^k|$$

$$+\sum_{|k|=m}\sup_{x,z\in\mathbf{R}^n,x\neq z}(|\partial^k F_p(t,x,b(t,x,\omega))/\partial x^k-\partial^k F_p(t,z,b(t,z,\omega))/\partial z^k|/|x-z|^\alpha)\in L^1_{loc}(dt)$$

and the same condition holds for each of G_j.Then $q_t^{-1}(x)$ exists for all $t\in[0,T]$ and is given by

$$dq_t^{-1}(x)=(\partial q_t/\partial x)^{-1}(q_t^{-1}(x))[F_p(t,x,b(t,x,\omega))dt+G_{j,p}(t,x,b(t,x,\omega))\circ dw^j \qquad (7)$$

An example of conservation law that will ensure global solution is $s_t(x)=c_X(q_t(x))$ in which case the Lagrangian submanifold $L=\{(q,s)\in\mathbf{R}^{2n}|\ s=c_X(q)\}$ is invariant for the stochastic characteristic system (3),(5).Note that (3),(5) is initialized on L as $s_0(x)=c_X(q_0(x))$. Non-random conservation laws for stochastic Hamiltonian systems were introduced by Bismut [2] .We can state now our main result.Define C_b^m as the class of continous functions of $(t,x,p)\in[0,T]\times\mathbf{R}^n\times\mathbf{R}^n$ with continous in(t,x,p) bounded derivatives up to order m with respect to (x,p) for each t.

<u>Theorem 1</u> Let F be C_b^{m+2} ,G_j be C_b^{m+3} and c have continous bounded derivatives up to order m+1,$m\ge3$. Assume there exists an adapted random field $b(t,x,\omega)$ with differential given by

$$db(t,x,\omega)=b^1(t,x,\omega)dt+b_j^2(t,x,\omega)\circ dw^j\ ,\ b(0,x,\omega)=c_X(x)$$

satisfying

(i) $b(t,\cdot,\omega)$ is C^{m+1} for $t\in[0,T]$ a.s. and $|\partial^k b(t,x,\omega)/\partial x^k|\le r(t,\omega)\in L^1_{loc}(dt)$ a.s. $|k|=m+1$

(ii) $b_i^1(t,x,\omega)+\{F(t,x,p),p_i-b_i(t,x,\omega)\}|_{L_t(\omega)}=0$

(iii) $b_{ij}^2(t,x,\omega)+\{G_j(t,x,p),p_i-b_i(t,x,\omega)\}|_{L_t(\omega)}=0$

for i,j=1,...,n and $(x,p)\in L_t(\omega)=\{(x,p)\in\mathbf{R}^{2d}|p=b(t,x,\omega)\}$ and

$$\{g(t,x,p),h(t,x,p)\}=(\partial h/\partial p_i)(\partial g/\partial x_i)-(\partial h/\partial x_i)(\partial g/\partial p_i)$$

is the Poisson bracket for each t in (x,p) of g,h.
Then there exists a unique global $C^{m,\beta}$ ($\beta>0$) solution of the SPDE (1) given by $v(t,x,\omega)=r_t^*\circ q_t^{*,-1}(x)$ where $r_t^*(x),q_t^*(x)$ are the flows of(3) for $s_t(x)=b(t,q_t(x),\omega)$ and respectively of (6) with $q_t^{*,-1}(x)$ given by (7).
<u>Proof</u> We have to show that the unique nonexploding solution of (3),(5) is $(q_t^*(x),b(t,q_t^*(x),\omega))$. Then under our assumptions condition (C) is satisfied and $q_t^*(x)$ is a global flow of diffeomorphisms so the stochastic characteristics representation of the solution will be global.Indeed we show that $b(t,q_t^*,\omega)$ satisfies (5) by Ito rule

$$db_j(t,q_t^*,\omega)=-F_{x_j}(t,q_t^*,b(t,q_t^*,\omega))dt-G_{l,x_j}(t,q_t^*,b(t,q_t^*,\omega))\circ dw^l$$

$$=b_j^1(t,q_t^*,\omega)dt+b_{jl}^2(t,q_t^*,\omega)\circ dw^l+b_{j,x_i}(t,q_t^*,\omega)[F_{p_i}(t,q_t^*,b(t,q_t^*,\omega))dt+G_{l,p_i}(t,q_t^*,b(t,q_t^*,\omega))\circ dw^l]$$

The above equality holds as it amounts after groupping terms to

$$[b_j^1(t,x,\omega)+\{F(t,x,p),p_j-b_j(t,x,\omega)\}]|_{x=q_t^*,p=b(t,q_t^*,\omega)}\times dt$$

$$+\ [b_{jl}^2(t,x,\omega)+\{G_l(t,x,p),p_j-b_j(t,x,\omega)\}]|_{x=q_t^*,p=b(t,q_t^*,\omega)}\circ dw^l=0$$

and this is true due to (ii),(iii) above as $(q_t^*,b(t,q_t^*,\omega))\in L_t(\omega)$ a.s.

3. Application: global solution of Hamilton Jacobi Bellman SPDE of anticipative control

As an application of our result to a particular SPDE where it is necessary to obtain a global solution we consider the anticipative optimal control problem (P^0) introduced in [5]

$$dx_t=f_0(x_t,u_t)dt+g(x_t)\circ dw_t,\ x_0=x^*\in\mathbb{R}^n,\ t\in[0,T] \tag{8}$$
$$\inf_{u\in\mathcal{A}} E[\theta(x_T)] \tag{9}$$

where $<w_t,\{\mathcal{F}_t\},\Omega,\mathcal{F},P>$ is an n-dimensional Brownian motion , $t\in[0,T]$, $g(x)$ is an $n\times n$ matrix of functions with columns $g_1(x),\dots,g_n(x)$ and $f_0(x,u)=f(x,u)-(1/2)g_{i,x}g_i(x)$ (summation convention is used throughout and as before $g_{i,x}$ denotes the Jacobian of g_i). $\mathcal{A}$ denotes the class of admissible $\mathcal{B}[0,T]\times\mathcal{F}$ measurable (possibly anticipative) controls to be defined exactly later.(8) is defined by means of the generalized Stratonovich integral of anticipative calculus [10] as the drift is anticipating.We introduce next Sobolev spaces over Wiener space and other elements of anticipative calculus required here after [10] where an important existence and uniqueness theorem is proved for SDE with anticipating drift.

Definition 1 The generalized Stratonovich integral is defined to be the limit in probability for any $t\in[0,T]$ of the sequence

$$\sum_{i=1}^{n}\sum_{l=0}^{2^m-1}2^m(w_{i,(l+1)2^{-m}}-w_{i,l2^{-m}})\int_{l2^{-m}\wedge t}^{(l+1)2^{-m}\wedge t}h_{i,s}ds$$

for any n -dimensional process $h_s(\omega)$ for which the limit exists($h_{i,s}$ and $w_{i,s}$ are the i-th components of the respective processes) .The process h_s is then called Stratonovich integrable and the limit of the above sequence is denoted

$$\int_0^t h_s\circ dw_s$$

$\mathcal{S}$ is defined next to be the dense subset of $L^2(\Omega,\mathcal{F},P)$ consisting of random variables

$F=j(I_{i_1}(m_1),...,I_{i_n}(m_n))$; $j\in C_b^\infty(\mathbb{R}^n)$; $m_i(t)\in L^2(\mathbb{R}_+)$; $i_1,...,i_n\in\{1,...,n\}$; $I_{i_l}(m_l)=\int_0^\infty m_l(t)dw_{i_l,t}$. $\mathbb{D}^{i,1,p}$ denotes the closure of $\mathcal{S}$ with respect to the norm $\| F \|_{i,1,p}=\| F \|_p+\| \| D^i F \|_{L^2(\mathbb{R}_+)}\|_p$, $p\geq 2$ where the Wiener derivative w.r.t. w_i at time t is $D_t^i F=\sum_{\{l;i_l=i\}} j_{x_l}(I_{i_1}(m_1),...,I_{i_n}(m_n))m_l(t)$ and $\mathbb{D}^{1,p}=\bigcap_{i=1}^n \mathbb{D}^{i,1,p}$, $\mathbb{L}^{1,p}=L^p_{loc}(\mathbb{R}_+,dt;\mathbb{D}^{1,p})$.We will use D_tF as notation for the vector of processes D_t^i.$\mathbb{L}_C^{1,p}$ is the set of processes h_s in $\mathbb{L}^{1,p}$such that for any T>0 the set of functions $\{s\rightarrow D_t^i h_s$, $s\in[0,T]-\{t\}\}_{t\in[0,T]}$ is equicontinous for all i=1,...,n with values in $L^p(\Omega)$ and $\operatorname{ess\,sup}_{(s,t)\in[0,T]^2} E(|D_s^i h_t|^p)<\infty$,for any T>0,i=1,...,n. The localization of $\mathbb{L}_C^1=\bigcap_{p\geq 2}\mathbb{L}_C^{1,p}$ denoted $\mathbb{L}_C^{1,loc}$ is the set of measurable processes h such that for each T>0 there exist a sequence $\{r_n\}_n$ in $\bigcap_{p\geq 2}\mathbb{D}^{1,p}$ such that $\{r_n=1\}\uparrow\Omega$ a.s., $1_{[0,t]}r_n h\in\bigcap_{p\geq 2}\mathbb{L}_C^{1,p}$ and $r_n D\cdot h\cdot\in\bigcap_{p\geq 2} L^p(\Omega;\, L^2([0,T]^2))$ for every n (r_n depends on T).$\mathbb{L}^{1,p,loc}$,the localization of $\mathbb{L}^{1,p}$ is defined similarly.Also all these spaces can be defined in the same way for $L^2(\mathbb{R}^n;\exp(-|x|^2)dx)$- valued stochastic processes in which case for example the notation $\mathbb{L}^{1,p}(L^2(\mathbb{R}^n;\exp(-|x|^2)dx)$ will be used.It will be assumed in this section (A) f is bounded and C_b^{m+2},g is bounded and C_b^{m+3}, θ is C_b^{m+1}($m\geq 3$).We define $\mathcal{A}$ to be the class of $\mathcal{B}[0,T]\times\mathcal{F}$-measurable control functions with values in the compact $\mathcal{U}$ of $\mathbb{R}^m$ such that under the previous assumptions on f,g the anticipative controlled SDE (8) has a unique solution $x_t\in\mathbb{L}_C^{1,loc}$.Applying the existence and uniqueness theorem proved by Ocone and Pardoux in [10] for uncontrolled SDE to our case we get that this class contains in particular open loop measurable (possibly anticipating) controls $u(t,\omega)$ such that $|D_s^i u(t,\omega)|\leq M$ and feedback anticipative controls $u(t,x,\omega)$,C^2 in x (t,ω) a.e.,with Wiener derivatives of u and u_x continous in x a.s. almost everywhere in time such that there exists $p\in\mathbb{N}$ and $K_{T,p}$ a positive constant such that

$$|D_s u(t,x,\omega)|+|D_s u_x(t,x,\omega)|+|u_x(t,x,\omega)|+|u_{xx}(t,x,\omega)|\leq K_{T,p}(1+|x|^p)\ ,\ (s,t,\omega,x)\in[0,T]^2\times\Omega\times\mathbb{R}^n \tag{10}$$

As proved in [10] this unique solution is given by the decomposition formula $x_t=z_t\circ y_t$ where

$$dz_t(x)=g(z_t(x))\circ dw_t, z_0(x)=x \tag{11}$$
$$dy_t/dt=(\partial z_t/\partial x)^{-1}(y_t)f_0(z_t\circ y_t)\ , y_0=x^* \tag{12}$$

The motivation to consider the stochastic optimal control without nonanticipativity requirement (P) comes from our previous works [5,6] where nonanticipative optimal control is reduced to a family of deterministic optimal control problems parametrized by $\omega\in\Omega$ with nonanticipativity introduced as an equality constraint by an extra integral cost term using Lagrange multiplier processes following ideas of Wets from stochastic programming [11].One can then consider this as a method of solving anticipative optimal control problems with controls allowed to anticipate.In [5] the solution of anticipative controlled SDE was taken by definition to be the decomposition formula whereas here an anticipative calculus approach is used ensuring such a solution exists uniquely in $\mathbb{L}_C^{1,loc}$.This will eliminate lengthy convergence arguments from the proof of the HJB SPDE replacing them by applications of the anticipative Ito Ventzell formula of [10] and will allow more general (random) conservation law conditions for global solutions of HJB SPDE of anticipative control which we will obtain. We would like to solve (8),(9) by solving first the family of problems $\{P^\omega\}_{\omega\in\Omega}$ over measurable controls in t(denote this class by $\mathcal{M}$) for each ω ,$u_\omega(.) \in\mathcal{M}$

$$\inf_{u_\omega(.)\in\mathcal{M}} [\theta(z_T(y_T)] \tag{13}$$

subject to (12) as controls only appear there with $z_t(.)$ being as before the flow of (11) .This way provided the infimum in (13) will be actually attained and it will be in $\mathcal{A}$(i.e.,measurable in ω with the required Wiener derivative properties), which will be shown to be the case here,we will have after averaging

$$\inf_{u\in\mathcal{A}} E[\theta(x_T)]=E[\inf_{u_\omega(.)\in\mathcal{M}} (\theta(z_T(y_T))]$$

Problems (12),(13) denoted (P^ω) are characterized by the value function

$$W(t,y)= \inf_{u_\omega(.)\in\mathcal{M}} \theta(z_T(y(T;t,y))$$

satisfying the dynamic programming equation (random HJB PDE):

$$W_t(t,y)+\min_{u\in\mathcal{U}}\{W_y(t,y)(\partial z_t/\partial x)^{-1}(y)f_o(z_t(y),u)\}=0;\ W(T,y)=\theta(z_T(y)) \tag{14}$$

W_t and W_y denote the respective gradients.Using the selection lemma of [1] for $\min_{u\in\mathcal{U}}\{af_o(b,u)\}$ it is possible to get an optimal selector function $\phi(a,b)$ such that the minimum above under the compacity assumption on $\mathcal{U}$ equals $f_o(b,\phi(a,b))$.(14) becomes

$$W_t+W_y(\partial z_t/\partial x)^{-1}(y)f_o(z_t(y),\phi(W_y(\partial z_t/\partial x)^{-1}(y),z_t(y)))=0,W(T,y)=\theta(z_T(y)) \tag{14'}$$

As the characteristics method will be used to get a global (global will mean in this section for all $t\in[0,T]$) unique $C^{1,2}$ solutions of the nonlinear PDE (14') as well as for the HJB SPDE to be obtained later as the SPDE for which (14') is the robust equation, it will be assumed that (B) ϕ is C_b^4 in both variables.Our next result characterizes anticipative control in terms of a HJB SPDE and gives existence and a representation of the value function and the optimal control.We make first some assumptions on the random conservation law $b(t,x,\omega)$ and its Stratonovich differential that will appear in the next theorem

(a) $b(t,x,\omega)$ is C^m in x for all t a.s. ,continous w.r.t. t for all x a.s. such that first and second order derivatives are bounded by constants a.s., the rest of derivatives up to order m are bounded by backward adapted processes in $L^1(dt)$ a.s. and for all $\epsilon>0$ $\exists$,$K_{t,\epsilon}(\omega)$ positive backward adapted in $\bigcap_{p\geq 1} L^p([0,T]\times\Omega)$ s.t. $|b(t,x,\omega)|\leq K_{t,\epsilon}(\omega)(1+|x|^{1-\epsilon})$

(b) Wiener derivatives $D_s b(t,x,\omega)$,$D_s b_x(t,x,\omega)$ are continous in x (s,t,ω) a.e. , have versions continous in t uniformly in s for $(s,t) \in[0,T]^2, s\neq t$ and $\exists$,p and C>0 s.t. $| D_s b(t,x,\omega)|+| D_s b_x(t,x,\omega)| \leq C(1+|x|^p)$ $\forall,(s,t,\omega,x)\in[0,T]^2\times\Omega\times\mathbb{R}^n$

(c) $b^1(t,x,\omega),b^2(t,x,\omega)$ are C^m in x $\forall$,t a.s. ,continous in t $\forall$,x a.s. and together with their first derivatives satisfy $| B(t,x,\omega) |\leq l_t(\omega)(1+|x|^q)$,$l_t(\omega)$ in $\bigcap_{p\geq 1} L^p([0,T]\times\Omega)$ and positive backward adapted with B replaced by b^1,b^2 and their derivatives

(d) (b) is satisfied by $b^1(t,x,\omega),b^2(t,x,\omega)$ with C replaced by a positive backward adapted process in $\bigcap_{p\geq 1} L^p([0,T]\times\Omega)$

For evaluating the cost of information on the future allowing us to solve the anticipative problem it will be assumed that the usual nonanticipative stochastic optimal control problem

$$\inf_{u\in\mathcal{N}} E[\theta(x(T))] \tag{15}$$

$$dx_t=f(x_t,u_t)dt + g(x_t)dw_t \text{ , } \mathcal{N}\text{- the class of } \mathcal{F}_t\text{-adapted controls} \tag{16}$$

has a feedback solution $u^*(t,x)$ for which the value function solves in the $C^{1,2}$ and bounded functions class the HJB PDE

$$\partial V^*/\partial t+\min_{u\in\mathcal{U}}\{V^*_x f(x,u)\}+(1/2)tr((\partial^2 V^*/\partial x^2)gg^T)=0 \text{ , } V^*(T,x)=\theta(x) \tag{17}$$

and the optimal trajectory of (16) starting at t=s from x will be denoted $x^*_s(x)$.

<u>Theorem 2</u> Under our previous assumptions on f,g,θ and ϕ (see (A),(B)) assume also that there exists a backward random field (adapted to the future increments of w) $b(t,x,\omega)$ with backward Stratonovich differential $db(t,x,\omega)=b^1(t,x,\omega)dt+b^2_l(t,x,\omega)\hat{\circ}dw^l_t, b(T,x,\omega)=\theta_x(x)$ ($\hat{\circ}$ is the usual backward Stratonovich differential [8]) satisfying (ii),(iii) of theorem 1 for $F(t,x,p)=F(x,p)=pf_o(x,\phi(p,x))$, $G_j(t,x,p)=G_j(x,p)=p_j g_{jl}(x)$ and (a)-(d) above.Then $V(t,x,\omega)=\inf_{u\in\mathcal{A}} E[\theta(x(T;t,x))]$,the value function of the anticipative optimal control problem (8),(9),is given by $V(t,x,\omega)=EV^o(t,x,\omega)$ where V^o is the unique global $C^{m,\alpha}$ solution (for some α>0) of the following backward HJB SPDE

$$dV^o+\min_{u\in\mathcal{U}}\{V^o_x f_o(x,u)\}dt +V^o_x g(x)\hat{\circ}dw_t=0 \text{ ,} V^o(T,x)=\theta(x) \tag{18}$$

There exists an optimal anticipative control $u^o(t,x,\omega)=\phi(b(t,x,\omega),x)$ which is admissible and the optimal trajectory x^o_t is the unique a.s. continous solution in $\mathbb{L}^{1,loc}_C$ of the anticipative SDE

$$dx^o_t=f_o(x^o_t,\phi(b(t,x^o_t,\omega),x^o_t))dt+g(x^o_t)\circ dw_t \text{ ,} x^o_0=x^* \tag{19}$$

The cost of information on the future is

$$\Delta(t,x)=V^*(t,x)\text{-}EV^o(t,x)=\int_t^T E[b(s,x^*_s(x),\omega)(f(x^*_s(x),u^*(s,x^*_s(x)))$$

$$-f(x^*_s(x),\phi(b(s,x^*_s(x),\omega),x^*_s(x))]ds \tag{20}$$

<u>Proof</u> Consider first the pathwise problems (P^ω) and (14') their random HJB PDE expressed with the optimal selector ϕ .The characteristic system is

$$dQ_t/dt=F_S(z_t(Q_t),S_t(\partial z_t)^{-1}(Q_t)) \ , \ Q_T=y \tag{21}$$

$$dR_t/dt=-(F(z_t(Q_t),S_t(\partial z_t)^{-1}(Q_t))-F_S(z_t(Q_t),S_t(\partial z_t)^{-1}(Q_t))S_t), \ R_T=\theta\circ z_T(y) \tag{22}$$

$$dS_t/dt=-F_Q(z_t(Q_t),S_t(\partial z_t)^{-1}(Q_t)) \ ,S_T=\theta_x(z_T(y))\partial z_T(y) \tag{23}$$

where we used the abreviated notation for the Jacobian of the flow $z_t(y)$,∂z_t , we omitted y from the flows notation $Q_t(y)$ and where $F(x,p)$ is defined in the statement of theorem 2 .Vectorial-matrix notation was used with $F_Q=F_x(\partial z_t)+F_p(\partial(\partial z_t)^{-1})^T S_t, F_S=F_p(\partial z_t)^{-1}$. $W(t,y)$ is given only locally by the characteristics formula $W(t,y)=R_t\circ Q_t^{-1}(y)$ but under the assumptions (ii),(iii) ,(a)-(d) it follows that (21),(22) admits the unique nonexploding solution $(Q_t^*,S_t^*=b(t,z_t(Q_t^*),\omega)(\partial z_t)(Q_t^*))$ due to the existence of a random conservation law. This is proved by applying Ito Ventzell anticipative formula of [10] to S_t^* and $b(t,z_t,\omega)$ as for example z_t is past adapted and b is future (backwards) adapted . We can understand now why it was necessary to make the Wiener derivative and growth assumptions for the random conservation law . The sublinear growth of (a) is essential for nonexplosion together with the estimates of [10] $|(\partial z_t)^{-1}(y)|\leq k(\delta,\omega)(1+|x|^2)^\delta$ while the continuity and polynomial growth on b,b^1,b^2, their derivatives w.r.t. x and the Wiener derivatives of these are needed for the anticipative Ito rule conditions .We get for example as required $b^2\in\mathbb{L}^{1,4}(L^2(\mathbb{R}^n;\ \exp(-|x|^2)dx))$ etc.By the verification theorem [,p.87] it is shown now that $U^o(t,y,\omega)=\phi(W_y(\partial z_t)^{-1}(y),y)$ is the optimal control solution of (P^ω) . The next step is to show that

$$dV^o+V^o_x f_o(x,\phi(V^o_x,x))dt+V^o_x g(x)\hat{\circ}dw_t=0 \ , \ V^o(T,x)=\theta(x) \tag{24}$$

has a global $C^{m,\alpha}$ solution . Let us remark that although (24) does not satisfy the assumptions of theorem 1 still the existence of a random conservation law decouples the first characteristic from the third one and we have

$$dq_t=F_p(q_t,b(t,q_t,\omega))dt+g(q_t)\hat{\circ}dw_t \ ,q_T=x \tag{25}$$

$$dr_t/dt=-(F(q_t,b(t,q_t\omega))-F_p(q_t,b(t,q_t,\omega))b(t,q_t,\omega)) \ , \ r_T=\theta(x)$$

and under our assumptions on f,g,b the flow of (25) is a global flow of $C^{m,\alpha}$ diffeomorphisms as condition (C) of section 2 is satisfied and $V^o(t,x,\omega)=r_t\circ q_t^{-1}(x)$ is the global solution.This was also pointed out by Kunita for semi-linear SPDE [10,p.294] where although the coefficients have non-Holderian derivatives due to semi-linearity still the first characteristic is independent of the third and as the nonlinear part of coefficients belong to the appropriate Holder spaces one gets the required global diffeomorphism property for the flow.To prove that the value function $V^o(t,x,\omega)=\inf_{u\in\mathcal{A}}\theta(x(T;t,x))$ satisfies (24) we have to show that (14') is the robust equation for (24) and we have $W(t,z_t^{-1}(x))=V^o(t,x,\omega)$ .We apply again Ito Ventzell anticipative formula of [10] to $W(t,y)$ (with the differential given by (14') with $W_y(t,y)=b(t,z_t(y),\omega)(\partial z_t)(y)$ which holds because of the characteristics formula for gradients of solutions and the existence of a conservation law) and $z_t^{-1}(x)$

(with $dz_t^{-1}(x)=-(\partial z_t)^{-1}(z_t^{-1}(x))g(x)\circ dw_t$) and $D_s^i z_t^{-1}(x)=-[\partial z_s(z_t^{-1}(x))]^{-1}g_i(z_s z_t^{-1}(x))$ having the polynomial growth proved in [10]).We get using the generalized Stratonovich integral of definition 1 :

$$dW(t,z_t^{-1}(x))=-[W_y(\partial z_t)^{-1}(y)f_0(z_t(y),\phi(W_y(\partial z_t)^{-1}(y),z_t(y)))]|_{y=z_t^{-1}(x)}\times dt$$

$$-W_y(\partial z_t)^{-1}(z_t^{-1}(x))g(x)\circ dw_t$$

Using the uniqueness of the solution of (24) it turns out that as $W(t,z_t^{-1}(x))$ solves that equation the value function $V^o(t,x,\omega)$ is indeed the unique global $C^{m.\alpha}$ solution of (24) interpreted in the backward Stratonovich sense (the generalized Stratonovich integral coincides with the backward one for the present case).The optimal control is feedback anticipative $u^o(t,x,\omega)=\phi(b(t,x,\omega),x)$ and due to (a)-(d) it satisfies the conditions of the existence and uniqueness of [10] yielding a unique optimal anticipative solution in $\mathbb{L}_C^{1,loc}$.The cost difference formula is obtained as in our paper [5] by averaging (24) ,interchanging expectation with integration and derivation (for this [9 ,p.308] is applied using $V_x^o=b$ and the growth from (a)-(d)) and subtracting it from (17) to get

$$\partial\Delta/\partial t+\Delta_x f(x,u^*(t,x))+(1/2)tr(\Delta_{xx}gg^T(x))+E[V_x^o(f(x,u^*(t,x))-f(x,\phi(V_x^o,x)))]=0,\ \Delta(T,x)=\theta(x)$$

for which a probabilistic representation of the solution under our assumptions yields the formula (20). For the case in which the Lagrangian submanifold is invariant for the stochastic characteristic system see [5] .

4.Example: Anticipative LQG

Consider the anticipative LQG problem (first considered in [3],[4] using extension by continuity):

$$dx_t=(Ax_t+Bu_t)dt+Cdw_t$$

$$\inf_{u\in\mathcal{A}} E[\int_0^T (x_t^T Mx_t+u_t^T Ru_t)dt+x_T^T Nx_T]\ ;\ M\ ,\ N\geq 0\ \ ,\ R>0\ .$$

Using the extension of our results to the case with integral cost term we obtain the HJB SPDE of LQG which is a quadratic nonlinear SPDE

$$dV^o+[V_x^o Ax-(1/4)V_x^o BR^{-1}B^T(V_x^o)^T+x^T Mx]dt+V_x^o C\hat{\circ}dw_t=0$$

$$V^o(T,x)=x^T Nx$$

and the optimal anticipative control in selector form $u^o(t,x,\omega)=-(1/2)R^{-1}B^T V_x^o$.The characteristics are(Ito backward differential notation is used [8]):

$$dq_t=[Aq_t-(1/2)BR^{-1}B^T s_t]dt-C\hat{d}w_t\ \ ;\ q_T=x$$

$$ds_t/dt=-2Mq_t-A^Ts_t \; ; \; s_T= 2Nx$$

There exists a random conservation law $s_t(x)=2P_tq_t(x)+2\bar{b}_t(\omega)$ where

$$-dP_t/dt=P_tA+A^TP_t+M-P_tBR^{-1}B^TP_t \; , \; P_T=N$$

$$d\bar{b}_t(\omega)=b_1(t,\omega)dt+b_2(t,\omega)\hat{\circ}dw_t \; , \; \bar{b}_T(\omega)= 0$$

$$-2(dP_t/dt)q-2b_1(t,\omega)+ \{s-2P_tq-2\bar{b}_t,s^TAq-(1/4)s^TBR^{-1}B^Ts+q^TMq\}|_{(q,s)\in L_t(\omega)}=0$$

$$-2b_2(t,\omega)+\{s-2P_tq-2\bar{b}_t,s^TC\}|_{(q,s)\in L_t(\omega)}=0$$

$$L_t(\omega)=\{(q,s)\in\mathbb{R}^{2n}|\; s=2P_tq+2\bar{b}_t(\omega)\}$$

We can comment again that although the coefficients of the SPDE and SDE are not bounded with bounded derivatives (as required by theorems 1 and 2) the existence of an affine conservation law leading to an affine first characteristic equation ,decoupled from the third one, yields a global solution . This shows that our results as remarked earlier and Kunita's characteristic method can be generalized if a random conservation law exists and the decoupled characteristic system is such that the first characteristic generates a global flow of diffeomorphisms . Also the initial condition can be allowed to have just Holder (and not bounded) derivatives and $\mathcal{U}=\mathbb{R}^m$ as quadraticity (convexity) is used to get the minimizer and not selection lemma.The value function and the optimal anticipative control are :

$$V^o(t,x)=x^TP_tx+2\bar{b}_t(\omega)x+c_t(\omega) \quad , \; u^o(t,x,\omega)=-R^{-1}B^T(P_tx+\bar{b}_t(\omega))$$

$$d\bar{b}_t=-(A^T- P_tBR^{-1}B^T)\bar{b}_tdt - P_tC\hat{d}w_t \; , \; \bar{b}_T=0$$

$$dc_t=\bar{b}_t^TBR^{-1}B^T\bar{b}_tdt-2b_t^TC\hat{d}w_t \; , \; c_T=0$$

We recover the results of [3],[4] . It is interesting to see that Riccati matrix differential equation appears in the conservation law formula.

Acknowledgement

We thank Boris Rozovskii and Bob Anderson for their hospitality and support which made possible the presentation of this paper at the Charlotte SPDE conference . The many fruitful discussions and email dialogue with Dan Ocone and David Nualart's lectures and help in understanding anticipative calculus are gratefully acknowledged. We also thank Hiroshi Kunita for recent discussions in Haifa on the stochastic characteristics method .

References

[1] V. BENES, "Existence of optimal stochastic control laws", SIAM J. Contr. 9(1971), pp. 446-475.

[2] J.M. BISMUT, Mecanique Aleatoire, Lecture Notes in Mathematics 866, Springer-Verlag, Berlin, 1981.

[3] M.H.A. DAVIS, "Anticipative LQG Control", IMA J Contr. & Inf. 6(1989), pp. 259-265.

[4] M.H.A.DAVIS, "Anticipative LQG Control II", Applied Stochastic Analysis M.H.A. Davis and R.J. Elliott eds., Stochastic Monographs 5, Gordon and Breach, London, 1990.

[5] M.H.A. DAVIS and G. BURSTEIN, "A deterministic approach to stochastic optimal control with application to anticipative control", submitted to Stochastics and Stochastics Reports, 1991.

[6] M.H.A. DAVIS, M.A.H. DEMPSTER and R.J. ELLIOTT,"On the value of information in controlled diffusion processes", in Stochastic Analysis, E. Mayer-Wolf et al eds., Academic Press, Orlando, 1991.

[7] W.H.FLEMING and R.W. RISHEL, Deterministic and Stochastic Optimal Control , Applications of Mathematics 1, Springer-Verlag, Berlin 1975.

[8] H.KUNITA, "Stochastic differential equations and stochastic flows of diffeomorphisms", in Ecole d'Ete de Probabilite de Saint-Flour XII, 1982, P.L. Hennequin ed., Lecture Notes in Mathematics 1097, Springer-Verlag, Berlin, 1984, pp. 144-303.

[9] H. KUNITA, Stochastic Flows and Stochastic Differential Equations, Cambridge Studies in Advanced Mathematics 24, Cambridge University Press, Cambridge 1990.

[10] D. OCONE and E. PARDOUX, "A generalized Ito-Ventzell formula. Application to a class of anticipating stochastic differential equations", Ann. Inst. Henri Poincare 25(1989), pp.39-71.

[11] R.J.B. WETS,"On the relation between stochastic and deterministic optimization", in Control Theory, Numerical Methods and Computer Systems Modelling, Lecture Notes in Economics and Mathematical Systems 107, Springer-Verlag, Berlin, 1975, pp. 350-361.

STOCHASTIC CALCULUS WITH ANTICIPATION AND SHIFT TRANSFORMATIONS OF WIENER'S MEASURE

O. ENCHEV

Mathematics Dept., Boston Univ., Boston, Massachusetts, 02215

1. Introduction

In the focus of this lecture is the following class of transformations, which we call simply shifts

$$\omega \mapsto \mathcal{T}^h \omega \equiv \omega(\cdot) - \int_0^{(\cdot)} h_t(\omega)\, dt\ .$$

Here, ω denotes a Brownian sample path; that is, an element of the space $C[0,1]$ distributed according to the standard Wiener measure μ. Shift transformations play a crucial role in the study of many problems in physics (cf. [14]), signal processing (cf. [8], [10]), large deviations (cf. [4]) and other areas.

The central problem is to describe the distribution of the element $\mathcal{T}^h\omega$ in $C[0,1]$, or, more precisely, the measure $(\mathcal{T}^h)^{-1} \circ \mu$. The first who addressed this problem were Cameron and Martin [2], [3]. Under certain assumptions for the process (h_t) they showed that the measures $(\mathcal{T}^h)^{-1} \circ \mu$ and μ are mutually absolutely continuous with

$$\frac{d[(\mathcal{T}^h)^{-1} \circ \mu]}{d\mu}(\omega) = \exp\left[\int_0^1 h_t dW_t - \tfrac{1}{2}\int_0^1 h_t^2 dt\right] .$$

The first integral above is the Wiener-Itô stochastic integral with respect to the coordinate Brownian motion. As the last formula suggests, the developments in the study of the shift transformations have been strongly correlated with the developments in stochastic calculus. The Itô calculus and the martingale theory naturally lead to important extensions of Cameron-Martin's result in the works of Cameron-Graves [1], Girsanov [5], Shepp

Key words and phrases. abstract Wiener spaces, stochastic integrals with anticipating integrands, Gohberg-Krein factorization, absolutely continuous transformations of the Wiener measure.

1991 Mathematics Subject Classification. Primary 60B11, 60H05, 60H07; Secondary 28C20, 46G12, 47A68

Research partially supported by Seed Grant 958-Math at Boston University

[13] and others. The above formula was extended for a general class of nonanticipating (causal) processes (h_t) and this result entered the probabilistic literature as *"the theorem of Girsanov."*

A radically new idea was put forward by Ramer [12]. He studied transformations on a general abstract Wiener space (E, H, μ), which have the form $T = I + K$, for some nonlinear operator $K : E \mapsto H$. His approach was entirely independent from the martingale theory. Instead of nonanticipativity, he assumed that the operator K is Fréchet-differentiable in all directions from the Hilbert subspace H. He described the Radon-Nikodym derivative in terms of the Carelman-Fredholm determinant of the derivative $T' = I + K'$ and a new functional, which "resembled" the Wiener-Itô integral and was identified by some authors as Itô-Ramer integral. Later, Kusuoka [9] succeeded to relax some of the smoothness conditions in Ramer's work.

Almost at the same time Kabanov and Skorohod [7] developed a new concept of smoothness for Wiener functionals, which is essentially weaker than Fréchet-differentiability. Based on this concept, they constructed the so called extended (or Skorohod) stochastic integral, which is analogous to the Wiener-Itô integral, but may be defined also for anticipating integrands.

In this lecture I will attempt to explain how the Skorohod integral can be used for studying the shift T^h, in the case where the process h is some general, not necessarily causal, Skorohod integrand and for this type of shifts we will establish an analog of the usual Girsanov theorem. What motivated my initial interest to this study was the desire to establish a formula for the density $\frac{d[(T^h)^{-1} \circ \mu]}{d\mu}$, which does not involve the Carelman-Fredholm determinant, as in Ramer's work. This determinant is particularly difficult to handle, because it involves an infinite number of eigenvalues of a random integral operator. Since the density $\frac{d[(T^h)^{-1} \circ \mu]}{d\mu}$, which is often called Girsanov's density or Girsanov's exponent, appears explicitly in virtually every algorithm for signal processing, it is highly desirable to have an expression for this density which is a function of the length of the observation interval and in this respect Carelman-Fredholm determinant seems to be of a little help.

So, my plan is the following. First we will establish an integral representation for a certain class of determinants. Second, by using this representation for the Jacobians of finite dimensional transformations, we will establish an anticipative analog of the Girsanov theorem for shifts T^h corresponding to smooth simple processes (h_t). The final stage will be to develop a special approximation for general Skorohod integrands by smooth simple integrands, and, by an appropriate limiting procedure, to extend the result for general Skorohod integrands h.

2. Notations

I will adopt all basic notations in Malliavin's calculus, as they were introduced by D. Nualart and E. Pardoux in their 1988 paper [11]; that is, our basic probability space is $(C[0,1],\mu)$, by $\mathcal{S}$ we denote the totality of all smooth cylinder Wiener functionals, by $\mathbb{D}^{2,1}$ we denote the class of all stochastically differentiable Wiener functionals, by $\mathbb{L}^{2,1}$ we denote the class of Skorohod integrands, and the stochastic derivative of $F \in \mathbb{D}^{2,1}$ we write as D_tF, $0 \le t \le T$. All these notations have already been used in several talks, so I will not go into more details, and for the same reason I skip the technical introduction to the Skorohod integral

$$\delta(h) = \int_0^1 h_t dW_t, \quad (h_t) \in \mathbb{L}^{2,1}.$$

We set

$$\Delta_i^n = [(i-1)2^n, i2^n), \quad 1 \le i \le 2^n, \quad n \ge 1.$$

3. Calculation of Certain Determinants

Let $K(s,t)$, $0 \le s,t \le 1$, be a real measurable function with

$$\int_0^1 \int_0^1 K(s,t)\, ds\, dt < 1 \tag{3.1}$$

and let

$$\mathcal{E}_t[K] \stackrel{\text{def}}{=} \sum_{m=1}^{\infty} \int_t^1 \dots \int_t^1 K(t,r_1)K(r_1,r_2)\dots K(r_{m-1},r_m)K(r_m,t)\, dr_1 \dots dr_m \qquad 0 \le t \le T. \tag{3.2}$$

Equivalently, one can write $\mathcal{E}_t[K]$ as $\int_t^1 K(t,r)V^+(r,t)dr$, V^+ being the right Volterra kernel in the factorization of the Fredholm operator with kernel $K(s,t)$. We use the term factorization exactly in the way it was introduced and studied by I. Gohberg and M. Krein [6].

Now let us assume that the above function K has the form

$$K(s,t) = \sum_{i,j=1}^{2^n} b_{i,j} 1_{\Delta_i^n}(t) 1_{\Delta_j^n}(s), \quad 0 \le s,t \le T, \quad b_{i,j} \in \mathbb{R}.$$

We remark that in this representation the number n can be taken arbitrary large, by simply writing each indicator $1_{\Delta_j^n}$ as a sum of indicators of smaller intervals. Let us define the following matrix

$$A(K,n) = \left(a_{i,j}^n \equiv b_{i,j}2^{-n}\right)_{i,j=1}^{2^n} .$$

Since $\sum_{i,j} \left|a_{i,j}^n\right|^2 = \int\int K^2 < 1$, the matrix $(I - A(K,n))$ is invertible, and the inverse can be written as

$$(I - A(K,n))^{-1} = (I + W^{+,n})(I + W^{-,n}) .$$

Here $W^{\pm,n} \equiv (w_{i,j}^{\pm,n})$ are triangular matrices, such that $w_{i,j}^{+,n} = 0$, for $i < j$, and $w_{i,j}^{-,n} = 0$, for $i \geq j$. The elements of $W^{+,n}$ satisfy the following relations, which actually determine $W^{+,n}$ and $W^{-,n}$ in a unique way:

$$w_{i,j}^{+,n} = a_{i,j}^n + \sum_{k=j}^{n} a_{i,k} w_{k,j}^{+,n}, \quad 1 \leq j \leq i \leq 2^n .$$

Thus we have the following expansion for the diagonal elements of $W^{+,n}$:

$$w_{i,i}^{+,n} = a_{i,i} + \sum_{m=1}^{\infty} \sum_{l_1,\ldots,l_m=i}^{n} a_{i,l_1} a_{l_1,l_2} \ldots a_{l_m,i} .$$

Obviously

$$\det(I - A(K,n)) = \frac{1}{(\det(I-A))^{-1}} = \exp\left[-\sum_{i=1}^{2^n} \ln\left(1 + w_{i,i}^{+,n}\right)\right] .$$

Here we have that $\lim_{n\to\infty} w_{i,i}^{+,n} = 0$, uniformly with respect to i, and so, for large n, $\ln\left(1 + w_{i,i}^{+,n}\right) = w_{i,i}^{+,n} + \mathcal{O}\left(\left|w_{i,i}^{+,n}\right|^2\right)$. This reasoning suggests the following result, which we will not prove.

Lemma 1. *For all sufficiently large n,*

$$\det(I - A(K,n)) = \exp\left[-\mathit{trace}\,(A(K,n)) - \int_0^1 \mathcal{E}_t[K]\,dt\right] . \quad \square$$

4. Girsanov Exponents

Assume that h is an integrand from the class $\mathbb{L}^{2,1}$, which satisfies the following condition

$$\int_0^1 \int_0^1 |D_s h_t|^2 \, ds \, dt < 1 \,, \quad \mu\text{-a.e. in } C[0,1] \,. \tag{4.1}$$

Then we can define the following measure

$$d\mu^h = R^h d\mu \,;$$

$$R^h = \exp\left[\int_0^1 h_t dW_t - \tfrac{1}{2}\int_0^1 h_t^2 dt - \int_0^1 \mathcal{E}_t\,[Dh]\; dt\right].$$

The first integral above is the Skorohod integral $\delta(h)$. $\mathcal{E}_t\,[Dh]$ has the same meaning as $\mathcal{E}_t[K]$, but for $K(s,t) = D_s h_t$, $0 \le s,t \le 1$. Note that $\mathcal{E}_t\,[Dh]$ is a stochastic process.

5. Girsanov's Theorem for Smooth Simple Shifts

Now we will study the shift T^h, in the case where h is a smooth simple process of the form

$$h_t = \sum_{i=1}^{2^n} f_i \, 1_{\Delta_i^n}(t), \quad f_i \in \mathcal{S} \,, \tag{5.1}$$

which, in addition, satisfies the following sharper form of condition (4.1)

$$\int_0^1 \int_0^1 |D_s h_t|^2 \, ds \, dt \le C < 1 \,, \quad \text{everywhere! in } C[0,1] \,. \tag{5.2}$$

In this case T^h may be regarded as a transformation in a finite-dimensional Euclidean space. So, let us consider the following integration scheme in $\mathbb{R}^d$. Let $f_i : \mathbb{R}^d \mapsto \mathbb{R}$, $1 \le i \le d$, be functions from $C_b^\infty\left(\mathbb{R}^d\right)$, such that

$$\sum_{i,j=1}^d \left|\frac{\partial f_i}{\partial x_j}\right|^2 < 1 \,.$$

For $F \in C_b^\infty\left(\mathbb{R}^d\right)$, consider the integral

$$\frac{1}{(\sqrt{2\pi})^d} \int_{\mathbb{R}^d} F\left(x_1 - f_1, \ldots, x_d - f_d\right)$$
$$\times \exp\left[\sum_{i=1}^d f_i x_i - \tfrac{1}{2}\sum_{i=1}^d |f_i|^2\right] \times \exp\left[-\tfrac{1}{2}\sum_{i=1}^d x_i^2\right]$$
$$\times \det\left[I - \left(\frac{\partial f_i}{\partial x_j}\right)_{i,j=1}^d\right] dx_1 \ldots dx_d \,.$$

By the following change of variables:

$$y_i = x_i - f_i(x_1, \ldots, x_d) \;\; 1 \le i \le d,$$

the same integral we can write as

$$\frac{1}{(\sqrt{2\pi})^d} \int_{\mathbb{R}^d} F(y_1, \ldots, y_d) \exp\left[-\frac{1}{2} \sum_{i=1}^{d} y_i^2\right] dy_1 \ldots dy_d \,.$$

This simple calculation, in fact, presents the very essence of the Girsanov theorem in all its forms. With similar, but somewhat more complicated computations, using our Lemma 1, one can prove the following result.

Theorem 1. *Let h be a smooth simple process of the type* (5.1), *which satisfies condition* (5.2). *Then, for every smooth functional $F \in \mathcal{S}$, we have*

$$\mathbb{E}_{\mu^h}\left\{F\left(T^h\right)\right\} = \mathbb{E}_{\mu}\left\{F\right\} . \;\; \square$$

In particular the last result shows that $\mathbb{E}_{\mu}\left\{R^h\right\} = 1$, for every smooth simple process of the form (5.1), which satisfies condition (5.2). It also implies that, for every Borel set $A \subset C[0,1]$,

$$\mu^h\left\{T^h \in A\right\} = \mu(A)\,,$$

and this explains the link with the standard Girsanov theorem.

6. The General Case

Now we turn to the case where h is some general integrand from the class $\mathbb{L}^{2,1}$, which satisfies condition (4.1). The next lemma presents the key step at this final stage. Unfortunately, the proof is too technical and we have to omit it.

Lemma 2. *Let $h \in \mathbb{L}^{2,1}$ and let condition* (4.1) *be met. Then there exists a sequence of smooth simple processes h^n, $n \ge 1$, such that $h^n \to h$ in $\mathbb{L}^{2,1}$, and, for every $n \ge 1$, h^n satisfies condition* (5.2) *(not necessarily with the same constant C).* $\square$

Here is our main result.

Theorem 2. *Let $h \in \mathbb{L}^{2,1}$ and let condition* (4.1) *be met. Assume that* $\mathbb{E}_{\mu}\left\{R^h\right\} = 1$. *Then, for every bounded Borel function $F : C[0,1] \mapsto \mathbb{R}$,*

$$\mathbb{E}_{\mu^h}\left\{F\left(T^h\right)\right\} = \mathbb{E}_{\mu}\left\{F\right\} .$$

Proof. Due to the monotone classes theorem, it is enough to consider the case $F \in \mathcal{S}$. Let $h^n, n \geq 1$, be a sequence of smooth simple processes, chosen as in Lemma 2. This sequence can be chosen also such that $R^{h^n} \to R^h$, μ-a.e.. Since $\mathbb{E}_\mu\left\{R^{h^n}\right\} = 1 = \mathbb{E}_\mu\left\{R^h\right\}$, it follows that $\left\{R^{h^n},\ n \geq 1\right\}$ is a uniformly integrable family. Therefore $\left\{F\left(T^{h^n}\right)R^{h^n},\ n \geq 1\right\}$ is also uniformly integrable. This yields

$$\mathbb{E}_{\mu^h}\left\{F\left(T^h\right)\right\} = \lim_{n\to\infty} \mathbb{E}_{\mu^{h^n}}\left\{F\left(T^{h^n}\right)\right\} = \mathbb{E}_\mu\{F\}\ . \quad \square$$

7. When is $\mathbb{E}_\mu\left\{R^h\right\}$ equal to 1 ?

The importance of this question is self-evident and I would like to thank D. Nualart for asking it. One possible answer can be found by almost literally repeating the argument of Chapter 7 in G. Kallianpur's book [8]. Namely, the following two conditions imply that $\mathbb{E}_\mu\left\{R^h\right\} = 1$:

(a) there exists some constant $C < 1$, such that,

$$\int_0^T \int_0^T |D_s h_t|^2 ds\, dt \leq C\,, \quad \mu\text{-a.e. in } C[0,1]\ ;$$

(b) there exists some constant $\alpha > 1$ such that

$$\mathbb{E}\left\{\exp\left[\frac{\alpha}{2}\int_0^T |h_t|^2\, dt\right]\right\} < \infty\,.$$

8. Acknowledgements

I would like to thank the Organizing Committee, and especially B. Rozovskii and R. Anderson, for the opportunity to participate at this conference and present my work.

References

1. R. H. Cameron and R. E. Graves, *Additive functionals on a space of continuous functions*, Trans. Amer. Math. Soc. **70** (1951), 160–176.
2. R. H. Cameron and W. T. Martin, *Transformation of Wiener integrals under translations*, Ann. Math. Stat. **45** (1944), 386–396.
3. ———, *The transformation of Wiener integrals by nonlinear transformations*, Trans. Amer. Math. Soc. **66** (1949), 253–283.
4. J.-D. Deuschel and D. W. Stroock, *Large deviations*, Acad. Press, Boston, 1989.
5. I. V. Girsanov, *On transforming a certain class of stochastic processes by absolutely continuous substitution of measures*, Theory Probab. Its Appl. **5** (1960), 285–301.
6. I. C. Gohberg and M. G. Krein, *Theory and applications of Volterra operators in Hilbert spaces*, Am. Math. Soc., Transl., vol. vol. 24, Am. Math. Soc., Providence, RI, 1970.

7. Yu. Kabanov and A. Skorohod [Ю. Кабанов и А. Скороход], Расширенные Стохастические Интегралы *[Extended Stochastic Integrals]*, Труды Школы-Семинара по Теории Случайных Процессов [Proc. of Workshop on Stochastic Processes] (Vilnius, Lithuania), 1975.

8. G. Kallianpur, *Stochastic filtering theory*, Springer-Verlag, New York, 1980.

9. Sh. Kusuoka, *The nonlinear transformation of Gaussian measure on Banach space and its absolute continuity (I)*, J. Fac. Sci. Univ. Tokyo, Sect. IA **29** (1982), 567–598.

10. R. Liptser and A. Shiriyayev, *Statistics of random processes*, vol. I&II, Springer-Verlag, New York, 1977.

11. D. Nualart and E. Pardoux, *Stochastic calculus with anticipating integrands*, Probab. Th. Rel. Fields **78** (1988), 535–581.

12. R. Ramer, *On nonlinear transformations of Gaussian measures*, J. Funct. Analysis **15** (1974), 166–187.

13. L. A. Shepp, *Radon-Nikodym derivatives of Gaussian measures*, Ann. Math. Stat. **37** (1966), 321–354.

14. B. Simon, *Functional integration and quantum physics*, Acad. Press, New York, 1979.

A PROPOS D'UN EXEMPLE D'ÉQUATION DIFFÉRENTIELLE STOCHASTIQUE EN DIMENSION INFINIE.

X. Fernique
Département de Mathématique, Institut de Recherche Mathématique Avancée,
Université Louis Pasteur, 7 rue René Descartes, F-67084 Strasbourg Cédex.

1. Introduction.

1.1 On se propose de présenter quelques réflexions pour la résolution des équations différentielles stochastiques en dimension infinie. Si dans de nombreux cas, les tentatives de solution restent très partielles, on dispose pourtant d'un cas où on peut résoudre et où on sait analyser les solutions. Ce cas développé ici est tellement simple qu'il est peut-être non significatif : les solutions seront en effet gaussiennes à valeurs dans des espaces de suites ; pourtant dans cette situation, les travaux antérieurs aux miens, utilisant les méthodes générales d'étude des SDE ou même les méthodes gaussiennes que j'avais mises en place, n'avaient pas été capables de fournir des solutions précises et complètes, même dans le cas utile de l'espace l_2. Nous utilisons donc ici les résultats de nos travaux concernant les suites indépendantes de fonctions aléatoires gaussiennes d'Ornstein-Uhlenbeck stationnaires pour étudier dans quels espaces de suites, on peut résoudre les équations linéaires de Langevin associées. Les arguments essentiellement nouveaux sont énoncés en **2.2** ; ils permettront de mettre en évidence une situation probablement nouvelle (exemple 2.5.5). Peut-être peut-on à partir de cet exemple particulier envisager des perspectives plus générales.

Le lecteur pourra noter que dans cette étude, on saura s'interdire : (a) tout usage d'intégrale stochastique, qui permettrait mal, à notre sens, de distinguer suffisamment fonction aléatoire et modification, (b) tout usage de fonction aléatoire séparable, dont l'existence serait trop rare dans les espaces de grande dimension. La nécessité de telles précautions sera nette en particulier en **2.5**.

1.2 Notations. On note E_0 l'espace $\mathbf{R}^{\mathbf{N}}$ qu'on munit de sa topologie et de sa tribu produits ; on note F l'espace des suites numériques à support fini qu'on identifie au dual topologique de E_0 ; on note $(e_n, n \in \mathbf{N})$ leur base canonique commune ; on note $(\lambda_n, n \in \mathbf{N})$ et

$(\sigma_n, n \in \mathbf{N})$ deux suites numériques positives. Λ et Σ désignent les opérateurs diagonaux que ces suites définissent dans E_0, ${}^t\Lambda$ et ${}^t\Sigma$ désignant les opérateurs transposés dans F.

Les fonctions aléatoires utilisées seront toutes sauf mention expresse, définies sur un espace d'épreuves $(\Omega, \mathbb{A}, \mathbf{P})$ complet. Soient $X = \{X(\omega, t), \omega \in \Omega, t \in T\}$ et $Y = \{Y(\omega, t), \omega \in \Omega, t \in T\}$ deux fonctions aléatoires sur un ensemble T, on rappelle qu'elles sont *modifications* l'une de l'autre si :

$$\forall\, t \in T, \mathbf{P}\{X(t) = Y(t)\} = 1 \;;$$

on rappelle aussi qu'elles sont *indistinguables* si :

$$\mathbf{P}\{\ \forall\, t \in T, X(t) = Y(t)\} = 1.$$

Enfin pour tout $\omega \in (\Omega, \mathbb{A}, \mathbf{P})$, la ω- *trajectoire* de X est l'application : $t \to X(\omega, t)$.

Soit $(W_n, n \in \mathbf{N})$ une suite de mouvements browniens réels normalisés et indépendants sur $\mathbf{R}^+$; on suppose que les W_n sont tous mesurables sur $\Omega \times \mathbf{R}^+$ et que toutes les trajectoires des W_n sont continues ; $(W_n, n \in \mathbf{N})$ définit donc un mouvement brownien W sur $\mathbf{R}^+$ à valeurs dans $\mathbf{R}^\mathbf{N}$, mesurable sur $\Omega \times \mathbf{R}^+$ et à trajectoires continues. On note enfin a une variable aléatoire à valeurs dans $\mathbf{R}^\mathbf{N}$.

Dans ces conditions, on forme l'équation différentielle stochastique :

1.2.1 $$dX = - \Lambda X\, dt + \Sigma\, dW, \quad X(0) = a, t \in \mathbf{R}^+, \text{ p.s.}$$

c'est-à-dire :

1.2.1 (i) $$\forall\, n \in \mathbf{N}, \quad dX_n = - \lambda_n X_n\, dt + \sigma_n\, dW_n, \quad X_n(0) = a_n, t \in \mathbf{R}^+, \text{ p.s.}$$

à entendre au sens suivant :

1.2.1 (ii) Il existe un ensemble presque sûr Ω_1 tel que :

$$\forall\, \omega \in \Omega_1, \forall\, n \in \mathbf{N}, \forall\, t \in \mathbf{R}^+, X_n(\omega,t) - a_n(\omega) = - \lambda_n \int_0^t X_n(\omega, u)\, du + \sigma_n W_n(\omega, t),$$

sous la condition technique :

1.2.1 (iii) $\forall\, \omega \in \Omega_1, \forall\, n \in \mathbf{N}$, $X_n(\omega)$ à valeurs dans $\mathbf{R}$, est localement intégrable sur $\mathbf{R}^+$.

1.3 La résolution dans $\mathbf{R}^\mathbf{N}$.

Les différentes équations 1.2.1(i) se résolvent séparément ; elles fournissent une solution X p.s. unique (i.e. sur Ω_1) de l'équation 1.2.1 :

1.3.1 (i) $$\forall\, \omega \in \Omega_1, \forall\, t \in \mathbf{R}^+, X(\omega,t) = \Sigma\, W(\omega,t) - \Sigma\, \Lambda \int_0^t e^{\Lambda(u - t)} W(\omega, u)\, du + e^{-\Lambda t} a(\omega),$$

vérifiant donc :

1.3.1 (ii) X à valeurs dans $\mathbf{R}^\mathbf{N}$ est mesurable sur $\Omega_1 \times \mathbf{R}^+$ et à trajectoires continues.

En particulier, il existe une solution stationnaire S et une seule qui est p.s. définie par :

1.3.2 $$\forall\, \omega \in \Omega, \forall\, t \in \mathbf{R}^+,\ S(\omega, t) = \Sigma\, W(\omega, t) - \Sigma\, \Lambda \int_{-\infty}^{t} e^{\Lambda(u-t)}\, W(\omega, u)\, du,$$

c'est une fonction aléatoire gaussienne et sa covariance $\Gamma = \Gamma_S$ vérifie pour tout couple (t, y), (s, z) d'éléments de $\mathbf{R}^+ \times F$:

1.3.3 $$\Gamma(t, y\, ;\, s, z) = \sum_{n=0}^{\infty} \sigma_n^2\, y_n\, z_n\, (2\lambda_n)^{-1}\, e^{-\lambda_n |t-s|}.$$

1.4 Dans tout ce travail, E désignera exclusivement l'un des espaces l_p, $p \in [1, \infty]$ ou l'espace c_0 munis de leurs topologies fortes ou affaiblies et de la tribu induite par celle de $\mathbf{R}^N$; excepté l_∞ fort, ce sont tous des espaces lusiniens et tous, même l_∞, sont des sous-ensembles boréliens de E_0. On présente la résolution dans E de l'équation 1.2.1 et on analyse cette solution. Le fait que l_∞ fort ne soit pas séparable n'introduira pas de difficulté. On privilégiera les solutions stationnaires bien que l'extension aux autres solutions soit facile. On utilisera pour cela divers résultats([1], [2], [3], [4], [5]) sur la régularité des fonctions d'Ornstein-Uhlenbeck.

2. La résolution de l'équation.

2.1 Pour résoudre l'équation 1.2.1 dans les espaces E cités ci-dessus, nous l'entendrons au sens suivant :

2.1.1(i) X est une fonction aléatoire sur $\mathbf{R}^+$ à valeurs dans E_0, p.s.à valeurs dans E,

$$\mathbf{P}\{\forall\, t \in \mathbf{R}^+, X(t) \in E\} = 1\ ;$$

2.1.1(ii) de plus X est solution de l'équation 1.2.1 et donc :

$$\mathbf{P}\{\ \forall\, y \in F, \forall\, t \in \mathbf{R}^+, d\langle X(t), y\rangle = -\langle \Lambda\, X(t), y\rangle\, dt + d\langle \Sigma\, W(t), y\rangle\} = 1.$$

On se propose (a_1) de discuter l'existence de telles solutions, (a_2) d'indiquer la régularité de leurs trajectoires dans E et aussi (b) d'étudier si la signification de l'équation peut être modifiée.

2.2 En fait les techniques dont nous disposons ne permettront pas de séparer les problèmes (a_1) et (a_2). Nous utiliserons en effet les remarques suivantes :

Soit X une fonction aléatoire sur $\mathbf{R}^+$ vérifiant l'équation 1.2.1; soit de plus E l'un des espaces cités en 1.4 .

2.2.1 Supposons pour commencer que X ait une modification X' à trajectoires continues dans E ; alors X et X' sont toutes deux p.s. à trajectoires continues dans E_0 de sorte qu'en fait elles sont indistinguables ; il en résulte que X vérifie en fait l'équation 2.1.1 et a de plus presque toutes ses trajectoires continues dans E.

2.2.2 Supposons maintenant que X ait seulement une modification X' à trajectoires localement bornées dans E ; supposons de plus que l'espace E ne soit pas c_0 ; alors pour tout y $\in$ F, $\langle X', y\rangle$ a une modification continue (cf. 1.3.1, ii) ; de plus pour tout $\varepsilon > 0$, il existe une partie compacte K de E pour la topologie affaiblie de E (cf. [1], 1.8) telle que :

$$P\{\forall\, t \in R^+,\ X(t) \in K\} > 1 - \varepsilon,$$

de sorte que X' a une modification X" à valeurs dans E et à trajectoires continues dans E pour la topologie affaiblie (cf. [1], 4.2.5) ; la remarque précédente s'applique alors et montre que X" et X sont indistinguables de sorte que X vérifie en fait l'équation 2.1.1 et a presque toutes ses trajectoires continues dans E pour la topologie affaiblie et a fortiori localement bornées dans E.

2.2.3 Supposons par contre que l'espace E soit c_0 et que X ait une modification à trajectoires localement bornées dans c_0 ; la remarque précédente montre donc que X est p.s. à trajectoires localement bornées et relativement compactes dans l_∞ affaibli ; ces trajectoires sont alors continues dans l_∞ affaibli. Supposons de plus que X n'ait pas de modifications à trajectoires continues dans c_0, même pour la topologie affaiblie ; alors aucune modification X' de X ne vérifie l'équation 2.1.1 dans c_0 ; sinon X' serait indistinguable de X, aurait p.s. ses trajectoires dans c_0 et continues pour la topologie affaiblie de l_∞, donc continues pour la topologie affaiblie de c_0 ; il y aurait donc contradiction.

2.3 Le problème (a) si $E = l_p$, $p \in [1, \infty[$.

Si E est l'un des espaces l_p, $p \in [1, \infty[$, on sait ([3], [5]) à quelles conditions la solution stationnaire S de l'équation 1.2.1 a une modification à trajectoires localement bornées ou continues. Les remarques 2.2.1 et 2.2.2 permettent donc d'énoncer :

Théorème : *Les quatre propriétés suivantes sont équivalentes :*

2.3.1 *La fonction aléatoire* S *est solution de l'équation* 2.1.1 *dans* l_p *et a presque toutes ses trajectoires continues dans* l_p *fort.*

2.3.2 *La fonction aléatoire* S *est solution de l'équation* 2.1.1 *dans* l_p *et a presque toutes ses trajectoires continues dans* l_p *affaibli.*

2.3.3 *La fonction aléatoire* S *est solution de l'équation* 2.1.1 *dans* l_p *et a presque toutes ces trajectoires localement bornées dans* l_p *fort ou affaibli.*

2.3.4 *Les conditions suivantes sont vérifiées :*

(i) *la série* $\sum_{n=0}^{\infty} [\frac{\sigma_n}{\sqrt{\lambda_n}}]^p$ *est convergente,*

(ii) *si* $p \in [1, 2[$, *l'intégrale* $\int_0^{\infty} [\sum \{ [\frac{\sigma_n}{\sqrt{\lambda_n}}]^r, \lambda_n > \exp x^2 \}]^{1-p/2} x^{p-1}\, dx$, $r = 2\, p/(2-p)$,

est convergente ; si $p \in [2, \infty[$, *l'intégrale* $\int_0^{\infty} [\ \sup\{ [\frac{\sigma_n}{\sqrt{\lambda_n}}]^p , \lambda_n > \exp x^2 \}]\ x^{p-1}\, dx$ *est convergente.*

2.4 Le problème (a) si E = l_∞.

Si E est l'espace l_∞, les résultats de [3], 4.1.3 et de [4] avec les mêmes remarques montrent que la situation est plus variée :

Théorème : *Les deux propriétés suivantes sont équivalentes :*

2.4.1 *La fonction aléatoire* S *est solution de l'équation* 2.1.1 *dans* l_∞ *et a presque toutes ses trajectoires continues dans* l_∞ *fort.*

2.4.2 *Les conditions suivantes sont vérifiées :*

(i) *il existe un nombre* $M > 0$ *tel que la série* $\sum \exp\{-M^2\lambda_n/\sigma_n^2\}$ *converge,*

(ii) *pour tout* $\varepsilon > 0$, *il existe un nombre* B *tel que la série* $\sum I_{\lambda_n > B} \exp\{-\varepsilon^2\lambda_n/\sigma_n^2\}$ *converge et que* $\sup_{\lambda_n > B} \sigma_n [\frac{\log^+\lambda_n}{\lambda_n}]^{1/2} \leq \varepsilon$.

De même, les trois propriétés suivantes sont équivalentes :

2.4.3 *La fonction aléatoire* S *est solution de l'équation* 2.1.1 *dans* l_∞ *et a presque toutes ses trajectoires localement bornées dans* l_∞ *fort ou affaibli.*

2.4.4 *La fonction aléatoire* S *est solution de l'équation* 2.1.1 *dans* l_∞ *et a presque toutes ses trajectoires continues dans* l_∞ *affaibli.*

2.4.5 *Les deux conditions suivantes sont vérifiées :*

(i) *il existe un nombre* M *tel que la série* $\sum \exp\{-M^2\lambda_n/\sigma_n^2\}$ *converge,*

(ii) $$\sup_{n \in \mathbf{N}} \sigma_n [\frac{\log^+\lambda_n}{\lambda_n}]^{1/2} < \infty .$$

2.5 Le problème (a) si E = c_0.

Si E est l'espace c_0, les résultats de [3], 4.1.2 avec les mêmes remarques donnent un résultat partiellement semblable que la remarque 2.2.3 complète de manière différente :

Théorème : *Les deux propriétés suivantes sont équivalentes :*

2.5.1 *La fonction aléatoire* S *est solution de l'équation* 2.1.1 *dans* c_0 *et a presque toutes ses trajectoires continues dans* c_0 *fort ou affaibli.*

2.5.2 *Les conditions suivantes sont vérifiées :*

(i) *Pour tout* $\varepsilon > 0$, *la série* $\sum \exp\{-\varepsilon^2\lambda_n/\sigma_n^2\}$ *converge,*

(ii) $$\lim_{n\to\infty} \sigma_n \left[\frac{\log^+\lambda_n}{\lambda_n}\right]^{1/2} = 0 .$$

Inversement si la propriété 2.5.2 *n'est pas vérifiée, alors pour toute modification* S' *de* S, *les deux propriétés suivantes sont incompatibles :*

2.5.3 S' *a presque toutes ses trajectoires dans* c_0,

2.5.4 $\mathbf{P}\{ \forall\, y \in F, \forall\, t \in \mathbf{R}^+, d\langle S'(t), y\rangle = -\langle \Lambda\, S'(t), y\rangle\, dt + d\langle \Sigma\, W(t), y\rangle\} = 1.$

2.5.5 **Exemple** : Le dernier énoncé permet de mettre en évidence des situations très singulières ; supposons par exemple que l'on ait simultanément :

(i) pour tout $\varepsilon > 0$, la série $\sum \exp\{-\varepsilon^2\lambda_n/\sigma_n^2\}$ converge,

(ii) $$\sup_{n\in\mathbf{N}} \sigma_n \left[\frac{\log^+\lambda_n}{\lambda_n}\right]^{1/2} < \infty ,$$

(iii) $$\lim_{n\to\infty} \sup \sigma_n \left[\frac{\log^+\lambda_n}{\lambda_n}\right]^{1/2} > 0 ;$$

alors les deux premières conditions assurent ([3], 4.1.2) que S qui vérifie l'équation 1.2.1, a une modification à valeurs dans c_0 ayant toutes ses trajectoires localement bornées dans cet espace ; par contre la remarque 2.2.3 montre que ni S, ni aucune de ses modifications, ne vérifient l'équation 2.1.1 dans c_0.

On peut rappeler d'ailleurs ([6], 5.3) que S n'a aucune modification séparable dans c_0.

2.6 Le problème (b).

2.6.1 Un exemple : Nous montrons ici par un exemple qu'il n'est pas possible en général de renforcer la signification de l'équation 2.1.1 sans modifier ses solutions. Nous supposons que $E = l_2$ et nous posons pour tout $n \in \mathbf{N}$, $\lambda_n = e^{3n}$, $\sigma_n = e^n$. Nous notons encore Σ et Λ les opérateurs non bornés associés à ces suites dans l_2. On constate alors immédiatement que la solution stationnaire S est solution de l'équation 2.1.1 dans l_2 où elle a presque toutes ses trajectoires continues. On constate aussi par contre qu'on a :

$$\forall\, t \in \mathbf{R}^+,\ \mathbf{P}\{\ \Lambda\, S(t) \in l_2\} = 0,\ \ \mathbf{P}\{\ \Sigma W(t) \in l_2\} = 0,$$

ce qui interdit dans ce cas les renforcements vectoriels de l'équation .

2.6.2 Nous revenons à la situation générale ; nous notons E l'un des espaces cités en 1.4 ; pour pouvoir intégrer sans ambiguité ([2], 3.5), nous supposons que E est lusinien et quasi-complet (i.e. E n'est pas l'espace c_0 affaibli, ni l'espace l_∞ fort).

Nous supposons que la solution stationnaire S de l'équation 1.2.1 vérifie simultanément les deux propriétés suivantes :

2.6.2 (i) $\quad\quad \forall\, t \in \mathbf{R}^+,\ \ \mathbf{P}\{\ S(t) \in E\} \ = \ 1,$

2.6.2 (ii) $\quad\quad \forall\, t \in \mathbf{R}^+,\ \ \mathbf{P}\{\Lambda\, S(t) \in E\} \ = \ 1.$

Ces deux hypothèses qui, pour les différents espaces E considérés se traduisent facilement en fonction des coefficients de Λ et Σ, semblent relativement faibles ; on va pourtant montrer qu'elles suffisent pour que S vérifie l'équation 1.2.1 dans E en un sens beaucoup plus fort.

Dans ces conditions en effet, puisque S est mesurable dans E_0, que Λ est continu dans E_0, que E est borélien dans E_0 et muni de la tribu induite, le théorème de Fubini montre que l'ensemble $\{\omega : u \rightarrow \Lambda S(\omega, u)$ mesurable dans E et fortement localement intégrable$\}$ est presque sûr de sorte que la fonction aléatoire $V : t \rightarrow \int_0^t \Lambda S(u)\, du$ est p.s.définie par intégration vectorielle forte dans E et p.s. à trajectoires continues dans E. Il en résulte en particulier que pour tout $t \in \mathbf{R}^+$, $\Sigma\, W(t)$ qui est gaussienne et dont la covariance se majore à partir de celles de S(t) et de V(t), a une modification W'(t) à valeurs dans E ; la fonction aléatoire : t $\rightarrow$ W'(t) est alors un processus de Wiener dans E et a une modification W'' à trajectoires continues dans E ; W'' est une modification de ΣW et la continuité de ΣW dans E_0 montre que ΣW et W'' sont indistinguables de sorte que ΣW a presque toutes ses trajectoires continues dans E ; la covariance de S se majore à partir des covariances de V et de ΣW de sorte que S a encore une modification S' à trajectoires continues dans E et la continuité de S dans E_0 montre que S, indistinguable de S', a aussi presque toutes ses trajectoires continues dans E. Finalement, en réglant trivialement le cas où E = c_0 affaibli, on peut énoncer :

Théorème : *On suppose que* E *n'est pas l'espace* l_∞ *fort et que les hypothèses* 2.6.2 (i) *et* (ii) *sont vérifiées, alors les deux propriétés suivantes sont aussi vérifiées :*

(i) $\quad \mathbf{P}\{\ t \rightarrow \overrightarrow{S(t)},\ t \rightarrow \overrightarrow{\Sigma W(t)},\ t \rightarrow \int_0^t \overrightarrow{\Lambda S(u)}\, du \ \ \text{continues dans } E\} = 1,$

(ii) $$P\{\forall\, t \in \mathbf{R}^+, \overrightarrow{S(t)} - \overrightarrow{S(0)} = -\int_0^t \overrightarrow{\Lambda S(u)}\, du + \overrightarrow{\Sigma W(t)} \quad \text{dans } E\} = 1.$$

Ce théorème montre donc en particulier que ce sont les singularités de l'opérateur Λ qui permettent à la fonction aléatoire S d'être régulière sans que Σ W le soit.

Conclusion : On espère avoir montré dans cet exposé et en particulier dans l'exemple 2.5.5, la nécessité de définir les équations différentielles stochastiques en dimension infinie de telle manière qu'on puisse distinguer avec soin les solutions et leurs modifications ; faute de quoi certaines singularités sont inaccessibles à l'analyse.

Références.

[1] X. Fernique Fonctions aléatoires à valeurs dans les espaces lusiniens, Expositiones Math., 8, 1990, 289-364.

[2] X. Fernique Régularité de fonctions aléatoires gaussiennes à valeurs vectorielles, Ann. Prob., 18, 1990, 1739-1745.

[3] X. Fernique Sur la régularité de certaines fonctions aléatoires d'Ornstein-Uhlenbeck, Ann. Instit. Henri Poincaré, 26, 3, 1990, 399-417.

[3] X. Fernique Régularité de fonctions aléatoires gaussiennes stationnaires, Probab. Th. Rel. Fields 88, 524-536 (1991).

[4] X. Fernique Régularité de fonctions aléatoires gaussiennes à valeurs dans l_∞, Probab. and Math. Statist, 13, 1, à paraître.

[5] X. Fernique Sur la régularité des fonctions aléatoires d'Ornstein-Uhlenbeck à valeurs dans l_p, $p \in [1, \infty[$. Ann. Prob., à paraître.

[6] X. Fernique Analyse de fonctions aléatoires gaussiennes stationnaires à valeurs vectorielles,Technical Report 331, Center for stochastic processes of U.N.C. at Chapel Hill, Mars 1991.

Xavier Fernique,
Département de Mathématique,
Institut de Recherche Mathématique Avancée,
Université Louis Pasteur, 7 rue René Descartes,
67084 Strasbourg Cédex, France.

STOCHASTIC EVOLUTION EQUATIONS WITH NON-COERCIVE MONOTONE OPERATORS

Franco Flandoli
Scuola Normale Superiore
Piazza dei Cavalieri 7, 56100 Pisa, Italy

1 Abstract Result

Let $E \subset V \subset H$ be three real vector spaces, E a separable Banach space, V a reflexive separable Banach space, and H a separable Hilbert space, with dense continuous injections, norms $||.||_E, ||.||_V, ||.||_H$, respectively, inner product $(.,.)$ in H, and dual pairing $< .,. >_E$ between E and its dual space E'. Let V' be the dual space of V, with norm $||.||_{V'}$; identifying H with its dual H', we can write the inclusions $E \subset V \subset H \subset V'$, and we shall still denote the dual pairing between V' and V by $(.,.)$.

Moreover, let $(\Omega, \mathcal{F}, \mathcal{F}_t, P)$ be a complete stochastic basis with right-continuous filtration $\mathcal{F}_t, t \geq 0$, and let $w(t)$ be a K-valued cylindrical Wiener process (K being a real separable Hilbert space).

We consider the following stochastic evolution equation in the evolution triple $V \subset H \subset V'$

$$\begin{cases} du(t) = [Au(t) + F(u(t))]dt + B(u(t))\, dw(t), & t \in [0,T], \\ u(0) = u_0, \end{cases} \tag{1}$$

under the following hypotheses.

H.1 — $A : D(A) \subset H \to H$ is the infinitesimal generator of a strongly continuous semigroup of contractions e^{tA} in H, $e^{tA}(E) \subset E$ and the restriction of e^{tA} to E is still a strongly continuous semigroup of contractions (we keep the notations e^{tA} and A for the semigroup and its generator in E, if no confusion may arise); in particular (cf. [8]),

$$(Au, u) \leq 0, \quad u \in D(A). \tag{2}$$

$$< Au, u^* >_E \leq 0, \quad u \in D(A), u^* \in \partial ||u||_E, \tag{3}$$

where $\partial ||u||_E$ is the subdifferential of the $E-$norm at u (see [1]).

H.2 — $F : V \to V'$ satisfies

$$(F(u) - F(v), u - v) \leq -\alpha ||u - v||_V^p \tag{4}$$

$$(F(u), u) \leq -\beta ||u||_V^p + \rho \tag{5}$$

$$||F(u)||_{V'}^q \leq \gamma ||u||_V^p + \rho, \tag{6}$$

where α and β are strictly positive constants, $p > 1$, q satisfies $\frac{1}{q} + \frac{1}{p} = 1$, $\gamma \geq 0, \rho \geq 0$.

H.3 — $F(E) \subset E$, F is locally Lipschitz in E and

$$< F(u), u^* >_E \leq 0, \quad u \in E, u^* \in \partial ||u||_E. \tag{7}$$

H.4 — The mapping $u(.) \to F(u(.))$ is continuous from $L^p_{\mathcal{P}}([0,T] \times \Omega; V) \cap L^2_{\mathcal{P}}(\Omega; C(0,T;H))$ to $L^1_{\mathcal{P}}([0,T] \times \Omega; V')_\sigma$ (given a Banach space X, we denote by X_σ the space X endowed with the weak topology; the other notations are introduced in subsection 1.1 below).

H.5 — $B : H \to L_2(K,H)$ is Lipschitz continuous ($L_2(K,H)$ denotes the space of Hilbert-Schmidt operators from K to H).

H.6 — There exists an operator $A_0 : D(A) \subset H \to H$ which generates a strongly continuous semigroup of contractions e^{tA_0} in V, H and V' (after appropriate restriction and extension), and commutes with A in the sense that $(\lambda - A_0)^{-1}$ commutes with $(\lambda - A)^{-1}$ for $\lambda > 0$.

H.7 — $D(A) \cap V$ is dense in H, $D(A^*) \cap V$ is dense in V, and $E \cap D(A)$ is dense in $V \cap D(A)$.

Theorem 1.1 *Under the assumptions (H.1-7), there exists a unique solution* $u \in L^2_{\mathcal{P}}(\Omega; C(0,T;H)) \cap L^p_{\mathcal{P}}([0,T] \times \Omega; V)$ *of equation (1), for all given* $u_0 \in L^2(\Omega, \mathcal{F}_0, P; H)$, *and the mapping* $u_0 \to u$ *is continuous from* $L^2(\Omega, \mathcal{F}_0, P; H)$ *to* $L^2_{\mathcal{P}}(\Omega; C(0,T;H)) \cap L^p_{\mathcal{P}}([0,T] \times \Omega; V)$.

Note that we say that u, with the above regularity, is a solution of (1) if the identity

$$(u(t), v) = (u_0, v) + \int_0^t ((u(s), A^* v) + (F(u(s)), v)) \, ds + \left(\int_0^t B(u(s)) \, dw(s), v \right) \tag{8}$$

is satisfied for all $v \in D(A^*) \cap V$ and $t \in [0,T]$.

Remark — This result put in abstract form and extends a preliminary result on a concrete reaction-diffusion equation, [3]. The method is also partially similar to an iterative method developed by [7]. The main fact here, with respect to the previous literature, is that we deal with non-coercive equations (of monotone type). For instance, Theorem 1.1 covers certain first order differential equations, wave equations, and Schroedinger equations, with monotone nonlinearities. These examples are given in section 3, whereas section 2 is devoted to the proof of Theorem 1.1. The use of more regular spaces (E in this approach) where the operators defining the system posses suitable dissipativity properties is conceptually similar to a method developed by [9] in the linear case, but from many other points ov view the two approaches are totally different. A more abstract version of this method, with other applications, is in preparation, [4]. As discussed in [4], alternative, or more general, assumptions can be given at the following three levels: i) joint monotonicity and coercivity of F and B, as in [5] (thus extending the lipschitz assumption (H.5)), relaxing also the uniform monotonicity assumed in (4); ii) other sufficient conditions for the solvability of the deterministic equation (9) below, without introducing a space E with properties (3) and (7); iii) other technical conditions, in place of (H.6), to perform computations with chain rule and Ito formula.

1.1 Notations

Given a separable Banach space X (norm $||.||_X$) and a real number $p \geq 1$, we denote by $L^p(\Omega, \mathcal{F}_0, P; X)$ the space of $\mathcal{F}_0$-measurable random variable $u : \Omega \to X$ such that $E||u||_X^p < \infty$; by $L^p_{\mathcal{P}}([0,T] \times \Omega; X)$ the space of X-valued progressively measurable (with respect to $\mathcal{F}_t$) processes $u(t,\omega)$ such that $E\int_0^T ||u(t)||_X^p < \infty$; and, finally, by $L^p_{\mathcal{P}}(\Omega; C(0,T;X))$ the space of X-valued progressively measurable and continuous processes $u(t,\omega)$ such that $E\sup_{0\leq t\leq T} ||u(t)||_X^p < \infty$.

Moreover, if $M \in L^2_{\mathcal{P}}(\Omega; C(0,T;X))$ is a martingale (X being an Hilbert space), we denote by $<< M >>_t$ the quadratic variational process associated to M (taking values in the space of nuclear operators), and by $tr << M >>_t$ the trace of $<< M >>_t$ (cf. [6], [7], for instance).

2 Proof of Theorem 1.1

Step 1. — Auxiliary random differential equation.

We start by considering the following differential equation

$$\frac{du}{dt}(t,\omega) = Au(t,\omega) + F(u(t,\omega)) + f(t,\omega), \quad u(0,\omega) = u_0(\omega) \tag{9}$$

with certain given stochastic data $u_0(\omega)$ and $f(t,\omega)$. Precisely, we deal with the mild form of (9) in E:

$$u(t,\omega) = e^{tA}u_0(\omega) + \int_0^t e^{(t-s)A}[F(u(s,\omega)) + f(s,\omega)]\, ds. \tag{10}$$

In the analysis of this equation e^{tA} and A are understood as operators in E.

Under the assumptions (H.1,3) we have:

(R.1) — let $\mathcal{G}_t$ be a complete filtration, with $\mathcal{F}_0 \subset \mathcal{G}_0$; let $u_0 : \Omega \to E$ be an $\mathcal{F}_0$-measurable random variable, and let $f : [0,T] \times \Omega \to E$ be a progressively measurable process with respect to $\mathcal{G}_t$, with $f(.,\omega) \in C([0,T];E)$ P-a.s.; then there exists a unique solution u of equation (10) in the space of processes progressively measurable with respect to $\mathcal{G}_t$, with $u(.,\omega) \in C([0,T];E)$ P-a.s.; moreover, u satisfies P-a.s.

$$(u(t),v) = (u_0,v) + \int_0^t ((u(s),A^*v) + (F(u(s)),v))\, ds + \int_0^t (f(s),v)\, ds \tag{11}$$

for all $v \in D(A^*) \cap V$ and $t \in [0,T]$.

To prove this claim, we first consider equation (10) pathwise. Let $\omega \in \Omega$ be fixed such that $f(.,\omega) \in C([0,T];E)$ and $u_0(\omega) \in E$. Since F is locally Lipschitz in E, by a standard contraction argument there exists a unique local solution of (10), i.e. there exists an interval $[0,\tau(\omega)]$ and a function $u(.,\omega) \in C([0,\tau(\omega)];E)$ which is the unique solution of (10) on $[0,\tau(\omega)]$ in this space. Iterating the local procedure in a classical way, we can extend the local solution to the unique maximal solution $u(.,\omega) \in C(I(\omega);E)$, where $I(\omega) \subset [0,T]$ is the maximal interval of existence. Let T^* be the supremum of

$I(\omega)$. Since the solution is obtained by an explicit iterative procedure starting from the data, one can see that for all $\tau \in [0,T)$ the function $u(t,\omega)1_{\{T^\bullet(\omega)>\tau\}} : [0,\tau] \times \Omega \to E$ is $\mathcal{B}(0,\tau) \times \mathcal{G}_\tau$-measurable. If we prove that there exists a constant $c(\omega) > 0$ such that P-a.s.

$$\|u(t,\omega)\|_E \leq c(\omega), \quad t \in I(\omega), \tag{12}$$

then $I(\omega) = [0,T]$, and $u(t,\omega) : [0,T] \times \Omega \to E$ is progressively measurable with respect to the filtration $\mathcal{G}_t$. Hence, our final aim is to prove the a priori bound (12). We omit to mention ω, which is fixed, in the rest of the proof.

Let $u_n \in C(I_n; E)$ be the maximal solution of the approximating equation

$$u_n(t) = e^{tA_n} u_0 + \int_0^t e^{(t-s)A_n}[F(u_n(s)) + f(s)]\, ds, \tag{13}$$

where $A_n = nA(n-A)^{-1}$ (the Josida approximations of A, cf. [8]). Since A_n are bounded operators in E, we have $u_n \in C^1(I_n; E)$ and

$$\frac{du_n}{dt}(t) = A_n u_n(t) + F(u_n(t)) + f(t), \quad u_n(0) = u_0.$$

From [1], Ch.II.5 and (II.2.3), for all $t \in I$ and $u^*_{t,n} \in \partial\|u_n(t)\|_E$ we have

$$\begin{aligned} \frac{d^-}{dt}\|u_n(t)\| \;&\leq\; <\frac{du_n}{dt}(t), u^*_{t,n}>_E \\ &= < A_n u_n(t) + F(u_n(t)) + f(t), u^*_{t,n} >_E \\ &\leq\; < f(t), u^*_{t,n} >_E \;\leq\; \|f(t)\|_E. \end{aligned}$$

Hence

$$\|u_n(t)\|_E \leq \|u_0\|_E + \int_0^t \|f(s)\|_E\, ds, \tag{14}$$

which shows that $I_n = [0,T]$ for all $n \in N$. By the contraction principle depending on a parameter we have $u_n \to u$ in $C([0,\tau]; E)$ for all $\tau \in I$, so that the a priori bound (12) follows from (14).

Step 2. — Preliminary energy inequalities.

We interrupt the main line of the proof to prove some basic energy inequalities.

(R.2) — If $u \in L^2_{\mathcal{P}}(\Omega; C(0,T;H)) \cap L^p_{\mathcal{P}}([0,T] \times \Omega; V)$ is a process satisfying the identity

$$(u(t),v) = (u_0,v) + \int_0^t (u(s), A^*v)\, ds + \int_0^t (F(s),v)\, ds + (M(t),v), \tag{15}$$

$t \in [0,T], v \in D(A^*) \cap V$, for some $u_0 \in L^2(\Omega, \mathcal{F}_0, P; H), F \in L^q_{\mathcal{P}}([0,T] \times \Omega; V') \cap L^2_{\mathcal{P}}([0,T] \times \Omega; H)$, and some martingale $M \in L^2_{\mathcal{P}}(\Omega; C(0,T;H))$ (with $M(0) = 0$), then u satisfies the following inequality

$$\left\{ \begin{aligned} &\|u(t)\|^2_H \leq \|u_0\|^2_H + 2\int_0^t (F(s), u(s))\, ds \\ &+2\int_0^t (u(s), dM(s)) + tr << M >>_t, \quad t \in [0,T]; \end{aligned} \right. \tag{16}$$

similarly, if $u \in C(0,T;H) \cap L^p(0,T;V)$ is a (deterministic) function satisfying the identity

$$(u(t),v) = (u_0,v) + \int_0^t (u(s),A^*v)\,ds + \int_0^t (F(s),v)\,ds, \tag{17}$$

$t \in [0,T], v \in D(A^*) \cap V$, for some $u_0 \in H$ and $F \in L^q(0,T;V') \cap L^2(0,T;H)$, then u satisfies the following inequality

$$||u(t)||_H^2 \le ||u_0||_H^2 + 2\int_0^t (F(s),u(s))\,ds, \quad t \in [0,T]. \tag{18}$$

We prove only (16) bacause the proof of (18) is similar and more elementary. Take $v = J_n^* z$ in (15), $z \in D(A^*) \cap V$, $J_n = n(n-A_0)^{-1}$ ($J_n^* z \in D(A^*) \cap V$ by assumption (II.6)). Let $u_n(t) = J_n u(t)$, $u_{0,n} = J_n u_0$, $F_n(t) = J_n F(t)$, $M_n(t) = J_n M(t)$. Then, by (15),

$$u_n(t) = u_{0,n} + \int_0^t [Au_n(s) + F_n(s)]\,ds + M_n(t), \tag{19}$$

as an identity in V' (note that $u_n(s) \in D(A)$). By the classical Ito formula of [7] or [5], and inequality (2), we obtain the inequality (16) with $u_n, u_{0,n}, F_n$, and M_n in place of u, u_0, F, and M. Then we prove (16) as $n \to \infty$, since J_n converges strongly to the identity in V, H and V' (by assumption (H.6)). The proof of (R.2) is complete.

Finally, localizing the martingale M and using a classical martingale inequality we obtain from (16)

$$E||u(t)||_H^2 \le E||u_0||_H^2 + 2E\int_0^t (F(s),u(s))\,ds + E\,tr << M >>_t, \quad t \in [0,T], \tag{20}$$

and

$$E\sup_{0\le t\le T} ||u(t)||_H^2 \le c\left(E||u_0||_H^2 + E\sup_{0\le t\le T}\int_0^t (F(s),u(s))\,ds + E\,tr << M >>_T\right), \tag{21}$$

for some constant $c > 0$ independent of u_0, F, and M.

Step 3. — Deterministic a priori estimates.

Let us consider now an extension of equation (9) (compare with (11)):

$$(u(t),v) = (u_0,v) + \int_0^t ((u(s),A^*v) + (F(u(s)),v))\,ds + (M(t,\omega),v), \tag{22}$$

$v \in D(A^*) \cap V, t \in [0,T]$, which formally corresponds to the differential equation

$$\frac{du}{dt}(t) = Au(t) + F(u(t)) + \frac{dM}{dt}(t), \quad u(0) = u_0. \tag{23}$$

The point here is that M is not assumed to be differentiable (it will be a martingale in the next step). We want to prove the following pointwise a priori estimate:

(R.3) — there exists a constant $c > 0$ such that the following two properties hold true:

i) if, for a given $\omega \in \Omega$, we have $u_0(\omega) \in H$, $M(.,\omega) \in C([0,T];H) \cap L^p(0,T;V) \cap L^2(0,T;D(A))$ with $M(0) = 0$, and if $u \in C([0,T];H) \cap L^p(0,T;V)$ satisfies equation (22), then

$$sup_{0 \le t \le T} ||u(t)||_H^2 + \int_0^T ||u(t)||_V^p \, dt$$

$$\le c \left(||u_0||_H^2 + sup_{0 \le t \le T} ||M(t)||_H^2 + \int_0^T ||AM(t)||_H^2 \, dt + \int_0^T ||M(t)||_V^p \, dt + 2\rho T \right); \tag{24}$$

ii) similarly, if $u_1(.), u_2(.)$ are solutions of equation (22) corresponding to initial data $u_{0,1}, u_{0,2}$ and righr-hand-sides $M_1(.), M_2(.)$, with the same regularity as above, then

$$sup_{0 \le t \le T} ||u_1(t) - u_2(t)||_H^2 + \int_0^T ||u_1(t) - u_2(t)||_V^p \, dt$$

$$\le c \left(||u_{0,1} - u_{0,2}||_H^2 + sup_{0 \le t \le T} ||M_1(t) - M_2(t)||_H^2 + \int_0^T ||A(M_1(t) - M_2(t))||_H^2 \, dt \right)$$

$$+ c \left(2\rho T + \int_0^T ||u_1(t)||_V^p \, dt + \int_0^T ||u_2(t)||_V^p \, dt \right)^{\frac{1}{q}} \times \left(\int_0^T ||M_1(t) - M_2(t)||_V^p \, dt \right)^{\frac{1}{p}}. \tag{25}$$

Let us prove (24). Let $v(t) = u(t) - M(t)$. Then (22) gives

$$(v(t), x) = (u_0, x) + \int_0^t ((v(s), A^* x) + (AM(s), x) + (F(u(s)), x)) \, ds \tag{26}$$

for all $x \in D(A^*) \cap V$. By (R.2) we have

$$||v(t)||_H^2 \le ||u_0||_H^2 + 2 \int_0^t ((AM(s), v(s)) + (F(u(s)), v(s))) \, ds,$$

whence

$$||v(t)||_H^2 \le 2 \int_0^t (F(u(s)), u(s)) \, ds + ||u_0||_H^2$$

$$+ \int_0^t ||AM(s)||_H^2 \, ds + \int_0^t ||v(s)||_H^2 \, ds + 2 \int_0^t ||F(u(s))||_{V'} ||M(s)||_V \, ds,$$

by the classical inequality $2ab \le a^2 + b^2$. Let us now use assumption (5) to bound the first term on the left-hand-side. Moreover, by the Young inequality, for all $\varepsilon > 0$ there exists a constant $c(\varepsilon) > 0$ such that $ab \le \varepsilon a^q + c(\varepsilon) b^p$ for all $a, b \ge 0$, with $\frac{1}{q} + \frac{1}{p} = 1$. Hence

$$||F(u(s))||_{V'} ||M(s)||_V \le \varepsilon ||F(u(s))||_{V'}^q + c(\varepsilon) ||M(s)||_V^p$$
$$\le \varepsilon \gamma ||u(s)||_V^p + \varepsilon \rho + c(\varepsilon) ||M(s)||_V^p$$

where we have used (6). These facts yield

$$\|v(t)\|_H^2 - \int_0^t \|v(s)\|_H^2\,ds + 2(\beta - \varepsilon\gamma)\int_0^t \|u(s)\|_V^p\,ds - 2\rho T$$

$$\le \|u_0\|_H^2 + \int_0^t \|AM(s)\|_H^2\,ds + 2\varepsilon\rho T + 2c(\varepsilon)\int_0^t \|M(s)\|_V^p\,ds.$$

Recalling that $v(t) = u(t) - M(t)$ and that $M \in C(0,T;H)$, from the last inequality and Gronwall Lemma we easily obtain (24) (for ε sufficiently small to have $(\beta - \varepsilon\gamma) > 0$).

Finally, let us prove (25). As above, let $v_i(t) = u_i(t) - h_i(t), i = 1,2$. From (R.2) we have

$$\|v_1(t) - v_2(t)\|_H^2 + 2\int_0^t (A(M_1(s) - M_2(s)), v_1(s) - v_2(s))\,ds$$

$$+2\int_0^t (F(u_1(s)) - F(u_2(s)), v_1(s) - v_2(s))\,ds \le \|u_{0,1} - u_{0,2}\|_H^2,$$

whence, arguing as above,

$$\|v_1(t) - v_2(t)\|_H^2 + \alpha\int_0^t \|u_1(s) - u_2(s)\|_V^p\,ds$$

$$\le \|u_{0,1} - u_{0,2}\|_H^2 + \int_0^t \|A(M_1(s) - M_2(s))\|_H^2\,ds + \int_0^t \|v_1(s) - v_2(s)\|_H^2\,ds$$

$$+2\int_0^t (F(u_1(s)) - F(u_2(s)), M_1(s) - M_2(s))\,ds.$$

By the Holder inequality, the last term is bounded from above by

$$c\left(\int_0^t (\|F(u_1(s))\|_{V'}^q + \|F(u_2(s))\|_{V'}^q)\,ds\right)^{\frac{1}{q}} \times \left(\int_0^t \|M_1(s) - M_2(s)\|_V^p\,ds\right)^{\frac{1}{p}}$$

$$\le c\left(2\rho T + \int_0^t (\gamma\|u_1(s)\|_V^p + \gamma\|u_2(s)\|_V^p)\,ds\right)^{\frac{1}{q}} \times \left(\int_0^t \|M_1(s) - M_2(s)\|_V^p\,ds\right)^{\frac{1}{p}}.$$

for soma constant $c > 0$. Using this inequality and Gronwall lemma, as in the first part of the proof, we finally obtain (25). The proof of (R.3) is complete.

Step 4. — Auxiliary stochastic equation.

Next step is to solve the stochastic equation

$$du(t) = [Au(t) + F(u(t))]dt + dM(t), \quad u(0) = u_0, \tag{27}$$

for any H-valued square integrable continuous martingale (with respect to the filtration $\mathcal{F}_t$), with $M(0) = 0$. Here we study equation (27) in the evolution triple $V \subset H \subset V'$, in contrast to equation (9), whereas the space E is used only in the regularization of $M(t)$, to use the result of step 1.

Definition 2.1 *Given $u_0 \in L^2(\Omega, \mathcal{F}_0, P; H)$ and a square integrable H-valued continuous martingale (with respect to the filtration $\mathcal{F}_t$), with $M(0) = 0$ We say that $u \in L^2_{\mathcal{P}}(\Omega; C(0,T;H)) \cap L^p_{\mathcal{P}}([0,T] \times \Omega; V)$ is a solution to equation (27) if it satisfies*

$$(u(t), v) = (u_0, v) + \int_0^t ((u(s), A^* v) + (F(u(s)), v)) \, ds + (M(t), v) \tag{28}$$

for all $v \in V \cap D(A^*)$.

With the help of the existence result of step 1 and the inequalities of steps 2 and 3, we can now prove the following result on equation (27):

(R.4) — if $u_0 \in L^2(\Omega, \mathcal{F}_0, P; H)$ and M is a square integrable H-valued continuous martingale (with respect to the filtration $\mathcal{F}_t$), with $M(0) = 0$, then there exists a unique solution u of equation (27) in $L^2_{\mathcal{P}}(\Omega; C(0,T;H)) \cap L^p_{\mathcal{P}}([0,T] \times \Omega; V)$. Moreover, if u_1 and u_2 are two solutions corresponding to data $u_{0,1}, M_1$ and $u_{0,2}, M_2$, respectively, with the previous regularity, the following inequalities hold:

$$E||u_1(t) - u_2(t)||_H^2 + 2\alpha E \int_0^t ||u_1(s) - u_2(s)||_V^p \, ds \leq E||u_{0,1} - u_{0,2}||_H^2 + Etr << M_1 - M_2 >>_t \tag{29}$$

and

$$E \sup_{0 \leq t \leq T} ||u_1(t) - u_2(t)||_H^2 \leq c \left(E||u_{0,1} - u_{0,2}||_H^2 + Etr << M_1 - M_2 >>_T \right), \tag{30}$$

for some constant $c > 0$ independent of u_0, F, and M.

The last claims of (R.4) (inequalities (29)-(30)) are strightforward consequences of (R.2) and of assumption (H.2). From (29) and (30) it then follows the uniqueness part of (R.4). Hence we have only to prove the existence claim of (R.4).

We first prove this result under the additional assumption that

$$M \in L^2_{\mathcal{P}}(\Omega; C(0,T; V \cap D(A))). \tag{31}$$

Using mollifiers with compact support of diameter tending to zero, and the density of $E \cap D(A)$ in $V \cap D(A)$ (assumption (H.7)), we can approximate M with a sequence of processes M_n (no more martingales) which are progressively measurable with respect to $\mathcal{F}_{t+\frac{1}{n}}$, with $M_n(.,\omega) \in C^1([0,T]; E \cap D(A))$ P-a.s., such that $M_n(.,\omega) \to M(.,\omega)$ in $C([0,T]; V \cap D(A))$ P-a.s. Then the solutions u_n of equation (28) corresponding to M_n (which exist by step 1, with $f = \frac{dM_n}{dt}$) converge to some u P-a.s. in $C([0,T]; H) \cap L^p(0,T;V)$, by the inequalities of step 3. u is progressively measurable with respect to $\mathcal{F}_t$ because u_n is progressively measurable with respect to $\mathcal{F}_{t+\frac{1}{n}}$ and $\mathcal{F}_t$ is right-continuous. Moreover, u satisfies equation (28) (we use assumption (H.4) to take the limit in the term with $F(u_n)$). Finally, $u \in L^2_{\mathcal{P}}(\Omega; C(0,T;H)) \cap L^p_{\mathcal{P}}([0,T] \times \Omega; V)$ by inequality (24), since $M \in L^2_{\mathcal{P}}(\Omega; C(0,T; V \cap D(A)))$. Therefore we have proved (R.4) under the additional assumption (31).

Finally, the general case follows from the previous one approximating the general martingale $M \in L^2_{\mathcal{P}}(\Omega; C(0,T;H))$ by a sequence of martingales $M_n \in L^2_{\mathcal{P}}(\Omega; C(0,T; V \cap$

$D(A)))$ ($V \cap D(A)$ is dense in H by assumption (H.7)). If u_n are the corresponding solutions of (27), then by (29) and (30) we see that u_n converges to some u in $L^2_{\mathcal{P}}(\Omega; C(0,T;H)) \cap L^p_{\mathcal{P}}([0,T] \times \Omega; V)$, and we conclude that u is a solution of (27) by the same limit argument used above. The proof of (R.4) is complete.

Step 5. — Solution to the original equation (1).

Let $\tau \in [0,T]$. Given $v \in L^2_{\mathcal{P}}(\Omega; C(0,\tau;H))$ (also less is sufficient), let $u \in L^2_{\mathcal{P}}(\Omega; C(0,\tau;H)) \cap L^p_{\mathcal{P}}([0,\tau] \times \Omega; V)$ be the unique solution of the stochastic equation

$$\begin{cases} du(t) = [Au(t) + F(u(t))]dt + B(v(t))\,dw(t), & t \in [0,\tau], \\ u(0) = u_0, \end{cases} \tag{32}$$

constructed in step 4 (indeed, $M(t) = \int_0^t B(v(s))\,dw(s)$ is a given martingale in $L^2_{\mathcal{P}}(\Omega; C(0,\tau;H))$). We have thus defined a mapping Γ_τ in $L^2_{\mathcal{P}}(\Omega; C(0,\tau;H))$, $\Gamma_\tau(v) = u$. Γ_τ is a contraction in $L^2_{\mathcal{P}}(\Omega; C(0,\tau;H))$, for sufficiently small τ, since from (30) we have:

$$E \sup_{0 \le t \le \tau} \|u_1(t) - u_2(t)\|_H^2 \le cE \int_0^\tau \|B(v_1(s)) - B(v_2(s))\|^2_{L_2(K,H)}\,ds, \tag{33}$$

and B is lipschitz continuous by (H.5) (here $u_j = \Gamma_\tau(v_j), j = 1,2$) . Hence there exists a unique fixed point of Γ_τ, i.e. a unique local solution to equation (1). From the previous estimate we also see that this local procedure can be repeated over interval of constant lenght; therefore, in a finite number of steps we obtain the global solution to (1). The last claim of the theorem on the continuous dependence of the solution on u_0 follows from inequalities (29) and (30) and the application of Gronwall Lemma. The proof of the Theorem is complete.

3 EXAMPLES

3.1 First Order Differential Equation

Consider the following first order differential equation in a bounded domain $D \subset R^d$ with smooth boundary ∂D

$$\begin{cases} du = \alpha(x)\nabla u\,dt - |u|^{p-2}u\,dt + g(u)\,dw(t) \\ u = 0 \quad \text{on } \partial D \\ u(0) = u_0, \end{cases} \tag{34}$$

where $w(t)$ is a one-dimensional Brownian motion (for simplicity), $\alpha(.) \in C^1(\overline{D}; R^3)$ is a vector field, and $g \subset C^1(R)$ has bounded first derivatives. To write this equation in the abstract form (1) we set $H = L^2(D)$, $V = L^p(D)$, $E = C_0(\overline{D}) = \{u \in C(\overline{D}); u = 0$ on $\partial D\}$. Moreover, let $A : D(A) \subset H \to H$ be the infinitesimal generator of the semigroup e^{tA} defined as $(e^{tA}u_0)(x) = u_0(\phi(t,x))$, where $\phi(t,x)$ is the characteristic flow defined by the equation $\frac{\partial\phi}{\partial t} + \alpha(\phi) = 0$, $\phi(0,x) = x$. With obvious definitions of F and B, and with the choice $A_0 = A$, the assumptions of Theorem 1.1 are satisfied.

3.2 Schroedinger Equation

Consider the following Schroedinger equation in one space dimension (for simplicity in the choice of E) on the space interval $x \in [0,1]$

$$\begin{cases} du = i\frac{\partial^2 u}{\partial x^2}\, dt - |u|^{p-2}u\, dt + g(u)\, dw(t) \\ u(t,0) = u(t,1) = 0 \\ u(0,x) = u_0(x), \quad x \in [0,1], \end{cases} \tag{35}$$

where $Re(g), Im(g) \in C^1(R)$ with bounded first derivatives. Here we rewrite the equation as a system in the real and immaginary parts of u. Then we take $H = L^2(0,1) \times L^2(0,1)$, $V = L^p(0,1) \times L^p(0,1)$, $E = H_0^1(0,1) \times H_0^1(0,1)$ (which is an algebra); $A : D(A) \subset H \to H$, $D(A) = D(\tilde{A}) \times D(\tilde{A})$ and $A(u_1,u_2) = (-\tilde{A}u_2, \tilde{A}u_1)$, where $D(\tilde{A}) = H^2(0,1) \times H_0^1(0,1)$ and $\tilde{A}u = \frac{\partial^2 u}{\partial x^2}$; and finally A_0 defined as $A_0(u_1,u_2) = (\tilde{A}u_1, \tilde{A}u_2)$. With obvious definitions of F and B, the assumptions of Theorem 1.1 are satisfied.

3.3 Wave Equation

Consider now the following wave equation in one space dimension on the space interval $x \in [0,1]$

$$\begin{cases} du_t = \frac{\partial^2 u}{\partial x^2}\, dt - |u_t|^{p-2}u_t\, dt + g(u_t)\, dw(t) \\ u(t,0) = u(t,1) = 0 \\ u(0,x) = u^0(x), u_t(0,x) = u^1(x), \quad x \in [0,1], \end{cases} \tag{36}$$

with g as in the first example. Here we rewrite (as usual) the equation as a system in the variables (u, u_t), and we take $H = H_0^1(0,1) \times L^2(0,1)$, $V = H_0^1(0,1) \times L^p(0,1)$, $E = D(\tilde{A}) \times H_0^1(0,1)$ (which is an algebra), where $\tilde{A}$ is defined in the previous example; $A : D(A) \subset H \to H$, $D(A) = D(\tilde{A}) \times H_0^1(0,1)$ and $A(u_1,u_2) = (u_2, \tilde{A}u_1)$; and finally A_0 defined as $A_0(u_1,u_2) = (\tilde{A}u_1, \tilde{A}u_2)$. With obvious definitions of F and B, the assumptions of Theorem 1.1 are satisfied, a part from conditions (4) and (5), which have to be weakened as follows:

$$(F(u) - F(v), u - v) \le -\alpha||u - v||_V^p + \lambda_1||u - v||_H^2 \tag{37}$$

$$(F(u), u) \le -\beta||u||_V^p + \rho + \lambda_2||u||_H^2 \tag{38}$$

for some constants $\lambda_1, \lambda_2 \in R$. The proof of Theorem 1.1 can be modified under these more general assumptions without introducing any conceptual novelty. Thus the result of Theorem 1.1 holds true also for the wave equation (36).

References

[1] G. Da Prato (1976) Applications Croissantes et Equations d'Evolutions dans les Espaces de Banach, Academic Press, New York.

[2] G. Da Prato, J. Zabczyk (1988) *A note on semilinear stochastic equations*, Diff. Int. Eq. 1, N. 2, 143-155.

[3] F. Flandoli (1991) *A stochastic reaction-diffusion equation with multiplicative noise*, Appl. Math. Lett. 4, 45-48.

[4] F. Flandoli *Nonlinear stochastic evolution equations with monotone operators*, in preparation.

[5] N. V. Krylov, B. L. Rozovskii (1981) *Stochastic evolution equations*, J. Sov. Math. **16**, 1233-1277.

[6] M. Metivier (1988) STOCHASTIC PARTIAL DIFFERENTIAL EQUATIONS IN INFINITE DIMENSIONAL SPACES, Quaderni Scuola Normale Superiore, Pisa.

[7] E. Pardoux (1975) *Equations aux derivees partielles stochastiques non lineaires monotones*, These, Paris Sud, Centre D'Orsay, n. 1556.

[8] A. Pazy (1983) SEMIGROUPS OF LINEAR OPERATORS AND APPLICATIONS TO PARTIAL DIFFERENTIAL EQUATIONS, Springer-Verlag, New York.

[9] B. L. Rozovskii (1990) STOCHASTIC EVOLUTION SYSTEMS, D. Riedel Publishing Company, Dordrecht-Boston.

EXISTENCE OF A SMOOTH DENSITY FOR THE FILTER IN NONLINEAR FILTERING ON MANIFOLDS.

Patrick FLORCHINGER
URA CNRS 399, Département de Mathématiques,
UFR MIM, Université de Metz, Ile du Saulcy,
F 57045 METZ Cedex, France.
&
INRIA LORRAINE, Projet Congé, METZ, France.

Summary : In this paper, we prove by means of Malliavin calculus on manifolds that under the Hörmander condition, the filter associated with a nonlinear filtering problem on Riemannian manifolds has a smooth density.

0. Introduction.

The purpose of this paper is to prove the existence of a smooth density for the filter associated with a nonlinear filtering problem on Riemannian manifolds.

This problem has been investigated, for nonlinear filtering problems on Euclidean spaces, by means of Malliavin calculus (for a detailed exposition of the Malliavin calculus, we refer to [12], [1], [15-16], [23], [13] and references there in) or by analytic techniques applied to stochastic partial differential equations (for the latter approach, we refer to [3], [17] and references there in).

Nonlinear filtering problems where the observation process evolves on a Riemannian manifold have been studied by T.Duncan [6] and M.Pontier and J.Szpirglas [18-19]. On the other hand, S.Ng and P.Caines [14] gave a general formulation of the nonlinear filtering problem when the system-process and the observation process are both with values on a Riemannian manifold. A Bayes formula for the conditional expectation of smooth functions of the system-process is proved. Furthermore, they proved that the density of the filter, provided it exists, solves a Zakai equation (c.f. [24]).

Here, by means of the Malliavin calculus on manifolds (c.f. [22], [2], [4]), we prove under the Hörmander condition on the coefficients of the system-process that the filter associated with a nonlinear filtering problem on Riemannian manifolds admits a smooth density.

This paper is divided in four sections organized as follows. In section one, we present some preliminary results about the Malliavin calculus on manifolds that will be needed in the sequel. Section two gives the model used to describe the nonlinear filtering problem when the processes take their values in a Riemannian manifold. In section three, an unnormalized filter linked with the initial filter by a Bayes formula is introduced. In section four, we prove by the Malliavin calculus, under a Hörmander condition, the existence of a smooth density for the filter associated to the problem considerer in the paper.

Finally, for any tool of differential geometry we refer the reader to [10] and [21]. Furthermore, notice that along the paper, in most cases, the summation sign is omitted for repeated indexes.

1. Some elements of Malliavin calculus on manifolds.

In this section, we recall some results on the stochastic calculus of variations on manifolds that we need in the sequel. We refer to [2], [22] or [4] for a complete presentation of the subject.

Denote by $\{w_t, t\in[0,T]\}$ a standard d-dimensional Wiener process on the canonical space $(\mathcal{W},\mathcal{F},P)$ (i.e. $\mathcal{W} = C\left([0,T];\mathbb{R}^d\right)$, P is the Wiener measure and $\mathcal{F}$ is the completion of the Borel σ-algebra of $\mathcal{W}$ with respect to P). For each $p\geq 1$, let $L^p(P)$ be the Banach space of p-th integrable random variables on the space $(\mathcal{W},\mathcal{F},P)$ endowed with the norm $\|f\|_p = \left(E\left(f^p\right)\right)^{1/p}$ where E stands for the expectation with respect to P.

Following [15-16] and [13], we say that a $\mathbb{R}$-valued random variable $F : \mathcal{W} \to \mathbb{R}$ is a smooth functional if it is expressed as $f\left(w_{t_1},\dots,w_{t_n}\right)$ for some $n\in\mathbb{N}^*$, $0\leq t_1,\dots,t_n\leq T$ where $f(x)=f\left(x^{11},\dots,x^{d1},\dots,x^{1n},\dots,x^{dn}\right)$ is in $C_b^\infty\left(\mathbb{R}^{dn}\right)$. In the following, the space of all smooth functionals will be denoted by $\mathcal{S}$.

The derivative of a smooth functional F can be defined as the d-dimensional stochastic process given by

$$D_t^j F = \sum_{i=1}^{n} \frac{\partial f}{\partial x^{ji}}\left(w_{t_1},\dots,w_{t_n}\right) 1_{[0,t_i]}(t) \tag{1.1}$$

$j\in\{1,\dots,d\}$, $t\in[0,T]$.

Moreover, denoting by H the Cameron-Martin subspace of the Wiener space $(\mathcal{W},\mathcal{F},P)$, the derivative DF can be regarded as a H-valued random variable. Hence, iterating formula (1.1), the N-derivative of a functional F in $\mathcal{S}$ will be the $H^{\otimes N}$-valued random variable $D^N F$ defined by

$$\left(D^N F\right)_{s_1,\dots,s_N}^{j_1,\dots,j_N} = D_{s_1}^{j_1}\dots D_{s_N}^{j_N} F \tag{1.2}$$

$$= \sum_{i_1,\dots,i_N=1}^{n} \frac{\partial^N f}{\partial x^{j_1 i_1}\dots\partial x^{j_N i_N}}\left(w_{t_1},\dots,w_{t_N}\right) 1_{[0,t_{i_1}]}(s_1)\dots 1_{[0,t_{i_N}]}(s_N)$$

On the other hand, for any integer $N\geq 1$ and real number $p>1$, introduce the norm $\|.\|_{p,N}$ on $\mathcal{S}$ by

$$\|F\|_{p,N} = \|F\|_p + \left\| \left\| D^N F \right\|_{H.S.} \right\|_p \tag{1.3}$$

where $\|.\|_{H.S.}$ denotes the Hilbert-Schmidt norm in $H^{\otimes N}$,

$$\|D^N F\|^2_{H.S.} = \sum_{j_1,\ldots,j_N=1}^{d} \int_{[0,1]^N} \left[\left(D^N F \right)^{j_1,\ldots,j_N}_{s_1,\ldots,s_N} \right]^2 ds_1 \ldots ds_N \tag{1.4}$$

Then, for any integer $N \geq 1$ and any real number $p>1$, denote by $\mathbb{D}_{p,N}$ the Banach space which is the completion of $\mathcal{S}$ in $L^p(P)$ with respect to the norm $\|.\|_{p,N}$.

Furthermore, notice that the derivation operator D is a closed linear operator defined in $\mathbb{D}_{2,1}$ with values in $L^2\left(\mathcal{W} \times [0,T]; \mathbb{R}^d\right)$ (c.f. [23], [20], [16]).

Moreover, introduce as in [26] and [18], the algebra $\mathbb{D}_\infty$ defined by

$$\mathbb{D}_\infty = \bigcap_{\substack{p\geq 1 \\ N\in\mathbb{N}^*}} \mathbb{D}_{p,N} \tag{1.5}$$

which is the space of all infinitely differentiable Wiener functionals in the sense of Malliavin.

Now, let us extend theses tools to manifold-valued Wiener functionals.

Let N be a σ-compact C^∞ manifold and, introduce

$$\mathbb{D}_\infty(N) = \left\{ G : \mathcal{W} \to N \text{ s.t. } F(G) \in \mathbb{D}_\infty \ \forall F \in C_0^\infty(N) \right\} \tag{1.6}$$

where $C_0^\infty(N)$ denotes the space of N-valued C^∞ functions with compact support.

Then, $\mathbb{D}_\infty(N)$ is the space of all "infinitely differentiable" N-valued Wiener functionals.

Next, we give some conditions under which the solution of a stochastic differential equation evolving on a manifold is infinitely differentiable in the sense of Malliavin.

Let M and N be two σ-compact connected Riemannian manifolds of dimensions m and n equipped with the Riemannian metrics g_M and g_N respectively.

Denote by $X_0, X_1, \ldots, X_d$, (d+1) C^∞-vector fields on the manifold N and by π a C^∞ mapping of N into M. Moreover, assume that the following conditions are fullfiled.

(C1) There exists a diffeomorphism ψ of N onto a closed submanifold $\tilde{N}$ of $\mathbb{R}^k$ and (d+1) C^∞-vector fields on $\mathbb{R}^k$, $\tilde{X}_0, \tilde{X}_1, \ldots, \tilde{X}_d$ such that

(1) $(i_*)_x \left(\psi_* X_j \right)_x = \left(\tilde{X}_j \right)_x$, $x \in \tilde{N}$, $j=0,\ldots,d$

where i denotes the inclusion mapping of N into $\mathbb{R}^k$.

(2) Set $\tilde{X}_j = \alpha^i_j \frac{\partial}{\partial x_i}$, $j=0,\ldots,d$.

Then, α^i_j, $i=1,\ldots,k$, $j=0,\ldots,d$ is of linear growth order and its derivatives of any order are bounded.

(C2) π is a proper mapping (i.e. for each compact subset K of M, the inverse image $\pi^{-1}(K)$ is a compact subset of N).

Let $(x_t)_{t\in[0,T]}$ be the N-valued stochastic process, solution of the following stochastic differential equation written in the sense of Stratonovitch (c.f. [9], [8], [5-7]),

$$\begin{cases} dx_t = X_0(x_t)dt + \sum_{i=1}^{d} X_i(x_t)\circ dw_t^i \\ x_0 \in M \end{cases} \tag{1.7}$$

and, $(\Phi_t)_{t\in[0,T]}$ be the stochastic flows of diffeomorphisms of N onto itself associated with the stochastic process $(x_t)_{t\in[0,T]}$ (c.f. [9], [8], [11]).

Introduce the M-valued stochastic process $(y_t)_{t\in[0,T]}$ defined for any t in [0,T] by $y_t = \pi(x_t)$. Then, applying the derivative operator D to the stochastic process $(y_t)_{t\in[0,T]}$, the following regularity result holds

Theorem 1.1 [22] : For any t in [0,T], y_t is in $\mathbb{D}_\infty(M)$ and, for every functions F in $C_0^\infty(M)$ and h in H, it holds

$$<DF(y_t)(\omega),h>_H = \left((\pi_*)_{x_t(\omega)} (X_{\omega,t,h})_{x_t(\omega)} \right) F$$

where $X_{\omega,t,h}$ is the random C^∞-vector field defined by

$$X_{\omega,t,h} = \int_0^t \sum_{k=1}^{d} \dot{h}^k \, (\Phi_t(\omega)\circ\Phi_s(\omega)^{-1})_* X_k \, ds .$$

In the following, we define the Malliavin covariance associated with the stochastic process $(y_t)_{t\in[0,T]}$.

With this aim, introduce the C^∞-tensor field B^0 on N of type (2,0) given by

$$B_x^0(u_1,u_2) = \sum_{i=1}^{d} u_1((X_i)_x)\, u_2((X_i)_x)$$

for any x in N and u_1, u_2 in T_x^*N.

Then, a non-negative definite symetric bilinear form $<<Dy_t,Dy_t>>(\omega)$ on $T^*_{y_t(\omega)}M$ defined by

$$<<Dy_t,Dy_t>>(\omega)(u_1,u_2) = \int_0^t (\pi_*)_{x_t(\omega)} \left\{ (\Phi_t(\omega)\circ\Phi_s(\omega)^{-1})_* B^0 \right\} (u_1,u_2)\, ds \tag{1.8}$$

for any u_1, u_2 in $T^*_{y_t(\omega)}M$ is called the Malliavin covariance for the stochastic process $(y_t)_{t\in[0,T]}$.

On the other hand, since $T^*_{y_t(\omega)}M$ is equipped with the inner product assigned by the Riemannian metric g_M, set

$$g^t(\omega) = \begin{cases} 1/\det(\langle\langle Dy_t, Dy_t\rangle\rangle(\omega)) & \text{if } \det(\langle\langle Dy_t, Dy_t\rangle\rangle(\omega))>0 \\ 0 & \text{otherwise} \end{cases}$$

In the rest of this section, we fix t in [0,T] and assume that condition

$$\text{(C)} \qquad f(y_t)\, g^t \in \bigcap_{p\in(1,\infty)} L^p(P) \qquad \text{for any f in } C_0^\infty(M)$$

is fullfiled.
In order to state a regularity result to the law of the stochastic process $(y_t)_{t\in[0,T]}$, let us first give an integration by parts formula which is an usefull tool of the Malliavin calculus.

Lemma 1.2 (Integration by parts formula) [22] **:** For any differential operator ∂ on M and ϕ in $C_0^\infty(M)$, there exist $p\in(1,\infty)$, $r\in\mathbb{N}$ and a continuous linear mapping $\xi : \mathbb{D}_{p,r} \to L^1(P)$ such that, for every f in $C_0^\infty(M)$ and G in $\mathbb{D}_{p,r}$

$$E\Big(\partial f(y_t)\, \phi(y_t)\, G\Big) = E\Big(f(y_t)\, \xi(G)\Big)$$

Then, the following result, similar to that obtained in the case of euclidean space, can be proved

Proposition 1.3 [22] **:** For each G in $\mathbb{D}_\infty$, the function $g_G(x) = \langle\hat{\delta}_G, G\rangle$ is C^∞ and, for any f in $C_0^\infty(M)$

$$E\Big(f(y_t)\, G\Big) = \int_M f(x)\, g_G(x)\, \nu(dx)$$

where δ_x is the Dirac's δ-function at x in M and ν is the Riemannian volume element on M. In fact, the function p defined by $p(x) = \langle\hat{\delta}_x, 1\rangle$ is the C^∞-density of the law of the stochastic process $(y_t)_{t\in[0,T]}$ with respect to ν.

Then, to prove the regularity of the law of the stochastic process $(y_t)_{t\in[0,T]}$, it suffices to check condition (C). To conclude this section, we recall the Hörmander condition (c.f. [1-2]) which is a sufficient condition that assumption (C) is fullfiled.

Notation 1.4 : In this paper, denote by

1) $\mathcal{L}(X_0,\ldots,X_d)$ the Lie algebra generated by the vector fields $X_0,\ldots,X_d$.

2) $\mathcal{I}(X_1,\ldots,X_d)$ the ideal generated by the vector fields $X_1,\ldots,X_d$ in the Lie algebra $\mathcal{L}(X_0,\ldots,X_d)$.

Hence, the following result holds

Theorem 1.5 [22] **:** Let $(x_t)_{t\in[0,T]}$ be the solution of the stochastic differential equation (1.7), the coefficients of which satisfy assumptions (C1) and (C2). For any t in [0,T], set $y_t = \pi(x_t)$. Then, for each t in [0,T], y_t satisfies assumption (C) provided condition

$$\text{(H)} \qquad (\pi_*)_x\, \mathcal{I}_x = T_{\pi(x)}M \qquad \text{for every x in N}$$

is fullfiled. In particular, if condition (H) is fullfiled, the law of the stochastic process $(y_t)_{t\in[0,T]}$ on M has a C^∞-density with respect to the Riemannian volume element.

2. The nonlinear filtering model.

In this section, we recall the model used in [14] by Ng-Caines to introduce a general formulation of nonlinear filtering problems in Riemannian manifolds.
Let $(\Omega,\mathcal{F},P)$ be a complete probability space and w, v two independent Wiener processes on this space of dimension d and d_1 respectively.

2.1 Modeling of the system-process.

Let M be a compact connected an oriented Riemannian manifold of dimension m equipped with the Riemannian metric g_M. Denote by O(M) the bundle of orthonormal frame over M and by p_M the projection of O(M) onto M. Let (x^i,e^i_j), i,j=1,...,d be local coordinates around (x,e) in O(M) and $\left\{\Gamma^q_{i\,l}\right\}$ the Christoffel symbols of the Riemannian connection on M compatible with the metric g_M. Introduce the family $\left\{L_1,\dots,L_d\right\}$ of canonical horizontal vector-fields on O(M) with respect to the Riemannian connection. Note that around (x,e) in O(M), L_j, j=1,....,d is expressed as

$$L_j = e^i_j\left(\frac{\partial}{\partial x^i} - \Gamma^q_{i\,l}\, e^l_p \frac{\partial}{\partial e^q_p}\right) \tag{2.1}$$

Let X_0 be a C^∞-vector field on M written as

$$X_0 = \alpha^i \frac{\partial}{\partial x^i} \tag{2.2}$$

in local coordinates. Denote by $\tilde{X}_0$ the horizontal lift of the vector field X_0 with respect to the connection $\left\{\Gamma^q_{i\,l}\right\}$. In local coordinates, $\tilde{X}_0$ is expressed as

$$\tilde{X}_0 = \alpha^i\left(\frac{\partial}{\partial x^i} - \Gamma^q_{i\,l}\, e^l_p \frac{\partial}{\partial e^q_p}\right) \tag{2.3}$$

Now, introduce the system-process $(x_t)_{t\in[0,T]}$ as the M-valued stochastic process defined by

$$x_t = p_M(r_t) \tag{2.4}$$

where $r_t = (x_t,e_t)$ is the solution of the following stochastic differential equation on O(M) written in the Stratonovitch sense (c.f. [9], [8])

$$\begin{cases} dr_t = \tilde{X}_0(r_t)\, dt + \sum_{i=1}^{d} L_i(r_t) \circ dw^i_t \\ \\ r_0 = (x_0,e_0) \in O(M) \end{cases} \tag{2.5}$$

To obtain the expression of (2.5) in local coordinates, we refer the reader to [14].

2.2 Modelling of the observation process.

Let N be a σ-compact connected Riemannian manifold of dimension d_1 with the Riemannian metric g_N. Denote by O(N) the bundle of orthonormal frame over N and by p_N the projection of O(N) onto N. Let (y^i, f^i_j), $i,j=1,\ldots,d_1$ be local coordinates around (y,f) in O(N) and $\{\gamma^q_{i1}\}$ the Christoffel symbols of the Riemannian connection on N compatible with the metric g_N. Denote by $\left\{H_1,\ldots,H_{d_1}\right\}$ the system of canonical horizontal vector-fields on O(N) with respect to the Riemannian connection. Note that around (y,f) in O(N), H_j, $j=1,\ldots,d_1$ is expressed as

$$H_j = f^i_j\left(\frac{\partial}{\partial y^i} - \gamma^q_{i1} f^1_p \frac{\partial}{\partial f^q_p}\right) \tag{2.6}$$

Introduce $h(x_t,y)$ a C^∞-bounded vector field on N written in local coordinates as

$$h(x_t,y) = h^i(x_t,y)\frac{\partial}{\partial y^i} \tag{2.7}$$

Let $\tilde{h}$ be the horizontal lift of the vector field $h(x_t,y)$ with respect to the connection $\{\gamma^q_{i1}\}$. Then, in local coordinates, $\tilde{h}$ is expressed as

$$\tilde{h} = h^i\left(\frac{\partial}{\partial y^i} - \gamma^q_{i1} f^1_p \frac{\partial}{\partial e^q_p}\right) \tag{2.8}$$

Let $(y_t)_{t\in[0,T]}$ be the observation process defined as the N-valued stochastic process given by

$$y_t = p_N(s_t) \tag{2.9}$$

where $s_t = (y_t, f_t)$ is the solution of the following stochastic differential equation on O(N) written in the Stratonovitch sense (c.f. [9], [8])

$$\begin{cases} ds_t = \tilde{h}(x_t,s_t)\,dt + \sum_{i=1}^{d_1} H_i(s_t)\circ dv^i_t, \\ s_0 \in O(N) \end{cases} \tag{2.10}$$

To obtain the expression of (2.10) in local coordinates, we refer the reader to [14].

2.3 The filter.

In this section, we define the filter associated with the system-process observation pair (x_t,y_t) with values in MxN.

First, as in [14], note from the expression of (2.10) in local coordinates that

$$\sigma(s_\tau, 0\le\tau\le t) = \sigma(y_\tau, 0\le\tau\le t) \tag{2.11}$$

which implies that it is equivalent to observe the stochastic process $(x_t)_{t\in[0,T]}$ throught $\sigma(s_\tau, 0\le\tau\le t)$ or $\sigma(y_\tau, 0\le\tau\le t)$.

Then, as usualy in the case of Euclidean space, let us state

Definition 2.1 : For any t in [0,T] and any function ψ in $C_0^\infty(M)$, denote by $\Pi_t\psi$ the filter associated with the system-process observation pair (x_t,y_t) given by ((2.5), (2.10)) defined by

$$\Pi_t\psi = E\left[\,\psi(x_t)\,/\,\mathcal{Y}_t\,\right] \tag{2.12}$$

where $\mathcal{Y}_t = \sigma(y_\tau, 0\le\tau\le t)$ and E stands for the expectation with respect to P.

3. A Bayes formula - The unnormalized filter.

In this section, we define by means of a change of probability measure an unnormalized filter linked with the filter Π_t by an abstract Bayes formula. In case of nonlinear filtering on Euclidean spaces, this new probability measure is called the reference probability measure (c.f. [17]).

Let $(\tilde\Omega,\tilde{\mathcal{F}},\tilde P)$ be an independent copy of the probability space $(\Omega,\mathcal{F},P)$. Introduce on the probability space $(\tilde\Omega,\tilde{\mathcal{F}},\tilde P)$ the M-valued stochastic process $(\tilde x_t)_{t\in[0,T]}$ which as the same probability law as the stochastic process $(x_t)_{t\in[0,T]}$. On the probability space $(\tilde\Omega,\tilde{\mathcal{F}},\tilde P)$, we may set

$$\tilde x_t = p_M(\tilde r_t) \tag{3.1}$$

where $\tilde r_t$ is the solution of the stochastic differential equation written in the Stratonovitch sense

$$\begin{cases} d\tilde r_t = \tilde X_0(\tilde r_t)\,dt + \sum_{i=1}^{d} L_i(\tilde r_t)\circ dw_t^i \\ \\ \tilde r_0 = r_0 \in O(M) \end{cases} \tag{3.2}$$

Next, introduce as usualy in nonlinear filtering the Girsanov exponential associated with the stochastic processes $(\tilde x_t)_{t\in[0,T]}$ and $(y_t)_{t\in[0,T]}$ by

$$\begin{aligned} \Lambda_t(\tilde x_t,y_t) = \exp\Bigg(&\int_0^t <h(\tilde x_s,y_s),dy_s>_{y_s} - \frac{1}{2}\int_0^t \sum_{j=1}^{d_1}\sum_{k=1}^{d_1} \gamma_{kj}^k(y_s)\,h^j(\tilde x_s,y_s)\,ds \\ &- \frac{1}{2}\int_0^t tr\left(\frac{\partial H}{\partial y}(\tilde x_s,y_s)\right) ds - \frac{1}{2}\int_0^t <h(\tilde x_s,y_s),h(\tilde x_s,y_s)>_{y_s} ds \Bigg) \qquad P\otimes\tilde P \text{ a.s.} \end{aligned} \tag{3.3}$$

where $H=(h^1,\ldots,h^{d_1})$ and $<.,.>_y$ denotes the scalar product in T_yN $(y\in N)$ induced by the Riemannian metric g_N.

Then, the unnormalized filter associated with the system-process observation pair can be defined by

Definition 3.1 : For any t in [0,T] and any function ψ in $C_0^\infty(M)$, denote by $\rho_t\psi$ the filter associated with the system-process observation pair defined by

$$\rho_t\psi = E_{\tilde{P}}\left[\psi(\tilde{x}_t)\,\Lambda_t(\tilde{x}_t,y_t)\right] \tag{3.4}$$

where $E_{\tilde{P}}$ stands for the expectation with respect to the probability law $\tilde{P}$.

Therefore, following the proof of the Girsanov formula in [8], it has been proved in [14],

Theorem 3.2 [14] **:** For any t in [0,T] and any function ψ in $C_0^\infty(M)$, it holds

$$\Pi_t\psi = \frac{\rho_t\psi}{\rho_t 1} \tag{3.5}$$

4. The main result.

In this section, we state by means of Malliavin calculus on manifolds, an existence result of a smooth density for the filter Π_t.
With this aim, introduce Hörmander's condition on the coefficients of equation (2.5) that we need to state our result.

Notation 4.1 : Denote by $\mathcal{I}(L_1,....,L_d)$ the ideal generated by the vector fields $L_1,....,L_d$ in the Lie algebra $\mathcal{L}(\tilde{X}_0,L_1,....,L_d)$.

Then, introduce condition

(H) $\qquad (p_{M*})_r\,\mathcal{I}_r = T_{p_M(r)}M \qquad$ for every r in O(M).

Therefore, we can state the main result of the paper,

Theorem 4.2 : Assume that condition (H) is fullfiled and that the vector field $\tilde{X}_0$ satisfies condition (C1) of section one. Then, for any t in [0,T], the filter Π_t admits a C^∞-density with respect to the Riemannian volume element on M.

Proof of Theorem 4.2 : As the filter Π_t is linked with the unnormalized filter ρ_t by the Bayes formula (3.5) (Theorem 3.2) it is equivalent to prove the existence of a C^∞-density for the filter Π_t or the unnormalized filter ρ_t.

In the sequel, we show how the Malliavin calculus on manifolds approach gives under Hörmander's condition (H) the existence of a regular density with respect to the Riemannian volume element.
In the view of the compactness of M, the bundle O(M) is compact. Hence, it is obvious that the vector fields $\tilde{X}_0,L_1,....,L_d$ and the map p_M satisfy conditions (C1) and (C2) stated in section one.

On the other hand, by the Malliavin calculus on manifolds (c.f. [22]) and by usual arguments of the Malliavin calculus on Euclidean spaces, the following result holds,

Proposition 4.3 : For any t in [0,T], $\Lambda_t(\tilde{x}_t,y_t)$ is in $\mathbb{D}_\infty$.

Furthermore, according with formula (3.4), we have

$$\rho_t\psi = E_{\tilde{P}}\left[\psi(\tilde{x}_t)\,\Lambda_t(\tilde{x}_t,y_t)\right].$$

Moreover, according with formula (3.1), for any t in [0,T], we have $\tilde{x}_t = p_M(\tilde{r}_t)$ where the stochastic process $(\tilde{r}_t)_{t\in[0,T]}$ solves the stochastic differential equation (3.2), the coefficients of which satisfies conditions (C1), (C2) and Hörmander's condition (H).
Thus, one can deduce from Proposition 4.3, the integration by parts formula, Proposition 1.3 and Theorem 1.5 that for any t in [0,T], there exists a C^∞-function p_t on M such that

$$\rho_t\psi = \int_M \psi(x)p_t(x)\upsilon(dx) \tag{4.1}$$

Hence, $p_t(x)$ is the C^∞-density of the unnormalized filter ρ_t.
This concludes the proof of Theorem 4.2.

References.

1. Bismut, J.M. : Martingales, the Malliavin calculus and hypoellipticity under general Hörmander conditions. Z. Wahrscheinlichkeitstheorie Verw. Gebiete **56** (1981) 469-505.
2. Bismut, J.M. : Large deviations and the Malliavin calculus. Progress in Mathematics **45**. Birkhaüser, Boston 1984.
3. Chaleyat-Maurel, M., Michel, D. : Hypoellipticity theorems and conditional laws. Z. Wahrscheinlichkeitstheorie Verw. Gebiete **65** (1984) 573-597.
4. Davis, M.H.A. : The Wiener space derivative for functionals of diffusion on manifolds. Nonlinearity **1** (1988) 241-251.
5. Duncan, T.E. : Stochastics systems in Riemannian manifolds. J. Optimization Theory Appl. **27** (1976) 399-426.
6. Duncan, T.E. : Some filtering results in Riemannian manifolds. Information and Control **35** (1977) 182-195.
7. Duncan, T.E. : Stochastic calculus on manifolds with applications. In : Baras, J.S., Mirelli, V. (eds.) Recent advances in stochastic calculus. (Progress in Automation and Information Systems pp 105-140) Berlin Heidelberg New York: Springer 1990.
8. Elworthy, K.D. : Stochastic differential equations on manifolds. London Mathematical Society. Lecture Note Series 70. Cambridge University Press 1982.
9. Ikeda, N., Watanabe, S. : Stochastic differential equations and diffusion processes. Amsterdam Oxford New York: North Holland, Tokyo: Kodansha 1981.
10. Kobayashi, S., Nomizu, K. : Fundations of differential geometry. Vol 1,2. Wiley-Interscience 1969.
11. Kunita, H. : Stochastic differential equations and stochastic flows of diffeomorphisms. In : P.L.Hennequin (ed.) Ecole d'Eté de Probabilités de Saint Flour XII (Lect. Notes Math., vol 1097, pp 144-303) Berlin, Heidelberg, New York : Springer 1984.
12. Malliavin, P. : Stochastic calculus of variations and hypoelliptic operators. Proccedings of the International Conference on Stochastic Differential Equations (Kyoto 1976), 195-263. Tokyo: Kinokuniya; New York: Wiley 1978.

13. Michel, D., Pardoux, E. : An introduction to Malliavin calculus and some of its applications. In : Baras, J.S., Mirelli, V. (eds.) Recent advances in stochastic calculus. (Progress in Automation and Information Systems pp 65-104) Berlin Heidelberg New York: Springer 1990.

14. Ng, S.K., Caines, P.E. : Nonlinear filtering in Riemannian manifolds. IMA J. Mathematical Control and Information **2** (1985) 25-36.

15. Nualart, D. : Non causal stochastic integral and calculus. In: Korezlioglu, H., Ustunel, A.S. (eds.) Stochastic analysis and related topics. (Lect. Notes Math., vol 1316, pp 80-129) Berlin Heidelberg New York: Springer 1988.

16. Ocone, D. : A guide to the stochastic calculus of variations. In: Korezlioglu, H., Ustunel, A.S. (eds.) Stochastic analysis and related topics. (Lect. Notes Math., vol 1316, pp 1-79) Berlin Heidelberg New York: Springer 1988.

17. Pardoux, E. : Filtrage non linéaire et équations aux dérivées partielles stochastiques associées. In: P.L.Hennequin (ed.) Ecole d'Eté de Probabilités de Saint Flour XIX (Lect. Notes Math.) Berlin, Heidelberg, New York : Springer (to appear).

18. Pontier, M., Szpirglas, J. : Filtrage non linéaire avec observation sur une variété. Stochastics **15** (1985) 121-148.

19. Pontier, M., Szpirglas, J. : Filtering with observation on a Riemannian symetric space. SIAM J. Control and Optimization **26** (3) (1988) 609-627.

20. Shigekawa, I. : Derivatives of Wiener functionals and absolute continuity of induced measures. J. Math. Kyoto Univ. **20** (2) (1980) 263-289.

21. Sternberg, S. : Lectures on differential geometry. Prentice-Hall 1965.

22. Taniguchi, S. : Malliavin's stochastic calculus of variations for manifold-valued Wiener functionals and its applications. Z. Wahrscheinlichkeitstheorie Verw. Gebiete **65** (1983) 269-290.

23. Watanabe, S. : Stochastic differential equations and Malliavin calculus. Tata Inst. of Fundamental Research, Bombay : Springer 1984.

24. Zakai, M. : On the optimal filtering of diffusion processes. Z. Wahrscheinlichkeitstheorie Verw. Gebiete **11** (1969) 230-243.

ON THE ITÔ FORMULA FOR TWO–PARAMETER MARTINGALES

by

Nikos Frangos, David Nualart and *Marta Sanz–Solé*

0. Introduction

In the reference [3] Föllmer has introduced a pathwise Itô calculus for functions of the Brownian motion. The basic idea which allows to construct this calculus is the fact that almost all the trajectories of the Brownian motion have a finite quadratic variation along a suitable decreasing sequence of partitions of [0,1].

The stochastic calculus for two–parameter processes has been developed from the basic work of Cairoli and Walsh [2]. In [8] a version of the Itô formula has been established for two-parameter continuous martingales bounded in L^4, assuming that the underlying two–parameter filtration verifies the conditional independence property (F4) introduced in [2]. Furthermore (see [7,10]), two–parameter continuous martingales possess different types of one and two–dimensional quadratic variations.

The purpose of this paper is to apply the ideas introduced by Föllmer in [3] in order to develop a pathwise Itô calculus for two–parameter continuous martingales. The organization of the paper is as follows. In Section 2 we establish a deterministic change of variable formula for paths $X_{s,t}$, $(s,t) \in [0,1]^2$ which have variations up to the fourth order along a fixed sequence of grids in $[0,1]^2$. In Section 3 we apply this result to prove an Itô formula for a two–parameter square integrable continuous martingale M analogous to that obtained in [8], but without requiring M to be bounded in L^4.

1. Notation and basic assumptions

The stochastic processes considered here are parametrized by $T = [0,1]^2$. The unit square is ordered by " $\leq$ ", which is the usual coordinatewise linear ordering. Usually the coordinates of $z \in T$ will be denoted by (s,t); rectangles in T by $J = J_1 \times J_2$, with J_i, $i = 1,2$, intervals in [0,1]. For $J = (z_1, z_2]$, we write $J^1 = (s_1, s_2] \times [0, t_1]$, $J^2 = [0, s_1] \times (t_1, t_2]$.

If f is a map from T to R the increment of f over a rectangle $J = (z_1, z_2]$, $z_i = (s_i, t_i)$, $i = 1,2$, is defined by

$$\Delta_J f = f(z_1) - f(s_2, t_1) - f(s_1, t_2) + f(z_2) .$$

(In the case $z_i \in R$, $\Delta_J f = f(z_2) - f(z_1)$). Let X be a real function on T, zero on the axes. We will write $X_{s,t}$ for $X(s,t)$. Let us consider two sequences of partitions of [0, 1], $\{\mathcal{S}_n^1, n \geq 1\}$ and $\{\mathcal{S}_m^2, m \geq 1\}$ whose mesh tends to zero as n and m tends to infinity, respectively, and let $\mathcal{S}_{n,m} = \mathcal{S}_n^1 \times \mathcal{S}_m^2$ be the grid of T defined by the product of $\mathcal{S}_n^1$ and $\mathcal{S}_m^2$. That means, if $\mathcal{S}_n^1 = \{0 = s_1^n \leq \ldots \leq s_{p_n}^n \leq s_{p_n+1}^n = 1\}$ and $\mathcal{S}_m^2 = \{0 = t_1^m \leq \ldots \leq t_{q_m} \leq t_{q_m+1}^m = 1\}$, the rectangles of $\mathcal{S}_{n,m}$ are given by $(s_i^n, s_{i+1}^n] \times (t_j^m, t_{j+1}^m]$, $i = 0, \ldots, p_n$, $j = 0, \ldots, q_m$. They will be denoted by J. To simplify the notation we will omit the superscripts n and m in the description of the partitions.

We assume the following conditions on X.

(H1) The sequence $\{ \sum_{s_i \in S_n^1} (X_{s_{i+1},t} - X_{s_i,t})^2 \, \delta_{s_i,t} \, , \; n \geq 1\}$ converges vaguely, for any $t \in [0,1]$, as n tends to infinity, to a positive measure on [0, 1] denoted by $d\langle X_{.,t}\rangle$.

(H2) The sequence $\{ \sum_{t_j \in S_m^2} (X_{s,t_{j+1}} - X_{s,t_j})^2 \, \delta_{s,t_j} \, , \; m \geq 1\}$ converges vaguely, for any $s \in [0,1]$, as m tends to infinity, to a positive measure on [0, 1] denoted by $d\langle X_{s,.}\rangle$.

(H3) The sequence of measures $\{ \sum_{J \in S_{n,m}} (\Delta_J(X))^2 \, \delta_{s_i t_j} \, , \; n, m \geq 1\}$ converges vaguely, as n, m tend to infinity, to a positive measure on T denoted by $d\langle X\rangle$.

(H4) The sequence $\{ \sum_{J \in S_{n,m}} (\Delta_{J^1}(X))^2 \, (\Delta_{J^2}(X))^2 \, \delta_{s_i,t_j} \, , \; n, m \geq 1\}$ converges vaguely, as n, m tend to infinity, to a positive measure on T denoted by $d\langle \tilde{X}\rangle$.

(H5) The sequence $\{ \sum_{J \in S_{n,m}} (\Delta_J(X)) \, (\Delta_{J^1}(X)) \, (\Delta_{J^2}(X)) \, \delta_{s_i,t_j} \, , \; n, m \geq 1\}$ converges vaguely, as n, m tend to infinity, to a signed measure on T denoted by $d\langle X \, , \tilde{X}\rangle$.

Following Föllmer ([3]), we will say that X possesses *variations up to the fourth order along the sequence* $\{S_{n,m} \, , \; n, m \geq 1\}$ if hypotheses (H1) through (H5) are satisfied.

2. Deterministic version of Itô's formula

The next theorem is the two-parameter version of the deterministic Itô's formula established in [3].

Theorem 2.1. Let X be a real continuous function on T, null on the axis and having variations up to the fourth order along some sequence $\{S_{n,m} \, , \; n, m \geq 1\}$. Let f be a real function of class C^4 such that $f(0) = 0$. Then

$$I_{s,t} := \lim_{n \to \infty} \lim_{m \to \infty} \sum_{s_i \in S_n^1 \cap [0,s]} \sum_{t_j \in S_m^2 \cap [0,t]} \{ f'(X_{s_i,t_j}) \Delta_J(X) + f''(X_{s_i,t_j}) \Delta_{J^1}(X) \Delta_{J^2}(X)\}$$

exists, and we have the following change of variables formula

$$\begin{aligned} f(X_{s,t}) = I_{s,t} &+ \frac{1}{2} \int_0^s f''(X_{x,t}) \, d\langle X_{.,t}\rangle_x + \frac{1}{2} \int_0^t f''(X_{s,y}) \, d\langle X_{s,.}\rangle_y \\ &- \frac{1}{2} \int_{R_{st}} f''(X_z) \, d\langle X\rangle_z - \int_{R_{st}} f'''(X_z) \, d\langle X, \tilde{X}\rangle_z \\ &- \frac{1}{4} \int_{R_{st}} f^{iv}(X_z) \, d\langle \tilde{X}\rangle_z \, , \end{aligned}$$

where R_{st} denotes the rectangle $[0, s] \times [0, t]$.

Proof. Using Taylor's expansion we obtain

$$\begin{aligned} (1) \qquad f(X_{1,1}) &= \sum_{s_i \in S_n^1} [f(X_{s_{i+1},1}) - f(X_{s_i,1})] \\ &= \sum_{s_i \in S_n^1} f'(X_{s_i,1}) \, (X_{s_{i+1},1} - X_{s_i,1}) + \frac{1}{2} \sum_{s_i \in S_n^1} f''(\overline{X}_i) \, (X_{s_{i+1},1} - X_{s_i,1})^2 \, , \end{aligned}$$

where $\overline{X}_i$ is an intermediate point between $X_{s_i,1}$ and $X_{s_{i+1},1}$. Hypothesis (H1) yields

(2) $$f(X_{1,1}) = \lim_{n\to\infty} \sum_{s_i\in S_n^1} f'(X_{s_i,1})(X_{s_{i+1},1} - X_{s_i,1}) + \frac{1}{2}\int_0^1 f''(X_{s,1})\, d\langle X_{\cdot,1}\rangle_s \,.$$

Set

$$A_{n,m} = \sum_{s_i\in S_n^1}\sum_{t_j\in S_m^2} f'(X_{s_i,t_j})\, \Delta_J(X)\,,$$

$$B_{n,m} = \sum_{s_i\in S_n^1}\sum_{t_j\in S_m^2} \left(f'(X_{s_i,1}) - f'(X_{s_i,t_j})\right) \Delta_J(X)\,.$$

We have

$$\sum_{s_i\in S_n^1} f'(X_{s_i,1})\,(X_{s_{i+1},1} - X_{s_i,1}) = A_{n,m} + B_{n,m}\,.$$

Furthermore $B_{n,m} = C_{n,m} + D_{n,m}$ with

$$C_{n,m} = \sum_{s_i\in S_n^1}\sum_{t_j\in S_m^2} \Delta_{J^2}\left(f'(X)\right)\Delta_J(X)\,,$$

and

$$D_{n,m} = \sum_{s_i\in S_n^1}\sum_{t_j\in S_m^2} \Delta_{J^2}\left(f'(X)\right)\,\Delta_{J^1}(X)\,,$$

Using Taylor's formula we can further develop $C_{n,m}$ as follows.

(3) $$\begin{aligned} C_{n,m} = &\sum_{s_i\in S_n^1}\sum_{t_j\in S_m^2} f''(X_{s_i,t_j})\,\Delta_{J^2}(X)\,\Delta_J(X) \\ &+\frac{1}{2}\sum_{s_i\in S_n^1}\sum_{t_j\in S_m^2} f'''(\overline{X}_{i,j})\,\left(\Delta_{J^2}(X)\right)^2\,\Delta_J(X), \end{aligned}$$

where $\overline{X}_{i,j}$ is an intermediate random point between $X_{s_i,t_{j+1}}$ and X_{s_i,t_j}. By hypothesis (H2) we have, for any $s_i \in S_n^1$,

(4) $$\lim_{m\to\infty} \sum_{t_j\in S_m^2} f'''(\overline{X}_{i,j})\,\left(\Delta_{J^2}(X)\right)^2\,\Delta_J(X) = 0\,.$$

Let us mention the following elementary identity: For any real number α_i, $i = 1,\ldots,4$,

(5) $$(\alpha_4 - \alpha_1)\,(\alpha_1 - \alpha_2 - \alpha_4 + \alpha_3) = \frac{1}{2}\left\{(\alpha_3 - \alpha_2)^2 - (\alpha_4 - \alpha_1)^2 - (\alpha_1 - \alpha_2 - \alpha_4 + \alpha_3)^2\right\}.$$

Consequently, the first term of the right hand side of (3) can be written as

$$\frac{1}{2}\Big\{\sum_{s_i\in S_n^1}\sum_{t_j\in S_m^2} f''(X_{s_i,t_j})\,\left[(X_{s_{i+1},t_{j+1}} - X_{s_{i+1},t_j})^2 - (\Delta_{J^2}(X))^2 - (\Delta_J(X))^2\right]\Big\}.$$

Hypotheses (H2) and (H3), together with (4) ensure

(6)

$$\lim_{n\to\infty}\lim_{m\to\infty} C_{n,m} = \frac{1}{2}\Big\{\lim_{n\to\infty}\sum_{s_i\in S_n^1}\int_0^1 f''(X_{s_i,y})\, d\langle X_{s_{i+1},\cdot}\rangle_y$$

$$-\int_T f''(X_{s,t})\, d\langle X\rangle_{s,t} - \lim_{n\to\infty} \sum_{s_i\in S_n^1} \int_0^1 f''(X_{s_i,y})\, d\langle X_{s_i,\cdot}\rangle_y \Big\}$$

$$= -\frac{1}{2}\int_T f''(X_{s,t})\, d\langle X\rangle_{s,t} + \frac{1}{2}\lim_{n\to\infty} \sum_{s_i\in S_n^1} \int_0^1 f''(X_{s_{i+1},y})\, d\langle X_{s_{i+1},\cdot}\rangle_y$$

$$-\frac{1}{2}\lim_{n\to\infty} \sum_{s_i\in S_n^1} \int_0^1 f''(X_{s_i,y})\, d\langle X_{s_i,\cdot}\rangle_y - \frac{1}{2}\lim_{n\to\infty} \sum_{s_i\in S_n^1} \int_0^1 [f''(X_{s_{i+1},y}) - f''(X_{s_i,y})]\, d\langle X_{s_{i+1},\cdot}\rangle_y$$

$$= -\frac{1}{2}\int_T f''(X_{s,t})\, d\langle X\rangle_{s,t} + \frac{1}{2}\int_0^1 f''(X_{1,y})\, d\langle X_{1,\cdot}\rangle_y$$

$$-\frac{1}{2}\lim_{n\to\infty} \sum_{s_i\in S_n^1} \int_0^1 [f''(X_{s_{i+1},y}) - f''(X_{s_i,y})]\, d\langle X_{s_{i+1},\cdot}\rangle_y\,.$$

We now take care of the term $D_{n,m}$. As for $C_{n,m}$ we use Taylor's expansion to obtain

$$\begin{aligned} D_{n,m} = &\sum_{s_i\in S_n^1}\sum_{t_j\in S_m^2} f''(X_{s_i,t_j})\, \Delta_{J^2}(X)\, \Delta_{J^1}(X) \\ &+\frac{1}{2}\sum_{s_i\in S_n^1}\sum_{t_j\in S_m^2} f'''(X_{s_i,t_j})\, (\Delta_{J^2}(X))^2\, \Delta_{J^1}(X) \\ &+\frac{1}{6}\sum_{s_i\in S_n^1}\sum_{t_j\in S_m^2} f^{iv}(\tilde{X}_{i,j})\, (\Delta_{J^2}(X))^3\, \Delta_{J^1}(X)\,, \end{aligned}$$

where $\tilde{X}_{i,j}$ is an intermediate point between X_{s_i,t_j} and $X_{s_i,t_{j+1}}$. Then, hypothesis (H2) yields

$$\begin{aligned} \lim_{n\to\infty}\lim_{m\to\infty} D_{n,m} = &\lim_{n\to\infty}\lim_{m\to\infty} \sum_{s_i\in S_n^1}\sum_{t_j\in S_m^2} f''(X_{s_i,t_j})\, \Delta_{J^2}(X)\, \Delta_{J^1}(X) \\ &+\frac{1}{2}\lim_{n\to\infty}\sum_{s_i\in S_n^1}\lim_{m\to\infty}\sum_{t_j\in S_m^2} f'''(X_{s_i,t_j})\, (\Delta_{J^2}(X))^2\, \Delta_{J^1}(X)\,. \end{aligned} \tag{7}$$

In order to finish the proof we have to identify the value of

$$\begin{aligned} L = \frac{1}{2}\Big\{ &\lim_{n\to\infty}\sum_{s_i\in S_n^1}\lim_{m\to\infty}\sum_{t_j\in S_m^2} f'''(X_{s_i,t_j})\, (\Delta_{J^2}(X))^2\, \Delta_{J^1}(X) \\ &-\lim_{n\to\infty}\sum_{s_i\in S_n^1}\int_0^1 [f''(X_{s_{i+1},y}) - f''(X_{s_i,y})]\, d\langle X_{s_{i+1},\cdot}\rangle_y \Big\}\,. \end{aligned} \tag{8}$$

Using hypothesis (H2) and the identity (5) the right hand side of (8) can be replaced by

$$\begin{aligned} &\frac{1}{2}\Big\{ \lim_{n\to\infty}\sum_{s_i\in S_n^1}\lim_{m\to\infty}\sum_{t_j\in S_m^2} [f'''(X_{s_i,t_j})\, (\Delta_{J^2}(X))^2\, \Delta_{J^1}(X) \\ &-\Delta_{J^1}(f''(X))\, (X_{s_{i+1},t_{j+1}} - X_{s_{i+1},t_j})^2] \Big\} \\ &= \frac{1}{2}\Big\{ \lim_{n\to\infty}\sum_{s_i\in S_n^1}\lim_{m\to\infty}\sum_{t_j\in S_m^2} [f'''(X_{s_i,t_j})\, (\Delta_{J^2}(X))^2\, \Delta_{J^1}(X) \\ &-\Delta_{J^1}(f''(X))\, \{2\Delta_{J^2}(X)\, \Delta_J(X) + (\Delta_{J^2}(X))^2 + (\Delta_J(X))^2\} \Big\}\,. \end{aligned} \tag{9}$$

Finally,

$$\Delta_{J^1}(f''(X)) = f'''(X_{s_i,t_j})\,((\Delta_{J^1}(X)) + \frac{1}{2}\, f^{iv}(X'_{i,j})\,(\Delta_{J^1}(X))^2 ,$$

for some random point $X'_{i,j}$ between X_{s_{i+1},t_j} and X_{s_i,t_j}.
Furthermore, by Schwarz's inequality

$$\sum_{(s_i,t_j)\in S_n^1\times S_m^2} f^{iv}(X'_{i,j})\,(\Delta_{J^1}(X))^2\,\Delta_{J^2}(X)\,\Delta_J(X)$$
$$\le C\Big\{\sum_{(s_i,t_j)\in S_n^1\times S_m^2} (\Delta_J(X))^2\Big\}^{1/2}\Big\{\sum_{(s_i,t_j)\in S_n^1\times S_m^2} (\Delta_{J^1}(X))^4\,(\Delta_{J^2}(X))^2\Big\}^{1/2} .$$

Consequently, due to (H1), (H3) and (H4)

$$\lim_{n\to\infty}\lim_{m\to\infty}\sum_{(s_i,t_j)\in S_n^1\times S_m^2} f^{iv}(X'_{i,j})\,(\Delta_{J^1}(X))^2\,\Delta_{J^2}(X)\,\Delta_{J^1}(X) = 0 ,$$

and

$$\lim_{n,m\to\infty}\sum_{s_i\in S_n^1}\sum_{t_j\in S_m^2}\Delta_{J^1}(f''(X))\,(\Delta_J(X))^2 = 0 .$$

Therefore

$$L = -\lim_{n,m\to\infty}\sum_{s_i\in S_n^1}\sum_{t_j\in S_m^2} f'''(X_{s_i,t_j})\,(\Delta_{J^1}(X))\,(\Delta_{J^2}(X))\,(\Delta_J(X))$$
$$-\frac{1}{4}\lim_{n,m\to\infty}\sum_{s_i\in S_n^1}\sum_{t_j\in S_m^2} f^{iv}(X'_{i,j})\,(\Delta_{J^1}(X))^2\,(\Delta_{J^2}(X))^2 ,$$

and hypotheses (H4) and (H5) ensure

$$L = -\int_T f'''(X_{s,t})\,d\langle X,\tilde X\rangle - \frac{1}{4}\int_T f^{iv}(X_{s,t})\,d\langle\tilde X\rangle . \tag{10}$$

¿From (2), (6), (7) and (10) it follows that

$$\lim_{n\to\infty}\lim_{m\to\infty}\sum_{s_i\in S_n^1}\sum_{t_j\in S_m^2}\{\,f'(X_{s_i,t_j})\,\Delta_J(X) + f''(X_{s_i,t_j})\,\Delta_{J^2}(X)\,\Delta_{J^1}(X)\,\}$$

exists, and the formula for $f(X_{1,1})$ holds. The proof of the theorem is complete. ■

3. An Itô formula

Let $(\Omega,\mathcal{F},P)$ be a complete probability space, $\{\mathcal{F}_z,\ z\in T\}$ an increasing family of sub σ–fields of $\mathcal{F}$. For any $z=(s,t)\in T$, the σ–fields $\mathcal{F}^1_{(s,t)}$ and $\mathcal{F}^2_{(s,t)}$ are defined to be $\mathcal{F}_{(s,1)}$ and $\mathcal{F}_{(1,t)}$ respectively. We assume that the usual conditions (F1) to (F4) of Cairoli–Walsh [2] are satisfied. For $p\ge 1$ let $\mathcal{M}^p_c$ be the class of all sample continuous martingales M null on the axis such that $E(|M_z|^p)<+\infty$ for any $z\in T$. We will denote by $M_{\cdot,t}$, $M_{s,\cdot}$, the one–parameter martingales $\{M_{(s,t)},\ \mathcal{F}_{(s,1)};\ s\in[0,1]\}$ and $\{M_{(s,t)},\ \mathcal{F}_{(1,t)};\ t\in[0,1]\}$, respectively.

Let $M\in\mathcal{M}^2_c$. In order to apply Theorem 2.1 we have to show that almost all paths of M satisfy hypotheses (H1) through (H5) along a suitable sequence of grids $\mathcal{S}_{n,m}$ independent of the path. Hypotheses (H1) through (H3) follow from Theorems 3.3 and 3.4 of [7] (see also [4], and [11] for not necessarily continuous martingales and just $L\log^+ L$–bounded). Hypotheses (H4) and (H5) hold true for uniformly bounded not necessarily continuous martingales (see Theorem 3.1 of [5] and the proof of Lemma 2.5 in [8]). In general we have the result stated in Proposition 3.3.

Lemma 3.1. Let M and N be two regular square integrable martingales. Then there exists a constant C, such that for any partition $\mathcal{S}$ of T we have

$$E\Big\{\big(\sum_{J\in\mathcal{S}}(\Delta_{J^1}(M))^2\,(\Delta_{J^2}(N))^2\big)^{1/2}\Big\}$$
$$\le C\Big\{E(M_{1,1}^2)\,E(N_{1,1}^2)\Big\}^{1/2}.$$

Proof. First notice that

$$\text{(11)}\qquad E\Big\{\big(\sum_{J\in\mathcal{S}}(\Delta_{J^1}(M))^2\,(\Delta_{J^2}(N))^2\big)^{1/2}\Big\}$$
$$\le E\Big\{\big(\sum_{J^1\in\mathcal{S}^1}\sup_{t\in[0,1]}(\Delta_{J^1}(M_{\cdot,t}))^2\sum_{J^2\in\mathcal{S}^2}\sup_{s\in[0,1]}(\Delta_{J^2}(N_{s,\cdot}))^2\big)^{1/2}\Big\}$$
$$\le\Big\{E\big(\sum_{J^1\in\mathcal{S}^1}\sup_{t\in[0,1]}(\Delta_{J^1}(M_{\cdot,t}))^2\big)\,E\big(\sum_{J^2\in\mathcal{S}^2}\sup_{s\in[0,1]}(\Delta_{J^2}(N_{s,\cdot}))^2\big)\Big\}^{1/2}$$
$$\le C\{E(M_{1,1}^2)\,E(N_{1,1}^2)\}^{1/2},$$

where the last inequality follows from Lemma 2.2 [9] (see also Lemma 1.1 [1] and Proposition 2.3 of [5]). ■

Lemma 3.2. Let M be a regular square integrable martingale, $\{\mathcal{S}_n=\mathcal{S}_n^1\times\mathcal{S}_n^2,\ n\ge 1\}$ a sequence of partitions of T whose mesh tends to zero as n tends to infinty. For any $m\in N$, set

$$M^m_{\cdot}=E\{(-m\vee M_{1,1})\wedge m/\mathcal{F}_{\cdot}\}.$$

Then the sequences

$$\text{(a)}\qquad \sum_{J\in\mathcal{S}_n}(\Delta_{J^1}(M^m))^2\,(\Delta_{J^2}(M^m))^2-\sum_{J\in\mathcal{S}_n}(\Delta_{J^1}(M))^2\,(\Delta_{J^2}(M))^2,\ m\ge 1,$$

and

$$\text{(b)}\qquad \sum_{J\in\mathcal{S}_n}(\Delta_J(M^m))(\Delta_{J^1}(M^m))(\Delta_{J^2}(M^m))-\sum_{J\in\mathcal{S}_n}(\Delta_J(M))(\Delta_{J^1}(M))(\Delta_{J^2}(M)),\ m\ge 1,$$

tend to zero as m tends to infinity in $L^0(\Omega,\mathcal{F},P)$, uniformly in n.

Proof. First observe that, for each $n,m\in N$

$$\text{(12)}\qquad \Big|\sum_{J\in\mathcal{S}_n}(\Delta_{J^1}(M^m))^2\,(\Delta_{J^2}(M^m))^2-\sum_{J\in\mathcal{S}_n}(\Delta_{J^1}(M))^2(\Delta_{J^2}(M))^2\Big|$$
$$\le\Big|\sum_{J\in\mathcal{S}_n}(\Delta_{J^1}(M^m))^2(\Delta_{J^2}(M^m))^2-\sum_{J\in\mathcal{S}_n}(\Delta_{J^1}(M^m))^2(\Delta_{J^2}(M))^2\Big|$$
$$+\Big|\sum_{J\in\mathcal{S}^n}(\Delta_{J^1}(M^m))^2(\Delta_{J^2}(M))^2-\sum_{J\in\mathcal{S}_n}(\Delta_{J^1}(M))^2(\Delta_{J^2}(M))^2\Big|.$$

Now

$$
\begin{aligned}
&\Big|\sum_{J\in\mathcal{S}_n}(\Delta_{J^1}(M^m))^2\left\{(\Delta_{J^2}(M))^2-(\Delta_{J^2}(M^m))^2\right\}\Big| \\
&\le \sum_{J\in\mathcal{S}_n}(\Delta_{J^1}(M^m))^2(\Delta_{J^2}(M-M^m))^2 \\
&+2\sum_{J\in\mathcal{S}_n}(\Delta_{J^1}(M^m))^2\,|(\Delta_{J^2}(M^m))|\,|\Delta_{J^2}(M-M^m)| \\
&\le \sum_{J\in\mathcal{S}_n}(\Delta_{J^1}(M^m))^2(\Delta_{J^2}(M-M^m))^2 \\
&+2\Big\{\sum_{J\in\mathcal{S}_n}(\Delta_{J^1}(M^m))^2(\Delta_{J^2}(M^m))^2\sum_{J\in\mathcal{S}_n}(\Delta_{J^1}(M^m))^2(\Delta_{J^2}(M-M^m))^2\Big\}^{1/2}
\end{aligned}
$$

Let

$$
\begin{aligned}
Y_{m,n} &= \sum_{J\in\mathcal{S}_n}(\Delta_{J^1}(M^m))^2(\Delta_{J^2}(M-M^m))^2, \\
Z_{m,n} &= \sum_{J\in\mathcal{S}_n}(\Delta_{J^1}(M^m))^2(\Delta_{J^2}(M^m))^2
\end{aligned}
$$

By Lemma 3.1 applied to the regular square integrable martingales M^m and $M-M^m$, we obtain the convergence of $\{Y_{m,n}, m \geq 0\}$, to 0 in L^0 as $m \to \infty$, uniformly in n. The same lemma applied to M^m ensures that $Z_{m,n}$ is a.s. bounded in m and n. Therefore the first term of the right hand side of (12) converges to 0 in L^0 as $m \to \infty$, uniformly in n. Similarly one can show that also the second term converges to zero. This finishes the proof concerning the sequence (a).

For the sequence in (b) we notice that

$$
\begin{aligned}
&\Big|\sum_{J\in\mathcal{S}_n}(\Delta_J(M^m))(\Delta_{J^1}(M^m))(\Delta_{J^2}(M^m))-\sum_{J\in\mathcal{S}_n}(\Delta_J(M))(\Delta_{J^1}(M))(\Delta_{J^2}(M))\Big| \\
&\le \alpha_{n,m}+\beta_{n,m}+\gamma_{n,m}
\end{aligned}
$$

with

$$
\begin{aligned}
\alpha_{n,m} &= \Big|\sum_{J\in\mathcal{S}_n}(\Delta_{J^1}(M)ig)(\Delta_{J^2}(M))(\Delta_J(M-M^m))\Big| \\
\beta_{n,m} &= \Big|\sum_{J\in\mathcal{S}_n}(\Delta_J(M^m))(\Delta_{J^2}(M))(\Delta_{J^1}(M-M^m))\Big|, \\
\gamma_{n,m} &= \Big|\sum_{J\in\mathcal{S}_n}(\Delta_J(M^m))(\Delta_{J^1}(M^m))(\Delta_{J^2}(M-M^m))\Big|
\end{aligned}
$$

by Schwarz's inequality

$$
\alpha_{n,m} \le \Big\{\sum_{J\in\mathcal{S}_n}(\Delta_J(M-M^m))^2\sum_{J\in\mathcal{S}_n}(\Delta_{J^1}(M))^2(\Delta_{J^2}(M))^2\Big\}^{1/2},
$$

and the desired convergence follows by Lemma 3.1. Analogous arguments apply to $\beta_{n,m}$ and $\gamma_{n,m}$. Hence the lemma is proved. ∎

Proposition 3.3. Let M be a square integrable regular martingale. Then hypotheses (H4) and (H5) hold true for almost all the paths of M.

Proof. For any $k,\ell,\ k',\ell',\ m \in N$ we have

$$
\begin{aligned}
&\Big| \sum_{J\in\mathcal{S}_{k,\ell}} (\Delta_{J^1}(M))^2 (\Delta_{J^2}(M))^2 - \sum_{J\in\mathcal{S}_{k',\ell'}} (\Delta_{J^1}(M))^2 (\Delta_{J^2}(M))^2 \Big| \\
&\leq \Big| \sum_{J\in\mathcal{S}_{k,\ell}} (\Delta_{J^1}(M))^2 (\Delta_{J^2}(M))^2 - \sum_{J\in\mathcal{S}_{k,\ell}} (\Delta_{J^1}(M^m))^2 (\Delta_{J^2}(M^m))^2 \Big| \\
&+ \Big| \sum_{J\in\mathcal{S}_{k,\ell}} (\Delta_{J^1}(M^m))^2 (\Delta_{J^2}(M^m))^2 - \sum_{J\in\mathcal{S}_{k',\ell'}} (\Delta_{J^1}(M^m))^2 (\Delta_{J^2}(M^m))^2 \Big| \\
&+ \Big| \sum_{J\in\mathcal{S}_{k',\ell'}} (\Delta_{J^1}(M^m))^2 (\Delta_{J^2}(M^m))^2 - \sum_{J\in\mathcal{S}_{k',\ell'}} (\Delta_{J^1}(M))^2 (\Delta_{J^2}(M)))^2 \Big|
\end{aligned}
$$

Apply Lemma 3.2 (a) to the first and last terms of the right hand side of the above inequality. The second one converges to zero as $k,\ell \to +\infty$ in L^2 (see Theorem 3.1 of [5] and the proof of Lemma 2.5 in [8]). Therefore, there exists a subsequence $\{\mathcal{S}_{k',\ell'}\}$ of $\{\mathcal{S}_{k,\ell}\}$ such that, for almost all $\omega \in \Omega$ and all rational points s,t, the sequence

$$
\sum_{J\in\mathcal{S}_{k',\ell'}\cap R_{s,t}} (\Delta_{J^1}(M))^2 \, (\Delta_{J^2}(M))^2
$$

converges to some limit $\langle \tilde{M} \rangle_{s,t}$. This implies (H4).

The arguments to check (H5) under the hypotheses of this proposition are similar to that for (H4). We have to use Lemma 3.2 (b), Theorem 3.1 of [5] and the ideas of the proof of Lemma 2.5 of [8]. ∎

We can finally state the Itô formula.

Theorem 3.4. Let $M \in \mathcal{M}_c^2$, and f be a real function satisfying the conditions of Theorem 2.1. Then for any sequence $\{\mathcal{S}_{n,m},\ n,m \geq n,m \geq 1\}$ of partitions of T whose mesh tends to zero

$$
P - \lim_{n\to\infty} \lim_{m\to\infty} \sum_{s_i\in\mathcal{S}_n^1\cap[0,s]} \ \sum_{t_j\in\mathcal{S}_m^2\cap[0,s]} f''(M_{s_i,t_j}) \Delta_{J^1}(M)\Delta_{J^2}(M),
$$

exists. We denote this limit by $\int_{R_{s,t}} f''(M_z)\, d\tilde{M}_z$.

Moreover

$$
\begin{aligned}
f(M_{s,t}) = &\int_{R_{st}} [f'(M_z)\, dM_z + f''(M_z)\, d\tilde{M}_z] \\
&+ \frac{1}{2}\int_0^s f''(M_{x,t})\, d\langle M_{\cdot,t}\rangle_x + \frac{1}{2}\int_0^t f''(M_{s,y})\, d\langle M_{s,\cdot}\rangle_y \\
&- \frac{1}{2}\int_{R_{st}} f''(M_z)\, d\langle M\rangle_z - \int_{R_{st}} f'''(M_z)\, d\langle M, \tilde{M}\rangle_z \\
&- \frac{1}{4}\int_{R_{st}} f^{iv}(M_z)\, d\langle \tilde{M}\rangle_z .
\end{aligned}
$$

Proof. The results proved in this section show that a.s. M satisfies the hypothesis of Theorem 2.1. Now we proceed as in [3], using the fact that the sequence

$$
\{ \sum_{J\in\mathcal{S}_{n,m}} f'(M_{s_i,t_j})\Delta_J(M)\},
$$

converges in probability to $\int_{R_{st}} f'(M_z) dM_z$ as $n,m \to \infty$.

Notice that in the above formula $\langle M, \tilde{M}\rangle_z$ and $\langle \tilde{M}\rangle_z$ denote random measures whose existence has been established in Proposition 3.3, and which can be identified as the quadratic variations corresponding to the martingale $\tilde{M}_z$ if M is bounded in L^4. ∎

References

[1] D. Bakry, Semimartingales à deux indices. Séminaire de Probabilités XVI, 1980/81. J. Azéma and M. Yor (Eds.). Lecture Notes in Math. 920, 355–369. Springer Verlag, 1981.

[2] R. Cairoli and J. B. Walsh, Stochastic integrals in the plane. Acta Math. 134, 111–183 (1975).

[3] H. Föllmer, Calcul d'Itô sans probabilités. Séminaire de Probabilités XV, 1979/80. J. Azéma and M. Yor (Eds.). Lecture Notes in Math. 850. Springer Verlag, 1981.

[4] P. Imkeller, The structure of two–parameter martingales and their quadratic quadratic variation. Lecture Notes in Math. 1308. Springer Verlag, 1986.

[5] P. Imkeller, Regulatory and integrator properties of variation processes of two parameter martingales with jumps. Preprint.

[6] S. Kalpazidou, A generalization of Burkholder's transform. Revue Roum. de Math. pures et appl. 30, 269–272 (1985).

[7] D. Nualart, On the quadratic variation of two–parameter continuous martingales. Annals of Probab. 12, 2, 445–457 (1984).

[8] D. Nualart, Une formule d'Itô pour les martingales continues à deux indices et quelques applications. Ann. Inst. H. Poincaré 20, 3 251–275 (1984).

[9] D. Nualart, M. Sanz and M. Zakai, On the relations between increasing functions associated with two–parameter continuous martingales. Stoch. Proc. and their Appl. 34, 99–119 (1990).

[10] M. Sanz, r–variations for two–parameter continuous martingales and Itô's formula. Stoch. Proc. and their Appl. 32, 69–92 (1989).

[11] N. Frangos and P. Imkeller, Quadratic variation for a class of $L\,L\log^+ L$–bounded two–parameter martingales. Annals of Probab. 15, 1097–1111 (1987).

Nikos Frangos
Department of Mathematics
Hofstra University
Hempstead NY 11550
U.S.A.

David Nualart and Marta Sanz–Solé
Facultat de Matemàtiques
Universitat de Barcelona
Gran Via 585
08007 BARCELONA
SPAIN

CENTRAL LIMIT THEOREM RESULTS
FOR A REACTION-DIFFUSION EQUATION WITH
FAST-OSCILLATING BOUNDARY PERTURBATIONS

BY MARK I. FREIDLIN * AND RICHARD B. SOWERS †

Department of Mathematics
University of Maryland
College Park, MD 20742

0. Introduction.

In this paper we consider some Central Limit theorems for stochastic reaction-diffusion equations (RDE's) of the form

$$\begin{aligned} \frac{\partial u^\epsilon}{\partial t} &= \Delta u^\epsilon + f(x, u^\epsilon) \\ u^\epsilon(0, \cdot) &= u_0 \\ \left.\frac{\partial u^\epsilon}{\partial \nu}\right|_{[0,T]\times S^1} &= \zeta^\epsilon \end{aligned} \tag{1}$$

where the space variable x takes values on the unit disc $D^2 := \{x \in I\!R^2 : \|x\| \leq 1\}$ and where the Neumann data ζ^ϵ is a properly normalized fast-oscillating random field on the boundary $[0, T] \times S^1$.

Infinite-dimensional evolution equations may be perturbed in many more natural ways than are possible with ordinary differential equations. Much work has been done on equations of the form

$$\begin{aligned} \frac{\partial u^\epsilon}{\partial t} &= \Delta u^\epsilon + f(u^\epsilon) + \zeta^\epsilon . \\ u^\epsilon(0, \cdot) &= u_0 \end{aligned} \tag{2}$$

for x in a one-dimensional manifold (see [2], [10] and [11]), but only recently have boundary perturbations been studied (see [3], [9] and [10]). Indeed, in a wide number of cases, such as for example several questions connected with nerve impulse propogation, the correct model for a physical phenomenon is (1) and not (2).

AMS 1985 subject classifications. Primary 60H15, Secondary 60F05, 60F17.

Key words and phrases. Central limit theorem, stochastic partial differential equations, random fields.

* The work of this author was supported NSF grant DMS 9106562.

† The work of this author was supported by a University of Maryland (College Park) Graduate School Dissertation Fellowship and partially by ONR grant N00014–91–J–1526. This author's present address is University of Southern California, Center for Applied Mathematical Sciences, 1042 W. 36th Place, DRB 155, Los Angeles, CA 90089–1113.

In this short note we shall investigate the behavior of u^ϵ of (1) where ζ^ϵ is a fast-oscillating (in some sense) random field. We of course expect that the limiting behavior will be given by the solution of the corresponding RDE

$$\frac{\partial u^0}{\partial t} = \Delta u^0 + f(x, u^0)$$
$$u^0(0,\cdot) = u_0 \qquad (3)$$
$$\left.\frac{\partial u^0}{\partial \nu}\right|_{[0,T]\times S^1} = \zeta^0$$

where ζ^0 is a (perhaps generalized) Gaussian field on $[0,T]\times S^1$ such that ζ^ϵ tends to ζ^0 in law. What is of interest here is the formulation of these results in the proper functional spaces; central limit theorems and other asymptotic results for randomly-perturbed processes must be understood as statements about probability measures on the appropriate function spaces. A more complete discussion of the motivation and goals of these types of questions may be found in [3]. We shall in this paper identify the functional spaces associated with the way in which the solution of (1) converges in law to the solution of (3).

The organization of this paper is as follows. In Section 1 we introduce two classes of rapidly-oscillating boundary perturbations which are natural models either of a white noise field on the boundary or of a field which has independent increments in the time direction but is smooth in the space direction. For both types of random fields, we define a Banach space in which to study convergence in law of the random boundary perturbation. In Section 2, we consider the RDE (1) with $f \equiv \mathbf{0}$; for this linear problem we write the solution as a linear transformation of the boundary perturbation; using the Banach space formalism introduced in Section 1, we then define Banach spaces which contain the solutions of (1) and in which the mapping of ζ^ϵ to u^ϵ is continuous. This allows us to conclude, using the convergence results of Section 1, the convergence in law of the solutions of (1) when $f \equiv \mathbf{0}$. Finally, in Section 3, we consider the fully nonlinear version of (1); it is easy to represent the solution of (1) as a continuous transformation of the solution of (1) with $f \equiv \mathbf{0}$; this allows us to immediately achieve the final goal of this paper; to understand the central limit theorem results for (1).

Two Classes of Boundary Perturbations.

In this section, we introduce two fairly general types of rapidly-oscillating boundary perturbations, and define the appropriate Banach spaces in which to study the asymptotics of the laws of these random fields.

To introduce these perturbations, we shall assume an underlying probability triple $(\Omega, \mathcal{F}, P)$ and an i.i.d. collection $\{\xi_i\}$ of real-valued second order stationary Markov processes indexed by $\mathbb{R}_+ := [0,\infty)$, and an i.i.d. collection $\{W_i\}$ of Wiener processes on $[0,T]$. Our general boundary perturbation shall be of the form

$$\zeta^\epsilon(t,x) := \sum_{i=0}^{\infty} c_{i,\epsilon}\phi_i(x)\epsilon^{-1}\xi_i(t\epsilon^{-2}) \qquad (t,x)\in[0,T]\times S^1 \quad (4)$$

for each $\epsilon > 0$, where $\{c_{i,\epsilon}\}$ are nonrandom constants such that $\sum_{i=1}^{\infty}|c_{i,\epsilon}| < \infty$, and where $\{\phi_i\}$ are the eigenfunctions of the Laplacian operator on S^1; i.e.,

$$\phi_i((\cos\theta, \sin\theta)) = (\cos i\theta, \sin i\theta)$$

for all θ in $[0, 2\pi)$ and all i. We assume that there is a $\{c_{i,0}\}$ such that $\lim_\epsilon c_{i,\epsilon} = c_{i,0}$. Under some natural statistical assumptions to be specified in the next paragraph, we then expect that ζ^ϵ will tend in law to the

Gaussian field ζ^0, where formally

$$\zeta^0(t,x) = \sum_{i=0}^{\infty} c_{i,0}\phi_i(x)\dot{W}_k(t). \tag{5}$$

If $c_{i,0} = 1$ for all i, then ζ^0 will be a white noise field on $[0,T] \times S^1$. If on the other hand $\lim_i c_{i,0} = 0$, this decay being fast enough, ζ^0 will be smooth in the x variable, but will have independent increments in the time variable. Our two classes of boundary perturbations, C_1, and C_2 will correspond to these two possibilities.

The exact statistical assumptions which we shall require of $\{\xi_0\}$ (and thus of $\{\xi_i\}$ for all $i = 0, 1, \ldots$, since the $\{\xi_i\}$ are i.i.d.) are the following. We assume that

(A.1) The process ξ is second-order stationary with $E\xi_0(0) = 0$ and $\int_0^\infty |E[\xi_0(t)\xi_0(0)]|dt < \infty$.

(A.2) There is a positive constant B such that $P\{|\xi_0(t)| \geq B\} = 0$ for all t in $I\!R_+$. Also, $\int_0^\infty E[\xi_0(t)\xi_0(t)dt = 1/2$.

To state the third requirement, let us define

$$\mathcal{F}_t := \sigma\{\xi_0(s) : s \leq t\} \quad \text{and} \quad \mathcal{F}^t := \sigma\{\xi_0(s) : s \geq t\}$$

for all t in $I\!R_+$. Let us also define the mixing coefficient α as

$$\begin{aligned}\alpha(\tau) := \sup\{E[\eta_1\eta_2] - E[\eta_1]E[\eta_2] : |\eta_1| \leq 1 \text{ and } |\eta_2| \leq 1 \ P\text{-a.s.},\\ \eta_1 \text{ is } \mathcal{F}_t\text{-measurable and } \eta_2 \text{ is } \mathcal{F}^{t+\tau}\text{-measurable},\\ \tau > 0\}\end{aligned}$$

for each $\tau > 0$. The third requirement on $\{\xi_0\}$ is that for some integer $k > 1$, we have

(A.3)$_k$ The integral $\int_0^\infty \tau^{k-1}\alpha(\tau)d\tau < \infty$.

Note that under these assumptions and the assumption that $\sum_{i=0}^\infty |c_{i,\epsilon}| < \infty$ for each $\epsilon > 0$, we have that

$$\sup_{(t,x)\in[0,T]\times S^1} |\zeta^\epsilon(t,x)| \leq B\epsilon^{-1} \sum_{i=0}^{\infty} |c_{i,\epsilon}| < \infty.$$

Also, these assumptions are sufficient to imply that $\tilde{\xi}_i^\epsilon$ tends to W_i weakly in $C^\gamma([0,T])$ for each $0 < \gamma < (k-1)/(2k)$, where

$$\tilde{\xi}_i^\epsilon(t) := \int_0^t \epsilon^{-1}\xi_i(s\epsilon^{-2})ds$$

for all i and all t in $[0,T]$ (see [3] and [8]).

The first class of boundary perturbations, which we shall denote by C_1, consists of boundary perturbations of the form (4) where we have

(C.1) that $\sup_{i,\epsilon\geq 0} |c_{i,\epsilon}| < \infty$.

This class obviously allows that $c_{i,0} = 1$ for all i, which gives us a white noise field as the limiting Gaussian boundary perturbations. This class thus admits that the limiting Gaussian field be a generalized function both in time and in space. Calculations will be somewhat easier with the second class of boundary perturbations, C_2, which consists of random fields of the form (4) where for some $0 < \beta \leq 1$, we have

(C.2)$_\beta$ that $\sup_{\epsilon\geq 0} \sum_{i=0}^\infty i^\beta |c_{i,\epsilon}| < \infty$.

We shall see that in this class of perturbations, the limiting Gaussian field will still maintain independent increments in the t direction, but will be smooth in the x direction; i.e., the limiting Gaussian field will be a generalized function in t but a regular function in x.

Recalling in more detail the results stated earlier, we have that the finite-dimensional distributions of $\tilde{\xi}_i$ converge to those of W_i for each i, and that for each $\gamma < (k-1)/(2k)$, there are nonnegative random variables $\{Y^{i,\epsilon,\gamma}\}$ with $\sup_{i,\epsilon\geq 0} E[Y^{i,\epsilon,\gamma}] < \infty$ for each γ and such that P-a.s.

$$|\tilde{\xi}_i^\epsilon(t) - \tilde{\xi}_i^\epsilon(s)| \leq Y^{i,\epsilon,\gamma}|t-s|^\gamma$$

for all $0 \leq s \leq t \leq T$ and all $\epsilon \geq 0$. Thus $\tilde{\xi}_i \Rightarrow W_i$ in $C^\gamma([0,T])$ for each $0 < \gamma < (k-1)/(2k)$ (here $\Rightarrow$ denotes convergence in law in the relevant Polish space, which in this case is the space of real-valued functions on $[0,T]$ which are Hölder-continuous of exponent γ).

We can immediately use these results to consider the limiting law of ζ^ϵ under the assumptions **(A.1)**–$\mathbf{(A.3)_k}$ and **(C.1)** or $\mathbf{(C.2)}_\beta$. We consider first boundary perturbations of the class C_2, since the analysis is somewhat easier. Set

$$\tilde{\zeta}^\epsilon(t,x) := \int_0^t \zeta^\epsilon(s,x)ds \tag{6}$$

for each $\epsilon \geq 0$ and each (t,x) in $[0,T]\times S^1$. By Prohorov's theorem to show the convergence of the laws of $\{\tilde{\zeta}^\epsilon\}$, we need to show that the laws are tight in the appropriate space and that all cluster points of these laws coincide. Let us first consider the tightness question. For each $0 < \gamma < (k-1)/(2k)$ and all (t,x) and (s,y) in $[0,T]\times S^1$, we have that

$$|\tilde{\zeta}^\epsilon(t,x) - \tilde{\zeta}^\epsilon(t,y) - \tilde{\zeta}^\epsilon(s,x) + \tilde{\zeta}^\epsilon(s,y)| \leq \sum_{i=0}^\infty |c_{i,\epsilon}||\phi_i(x)-\phi_i(y)|Y^{i,\epsilon,\gamma}|t-s|^\gamma.$$

Using the fact that $|\phi_i(x)-\phi_i(y)| \leq \min\{2, ir(x,y)\}$ for all i and all x and y (here $r(\cdot,\cdot)$ is the natural metric on S^1), we have that for each (t,x) and (s,y) in $[0,T]\times S^1$,

$$|\tilde{\zeta}^\epsilon(t,x) - \tilde{\zeta}^\epsilon(t,y) - \tilde{\zeta}^\epsilon(s,x) + \tilde{\zeta}^\epsilon(s,y)| \leq 2^{1-\beta}\sum_{i=0}^\infty i^\beta|c_{i,\epsilon}|Y^{i,\epsilon,\gamma}r^\beta(x,y)|t-s|^\gamma. \tag{7}$$

This indicates that we should define the Banach spaces $C_{\beta,\gamma}$ for each $0<\gamma<1$ and $0<\beta'<1$ as as the closure of $C^\infty([0,T]\times S^1)$ with respect to the norm

$$|||\varphi|||_{\beta,\gamma} := \sup_{(t,x)\in[0,T]\times S^1} |\varphi(t,x)| + \sup_{\substack{0\leq s<t\leq T\\ x,y\in S^1,\, x\neq y}} \frac{|\varphi(t,x)-\varphi(s,x)-\varphi(t,y)+\varphi(s,y)|}{r^\alpha(x,y)|t-s|^\gamma}$$

on $C^\infty([0,T]\times S^1)$. The norm $|||\varphi|||_{\beta,\gamma}$ measures the variation of φ over rectangles, in a way analogous to simple Hölder continuity. From the bound (7), we have that

$$E[|||\tilde{\zeta}^\epsilon|||_{\gamma,\beta}] \leq K\left(\sup_{\epsilon>0}\sum_{i=0}^\infty i^\beta|c_{i,\epsilon}|\right)\left(\sup_{i,\epsilon>0} E[Y^{i,\epsilon,\gamma}]\right)$$

for all $\epsilon > 0$, where K is some constant independent of $\epsilon > 0$. For each $0<\gamma<(k-1)/(2k)$. By standard embedding techniques and Arzela-Ascoli, clearly C_{γ_1,α_1} will be compactly embedded in C_{γ_2,α_2} whenever $\gamma_1 < \gamma_2$

and $\alpha_1 < \alpha_2$. This gives us the required tightness. It is easy to see, by projecting along a finite number of $\{\phi_i\}$, that all cluster points (in the weak topology) of the laws of $\{\tilde{\zeta}^\epsilon\}$ must be the law of the Gaussian random field

$$\tilde{\zeta}^0(t,x) := \sum_{i=0}^{\infty} c_{i,0}\phi_k(x)W_i(t).$$

This is enough to complete the proof of the following claim:

Proposition 1. *Assume that assumptions* **(A.1)**–**(A.3)**$_k$ *and* **(C.2)**$_\beta$ *hold. Then for all* $\gamma < (k-1)/(2k)$ *and all* $\beta' < \beta$, $\tilde{\zeta}^\epsilon \Rightarrow \tilde{\zeta}^0$ *in* $C_{\gamma,\beta'}$.

Now consider the assumptions **(A.1)**–**(A.3)**$_k$ and **(C.1)**. We expect the limiting law to be that of ζ^0 as in (5). As we earlier noted, the assumption **(C.1)** allows us to consider white noise on $[0,T]\times S^1$, so it is natural to now integrate ζ^ϵ in both variables. In analogy to (6), let us set

$$\bar{\zeta}^\epsilon(t,x) := \int_0^t \int_{y\in\mathcal{E}(x)} \zeta^\epsilon(s,y)dsdy$$

for all (t,x) in $[0,T]\times[0,2\pi]$, where $\mathcal{E}(x) := \{(\cos\theta,\sin\theta) : 0 \le \theta \le x\}$ for each x in $[0,2\pi)$. To show the tightness, we can use the procedures of [5] to show that for any (t,x) ands (s,y) in $[0,T]\times[0,2\pi]$,

$$E[|\bar{\zeta}^\epsilon(t,x) - \bar{\zeta}^\epsilon(t,y) - \bar{\zeta}^\epsilon(s,x) + \bar{\zeta}^\epsilon(s,y)|^{2k}] \le K|t-s|^k|x-y|^k \tag{8}$$

(see the appendix) where K is some constant independent of the (t,x) and (s,y). By using a natural extension of Garsia's estimate (see [4]), we then get that for each $0 < \gamma < (k-1)/(2k)$, there are nonnegative random variables $\{\bar{Y}^{\gamma,\epsilon}\}$ with $\sup_{\epsilon>0} E[\bar{Y}^{\gamma,\epsilon}] < \infty$ such that

$$|\bar{\zeta}^\epsilon(t,x) - \bar{\zeta}^\epsilon(t,y) - \bar{\zeta}^\epsilon(s,x) + \bar{\zeta}^\epsilon(s,y)| \le \bar{Y}^{\gamma,\epsilon}|t-s|^\gamma|x-y|^\gamma \tag{9}$$

for all (t,x) and (s,y) in $[0,T]\times[0,2\pi]$ (see Lemma 1 in the appendix). Once again, it is easy to show that all cluster points of the laws of $\{\bar{\zeta}^\epsilon\}$ must coincide with the law of

$$\bar{\zeta}^0(t,x) := \sum_{i=0}^{\infty} \left(\int_{y\in\mathcal{E}(x)} \phi_k(y)dy\right) W_i(t).$$

This gives us the result analogous to Proposition 1:

Proposition 2. *Assume that assumptions* **(A.1)**–**(A.3)**$_k$ *and* **(C.1)** *hold. Then for all* $0 < \gamma < (k-1)/(2k)$, $\bar{\zeta}^\epsilon \Rightarrow \bar{\zeta}^0$ *in* $C_{\gamma,\gamma}$.

The linear problem.

We can directly use the results of the last section to study the limiting laws of the linear problem associated with (1); let u_l^ϵ be given by

$$\frac{\partial u_l^\epsilon}{\partial t} = \Delta u_l^\epsilon$$
$$u_l^\epsilon(0,\cdot) = 0 \tag{10}$$
$$\left.\frac{\partial u_l^\epsilon}{\partial \nu}\right|_{[0,T]\times S^1} = \zeta^\epsilon.$$

If p is the Green's function for (10); i.e., the solution of the generalized problem

$$\frac{\partial p}{\partial t} = \Delta_x p$$
$$p(0,\cdot,y) = \delta_y$$
$$\left.\frac{\partial_x p}{\partial \eta}\right|_{[0,T]\times S^1} = 0,$$

(here the subscript x is added to the Laplacian and normal derivation operators to emphasize that these operators are acting on the x-dependence of the arguments, and here δ_y is the Dirac delta generalized function at y) then u_i^ϵ of (10) may be represented as

$$u_i^\epsilon(t,x) = \int_0^t \int_{S^1} p(t,x,y)\zeta^\epsilon(s,y)dsdy \tag{11}$$

for all (t,x) in $[0,T]\times D^2$ and all $\epsilon > 0$. (Under the assumption that $\sum_{i=0}^\infty |c_{i,\epsilon}| < \infty$ for each $\epsilon > 0$, ζ^ϵ is a well-defined function, so classical results of existence and uniqueness hold.) The goal is then to write (11) as $u^\epsilon = B_1(\tilde{\zeta}^\epsilon)$ for $\{\zeta^\epsilon\}$ in class C_1 and $u^\epsilon = B_2(\bar{\zeta}^\epsilon)$ for $\{\zeta^\epsilon\}$ in class C_2, where B_1 and B_2 are bounded linear operators from the appropriate boundary-function spaces as introduced in Section 1 to the appropriate solution space.

The basis for the necessary estimates is the following calculation; consider the problem

$$\frac{\partial v}{\partial t} = \Delta v$$
$$v(0,\cdot) = 0 \tag{12}$$
$$\left.\frac{\partial v}{\partial \nu}\right|_{[0,T]\times\partial H^2} = g$$

in the half space $H^2 := \{(x_1,x_2) : x_2 \geq 0\}$. Here g is any continuous and bounded function on $[0,T]\times\partial H^2$ For the problem (12), the solution is given by

$$v(t,x) = \int_0^t \int_{\mathbb{R}} p_G(t-s,x,(y,0))g(s,(y,0))dsdy \tag{13}$$

for all (t,x) in $[0,T]\times H^2$, where p_G is the heat kernel

$$p_G(t,x,y) = (4\pi t)^{-1}\exp\left(\frac{\|x-y\|^2}{4\pi t}\right)$$

for all $t > 0$ and all x and y in $\mathbb{R}^2$. Rewriting, we have that

$$\begin{aligned} v(t,(x_1,x_2)) &= -2\int_0^t \int_{\mathbb{R}} \frac{\partial p_G}{\partial t}(r,(x_1,x_2),(y,0))\left(\int_{t-r}^t g(s,(y,0))ds\right)drdy \\ &= -2\int_0^t \int_{\mathbb{R}} \frac{\partial p_G}{\partial t}(r,(x_1,x_2),(y,0)) \\ &\quad \cdot\left\{\int_{t-r}^t g(s,(y,0))ds - \int_{t-r}^t g(s,(x_1,0))ds\right\}drdy. \end{aligned} \tag{14}$$

Using the easily-verified fact that there is a positive constant K such that

$$\left|\frac{\partial p_G}{\partial t}(t,x,y)\right| \leq Kt^{-2}\exp\left(-\frac{\|x-y\|^2}{3t}\right)$$

for all $t > 0$ and all x and y in $\mathbb{R}^2$, we thus have that

$$|v(t,(x_1,x_2))| \leq 2K \int_0^t \int_{\mathbf{R}} r^{-2} \exp\left(-\frac{|x_1-y|^2}{3r}\right) \exp\left(-\frac{|x_2|^2}{3r}\right) \cdot \left| \int_{t-r}^t g(s,(y,0))ds - \int_{t-r}^t g(s,(x_1,0))ds \right| drdy. \tag{15}$$

We then get, after transferring the calculation (15) back to the manifold D^2, that if $0 < \gamma'' < \gamma < (k-1)/(2k)$ and $0 < \beta'' < \beta' < \beta$, there is a positive constant K such that

$$|u^\epsilon(t,x)| \leq \begin{cases} K|||\tilde{\zeta}^\epsilon|||_{\gamma',\beta'} & \text{if } 2\gamma'' + \beta'' - 1 \geq 0; \\ K|||\tilde{\zeta}^\epsilon|||_{\gamma,\beta'}(\text{dist}(x,S^1))^{-(2\gamma''+\beta''-1)} & \text{if } 2\gamma''+\beta''-1 < 0 \end{cases}$$

for all $\epsilon \geq 0$. Here $\text{dist}(x, S^1)$ is the distance between x and S^1 for any point x in D^2. The proper space in which to study the laws of $\{u^\epsilon\}$ for boundary perturbations of class C_2 is thus the collection of spaces C'_η, where C'_η is the closure of $C^\infty([0,T] \times D^2)$ in the norm

$$|||\varphi|||'_\eta := \begin{cases} \sup_{(t,x)\in[0,T]\times D^2} |\varphi(t,x)| & \text{if } \eta \geq 0 \\ \sup_{(t,x)\in[0,T]\times D^{2,\circ}} |\varphi(t,x)|(\text{dist}(x,S^1))^{-\eta} & \text{if } \eta < 0. \end{cases}$$

The calculations of (15) give us that for any $0 < \gamma'' < \gamma < (k-1)/(2k)$ and $0 < \beta'' < \beta' < \beta$, the mapping which takes $\tilde{\zeta}^\epsilon$ to u^ϵ is a continuous linear transformation from $C_{\gamma',\beta'}$ to $C'_{2\gamma''+\beta''-1-\eta}$, so the following result is true;

Proposition 3. *Under the assumptions* **(A.1)**–$\mathbf{(A.3)_k}$ *and* $\mathbf{(C.2)}_\beta$, *for each* $\beta' < \beta$ *and each* $\gamma < (k-1)/(2k)$, $u^\epsilon \Rightarrow u^0$ *in* $C'_{2\gamma+\beta'-1}$.

Similar calculations hold for boundary perturbations of class C_2. Instead of (15), we should rewrite (13) as

$$\begin{aligned} v(t,(x_1,x_2)) &= -2\int_0^t \int_{\mathbf{R}} \frac{\partial^2 p_G}{\partial t \partial y}(r,(x_1,x_2),(z,0)) \left(\int_{t-r}^t \int_{y\geq z} g(s,(y,0))dsdy \right) drdz \\ &= -2\int_0^t \int_{\mathbf{R}} \frac{\partial^2 p_G}{\partial t \partial y}(r,(x_1,x_2),(z,0)) \\ &\quad \cdot \left(\int_{t-r}^t \int_{y\geq z} g(s,(y,0))dsdy - \int_{t-r}^t \int_{y\geq x} g(s,(y,0))dsdy \right) drdz \end{aligned}$$

so that using the fact that there is a positive constant K such that

$$\left| \frac{\partial^2 p_G}{\partial t \partial y}(t,x,y) \right| \leq K t^{-5/2} \exp\left(-\frac{\|x-y\|^2}{3t}\right)$$

for all $t > 0$ and all x and y in $\mathbb{R}^2$, we thus have that

$$|v(t,(x_1,x_2))| \leq 2K \int_0^t \int_{\mathbf{R}} r^{-5/2} \exp\left(-\frac{|x_1-y|^2}{3r}\right) \exp\left(-\frac{|x_2|^2}{3r}\right) \cdot \left| \int_{t-r}^t \int_{y\geq z} g(s,(y,0))dsdy - \int_{t-r}^t \int_{y\geq x} g(s,(y,0))dsdy \right| drdz. \tag{16}$$

Transferring this calculation to D^2, we get that if $0 < \gamma' < \gamma < (k-1)/(2k)$, then there is a constant K such that

$$|u^\epsilon(t,x)| \le \begin{cases} K|||\tilde{\zeta}^\epsilon|||_{\gamma',\gamma'} & \text{if } 3\gamma'-2 \ge 0; \\ K|||\tilde{\zeta}^\epsilon|||_{\gamma,\beta'}(\text{dist}(x,S^1))^{-(3\gamma'-2)} & \text{if } 3\gamma'-2 < 0 \end{cases}$$

for all $\epsilon > 0$; i.e., for any $0 < \gamma' < \gamma < (k-1)/(2k)$ the mapping from $\bar{\zeta}^\epsilon$ to u^ϵ is a continuous linear transformation from $C_{\gamma,\gamma}$ to $C'_{3\gamma'-2}$. This gives us

Proposition 4. *Under assumptions* **(A.1)–(A.3)$_k$** *and* **(C.1)**, *for each* $0 < \gamma < (k-1)/(2k)$, $u^\epsilon \Rightarrow u^0$ *in* $C'_{3\gamma-2}$.

The nonlinear problem.

The only remaining task before us is to use the results of the last section, i.e., the results about the linear equation (10) to show the corresponding results for the nonlinear problem. This is easily accomplished by considering the mapping from $\varphi \mapsto \hat{B}(\varphi)$ defined by the integral equation

$$(\hat{B}(\varphi))(t,x) = \int_{D^2} p(t,x,y)u_0(y)dy + \varphi(t,x) + \int_0^t \int_{D^2} p(t-s,x,y)f(y,(\hat{B}(\varphi))(s,y))dsdy \tag{17}$$

for all (t,x) in $[0,T] \times D^2$ and all functions φ on $[0,T] \times D^2$ such that (17) is well-defined. For each γ in $I\!R$, the assumptions

(D.1)$_\gamma$ There is a constant $\bar{F}$ and two exponents δ_1 and $\delta_2 > 0$ satisfying $\delta_1 - \gamma\delta_2 < -1$, such that for all x in D^2 and all u in $I\!R$,

$$|f(x,u)| \le \bar{F}(1 + (\text{dist}(x,S^1))^{\delta_1}|u|^{\delta_2})$$

and

(D.2)$_\gamma$ There is a constant $\bar{f}$ such that for all x in D^2 and all u and v in $I\!R$,

$$|f(x,u) - f(x,v)| \le \bar{f}(\text{dist}(x,S^1))^\gamma |u-v|$$

are sufficient to ensure that $\hat{B}$ is a well-defined homeomorphism of C'_γ to itself (see [9] or [10]). Note that if $0 < \gamma_1 < \gamma_2$, then assumptions **(D.1)$_{\gamma_2}$** and **(D.2)$_{\gamma_2}$** imply assumptions **(D.1)$_{\gamma_1}$** and **(D.2)$_{\gamma_1}$**. This gives us the complete results about (1);

Theorem 1. *Assume that assumptions* **(A.1)–(A.3)$_k$** *and* $(C.1)_\beta$ *hold, and that assumption* $(D.1)_\gamma$ *and* $(D.2)_\gamma$ *hold for some* $0 < \gamma < 3(k-1)/(2k) - 2$. *Then* $u^\epsilon \Rightarrow u^0$ *in* C'_γ.

Theorem 2. *Assume that assumptions* **(A.1)–(A.3)$_k$** *and* **(C.2)$_\beta$** *hold, and that assumption* **(D.1)$_\gamma$** *and* **(D.2)$_\gamma$** *hold for some* $0 < \gamma < (k-1)/k+\beta-1$. *Then* $u^\epsilon \Rightarrow u^0$ *in* C'_γ.

The authors would like to thank professor Boris Rozovskii for inviting them to present this work at the International Conference on Stochastic Partial Differential Equations and Their Applications.

We now explain in greater detail how to get the estimates (8) and (9). Let us begin with by estimating

$$E\left[\left(\sum_{i=l_1}^{l_2} c_{i,\epsilon} \int_s^t \epsilon^{-1}\xi_i(r\epsilon^{-2})dr \int_{z\in\mathcal{E}(y)\sim\mathcal{E}(x)} \phi_i(z)dz\right)^{2k}\right] \tag{A.1}$$

for any fixed $0 \leq s < t \leq t \leq T$, $0 \leq x < y \leq 2\pi$ and $\epsilon > 0$, and any two nonnegative integers l_1 and l_2. For convenience, let us define

$$\Xi_i := c_{i,\epsilon} \int_s^t \epsilon^{-1}\xi_i(r\epsilon^{-2})dr \int_{z\in\mathcal{E}(y)\sim\mathcal{E}(x)} \phi_i(z)dz$$

for each $i = 0, 1, \ldots$. Using the notation $E := \{l_1, l_1+1, \cdots, l_2\}$, we can write

$$E\left[\left(\sum_{i=l_1}^{l_2} \Xi_i\right)^{2k}\right] = \sum_{(i_1,i_2,\cdots i_{2k})\in E^{2k}} E\left[\Xi_{i_1}\Xi_{i_2}\cdots\Xi_{i_{2k}}\right].$$

The expression on the right may be rearranged so that the indices are increasing; let us define for each positive integer p the set $E_p := \{(i_1, i_2, \cdots, i_p) \in E^p : i_1 \leq i_2 \cdots i_p\}$. Then

$$E\left[\left(\sum_{i=l_1}^{l_2} \Xi_i\right)^{2k}\right] = (2k)! \sum_{(i_1,i_2,\cdots i_{2k})\in E_{2k}} E\left[\Xi_{i_1}\Xi_{i_2}\cdots\Xi_{i_{2k}}\right]. \tag{A.2}$$

Since the $\{\xi_i\}$ are i.i.d. with zero mean, we have that $E\left[\Xi_{i_1}\Xi_{i_2}\cdots\Xi_{i_{2k}}\right] = 0$ for any $(i_1, i_2, \cdots i_{2k})$ in E_{2k} such that $i_{p-1} < i_p < i_{p+1}$ for some integer $1 \leq p < 2k$; thus the sum in (A.2) consists of terms with *groups* of at least two equal indices. With this thought, we have

$$E\left[\left(\sum_{i=l_1}^{l_2} \Xi_i\right)^{2k}\right] = (2k)! \sum_{n=1}^{2k} \sum_{\substack{(j_1,j_2,\cdots,j_n)\in \mathbb{I}^n \\ j_p\geq 2 \text{ all } p,\ \sum_{p=1}^n j_p = 2k}} \sum_{(i_1,i_2,\cdots i_n)\in E_n} E[(\Xi_{i_1})^{j_1}]E[(\Xi_{i_2})^{j_2}]\cdots E[(\Xi_{i_n})^{j_n}] \tag{A.3}$$

where $\mathbb{I} := \{1, 2, \ldots\}$. Consider now any such term $E[(\Xi_i)^p]$ for any nonnegative integers i and p with $p \geq 2$. We have that

$$E[(\Xi_i)^p] = c_{i,\epsilon}^p \langle \phi_i, \chi_A\rangle^p E\left[\left(\int_s^t \epsilon^{-1}\xi_i(r\epsilon^{-2})dr\right)^p\right].$$

By the results of Khas'minskii, we know that by assumption $(\mathbf{A.3})_\mathbf{k}$ implies that there is a constant K, which is independent of s, t, and ϵ, and which we may assume to be greater than 1, such that

$$E\left[\left(\int_s^t \epsilon^{-1}\xi_i(r\epsilon^{-2})dr\right)^p\right] \leq K|t-s|^{p/2}.$$

(Khas'minskii gave the result for p even in [5]; if p is odd, we may simply use Lyapunov's inequality–see [6], P. 34). With this estimate and Cauchy-Schwartz, we have that

$$E[(\Xi_i)^p] \leq K(\sup_{q,\epsilon} |c_{q,\epsilon}|^p)\langle \phi_i, \chi_A\rangle^2 \|\chi_A\|_{L^2(S^1)}^{p-2} |t-s|^{p/2}.$$

Inserting this into (A.3), we have (recall that we assumed that $K \geq 1$)

$$E\left[\left(\sum_{i=l_1}^{l_2} \Xi_i\right)^{2k}\right] \leq (2k)! K^{2k} \left(\sup_{p,\epsilon} |c_{p,\epsilon}|^{2k}\right) |t-s|^k$$

$$\cdot \sum_{n=1}^{2k} \|\chi_A\|_{L^2(S^1)}^{2k-2n} \sum_{\substack{(j_1,j_2,\cdots,j_n)\in \mathbb{I}^n \\ j_p\geq 2 \text{ all } p,\ \sum_{p=1}^n j_p = 2k}} \sum_{(i_1,i_2,\cdots i_n)\in E_n} \prod_{q=1}^n \langle \phi_{i_q}, \chi_A\rangle^2.$$

The innermost sum is bounded from above by

$$\sum_{(i_1,i_2,\cdots i_n)\in E_n}\prod_{q=1}^{n}\langle\phi_{i_q},\chi_A\rangle^2 \le \sum_{(i_1,i_2,\cdots i_n)\in E^n}\prod_{q=1}^{n}\langle\phi_{i_q},\chi_A\rangle^2 \le \left(\sum_{i\in E}\langle\phi_{i_q},\chi_A\rangle^2\right)^n$$

so since the cardinality of the set $\{(j_1,j_2,\cdots,j_n)\in I\!I^n : j_p \ge 2 \text{ all } p, \sum_{p=1}^{n} j_p = 2k\}$ is clearly less than $(2k-2)^n$, we have that

$$E\left[\left(\sum_{i=l_1}^{l_2} c_{i,\epsilon}\int_s^t \epsilon^{-1}\xi_i(r\epsilon^{-2})dr\int_{z\in\mathcal{E}(y)\sim\mathcal{E}(x)}\phi_i(z)dz\right)^{2k}\right]$$

$$\le (2k)!K^{2k}\left(\sup_{p,\epsilon}|c_{p,\epsilon}|\right)^{2k}(2k)(2k-2)^{2k}|t-s|^k\|\chi_A\|_{L^2(S^1)}^{2k-2}\left(\sum_{i=l_1}^{l_2}\langle\phi_i,\chi_A\rangle^2\right)$$

$$\le \tilde{K}|t-s|^k\|\chi_A\|_{L^2(S^1)}^{2k-2}\left(\sum_{i=l_1}^{l_2}\langle\phi_i,\chi_A\rangle^2\right).$$

It is not difficult to see from this that we may pass to the limit $l_1 = 0$ and $l_2 \to \infty$ in (A.1), and that (8) must hold.

Now we should indicate why (8) implies (9). We shall do this by adapting Garsia's celebrated result (see [1], [4], and [11]). Let $\{n_i; i = 1,2,\ldots,l\}$ be a collection of positive integers, and let C_i be the unit cube in $I\!R^{n_i}$. Let Ψ be a positive, convex, and continuous function on $[0,\infty)$ with $\lim_{x\to\infty}\Psi(x) = \infty$. Let for each $i = 1,2,\ldots,l\}$, p_i be a nonnegative and increasing function on $[0,\infty)$ with $p_i(0) = 0$. Let $\|x\|$ be the Euclidean distance of x, for any x in $\bigcup_{i=1}^{l} I\!R^{n_i}$. Finally, if φ is a mapping from $\bar{C} := C_{n_1}\times C_{n_2}\cdots C_{n_l}$ to $I\!R$, and $\bar{x}^1 = (x_1^1, x_2^1,\ldots,x_l^1)$ and $\bar{x}^2 = (x_1^2,x_2^2,\ldots,x_l^2)$ are in $\bar{C}$, the define by an abuse of notation the variation of φ over the rectangle $[\bar{x}^1,\bar{x}^2]$ as

$$\varphi([\bar{x}^1,\bar{x}^2]) := \sum_{(i_1,i_2,\ldots,i_l)\in\{1,2\}^l}(-1)^{|\{j:i_j=2\}|}\varphi(x_1^{i_1},x_2^{i_2},\ldots,x_l^{i_l}) \tag{A.4}$$

where for any set A, $|A|$ is its cardinality. Our theorem is

Lemma 1. *If φ is a measurable function on $\bar{C}$ such that*

$$B := \int_{\bar{x}^1=(x_1^1,x_2^1,\ldots,x_l^1)\in\bar{C}}\int_{\bar{x}^2=(x_1^2,x_2^2,\ldots,x_l^2)\in\bar{C}}\Psi\left(\frac{|\varphi([\bar{x}^1,\bar{x}^2])|}{\prod_{i=1}^{l}p_i(\|x_i^2-x_i^1\|/\sqrt{n_i})}\right)d\bar{x}^1d\bar{x}^2$$

is finite, then there is a subset K of $\bar{C}$ of Lebesgue measure zero such that if $\bar{x}^1 = (x_1^1,x_2^1,\ldots,x_l^1)$ and $\bar{x}^2 = (x_1^2,x_2^2,\ldots,x_l^2)$ are in $\bar{C}\sim K$, then

$$|\varphi([\bar{x}^1,\bar{x}^2])| \le 2^l\cdot 4\int_{\substack{0\le u_i\le\|x_i^2-x_i^1\|\\ i=1,2,\ldots,l}}\Psi^{-1}\left(\frac{B}{\prod_{i=1}^{l}u^{2n_i}}\right)dp_1(u_1)dp_2(u_2)\cdots dp_l(u_l).$$

Proof. For any rectangle Q in $\bigcup_{i=1}^{l} I\!R^{n_i}$, let $|Q|$ be its volume and $e(Q)$ the length of its edge. Let us consider a collection of rectangles $\{Q_i^1; i = 1,2,\ldots,l\}$ with Q_i^1 contained in C_i for each $i = 1,2,\ldots$. We allow the possibility of a degenerate rectangle, i.e., a single point. Let us define, by another abuse of notation, the averaged value of φ as

$$\varphi(Q_1, Q_2, \ldots, Q_l) = \left(\prod_{i=1}^{l} \frac{1}{|Q_i|}\right) \int_{\substack{x_i \in Q_i \\ i=1,2,\ldots,l}} \varphi(x_1, x_2, \ldots, x_l) dx_1 dx_2 \ldots dx_l,$$

with averaginng being replaced by simple evaluation for any indices for which the rectangles are degenerate. Let us now two such collection of rectangles $\{Q_i^1; i = 1,2,\ldots,l\}$ and $\{Q_i^2; i = 1,2,\ldots,l\}$ with $Q_i^1 \supset Q_i^2$ and $p_i(e(Q_i^2)) = \frac{1}{2}p_i(e(Q_i^1))$ for all $i = 1,2,\ldots,l$. The techniques of Garsia (see [1], [4], or [11]) then give us that

$$\left| \sum_{(j_1,j_2,\ldots,j_l)\in\{1,2\}^l} (-1)^{|\{r:j_r=2\}|} \varphi(Q_1^{j_1}, Q_2^{j_2}, \cdots, Q_l^{j_l}) \right|$$

$$\leq 4 \int_{\substack{e(Q_i^1)\leq u_i \leq e(Q_i^2) \\ i=1,2,\ldots,l}} \Psi^{-1}\left(\frac{B}{\prod_{i=1}^{l} u^{2n_i}}\right) dp_1(u_1) dp_2(u_2) \cdots dp_l(u_l). \tag{A.5}$$

Consider now a third such collection of rectangles $\{Q_i^1; i = 1,2,\ldots,l\}$ with with $Q_i^2 \supset Q_i^3$ and $p_i(e(Q_i^3)) = \frac{1}{2}p_i(e(Q_i^2))$ for all $i = 1,2,\ldots,l$. Consider the sum

$$\sum_{(k_1,k_2,\ldots,k_l)\in\{1,3\}^l} (-1)^{|\{q:k_q=3\}|} \sum_{(j_1,j_2,\ldots,j_l)\in I(k_1,k_2,\ldots,k_l)} (-1)^{|\{r:j_r=2\}|} \varphi(Q_1^{j_1}, Q_2^{j_2}, \ldots, Q_l^{j_l}) \tag{A.6}$$

where $I(k_1, k_2, \ldots, k_l) = \{k_1, 2\} \times \{k_2, 2\} \cdots \{k_l, 2\}$ for all $(k_1, k_2, \ldots, k_l)$ in $\{1,3\}^l$ (we are adding up terms of the type (A.5)). By rewriting (A.6) as

$$\sum_{(i_1,i_2,\ldots,i_l)\in\{1,2,3\}^l} (-1)^{|\{r:i_r=2\}|} \varphi(Q_1^{i_1}, Q_2^{i_2}, \ldots, Q_l^{i_l})$$

$$\sum_{(k_1,k_2,\ldots,k_l)\in\{1,3\}^l} \sum_{(j_1,j_2,\ldots,j_l)\in I(k_1,k_2,\ldots,k_l)} \chi_{\{(j_1,j_2,\ldots,j_l)=(i_1,i_2,\ldots,i_l)\}} (-1)^{|\{q:j_q=3\}|}$$

and considering each index $(i_1, i_2, \ldots, i_l)$ individually, we can show that in fact (A.6) reduces to

$$\sum_{(j_1,j_2,\ldots,j_l)\in\{1,3\}^l} (-1)^{|\{j:i_j=3\}|} \varphi(Q_1^{j_1}, Q_2^{j_2}, \cdots, Q_l^{j_l}). \tag{A.7}$$

Thus by summing up the bounds as in (A.5) according to (A.6), we can see that

$$\left| \sum_{(j_1,j_2,\ldots,j_l)\in\{1,3\}^l} (-1)^{|\{r:j_r=3\}|} \varphi(Q_1^{j_1}, Q_2^{j_2}, \cdots, Q_l^{j_l}) \right|$$

$$\leq 4 \int_{\substack{e(Q_i^1)\leq u_i \leq e(Q_i^3) \\ i=1,2,\ldots,l}} \Psi^{-1}\left(\frac{B}{\prod_{i=1}^{l} u^{2n_i}}\right) dp_1(u_1) dp_2(u_2) \cdots dp_l(u_l).$$

By taking sequences of rectangles decreasing to $\bar{x}^1 = (x_1^1, x_2^1, \ldots, x_l^1)$ and $\bar{x}^2 = (x_1^2, x_2^2, \ldots, x_l^2)$ and by again using the calculations (A.6)–(A.7), we may, as in the proof of Garsia's theorem, conclude the result. ∎

Using Lemma 1 we can easily derive some classical moduli of continuity (see [7] and the calculations of [11]).

References.

[1] Adler, R.J. (1981). *The Geometry of Random Fields.* Wiley, New York.

[2] Freidlin, M.I. (1988). Random perturbations of reaction-diffusion equations: the quasi-deterministic approach. *Trans. Amer. Math. Soc.* **305**, 665–697.

[3] Freidlin, M.I. and Wentzell, A.D. Reaction-diffusion equations with randomly-perturbed boundary conditions. To appear.

[4] Garsia, A. (1972). Continuity properties of Gaussian processes with multidimensional time parameter. *Proc. Sixth Berkeley Symp. Math. Stat. Prob.*, **2**, 369–374, Univ. of California Press, Berkeley.

[5] Khas'minskii, R. Z. (1966). On stochastic processes defined by differential equations with a small parameter. *Theory of Probability and Its Applications* **9**, No 2, 211–228.

[6] Laha, R. G. and Rohatgi, V. K. (1979) *Probability Theory.* Wiley, New York.

[7] Orey, S. and Pruitt, W. (1973) Sample functions of the N-parameter Wiener process. *Annals of Probability* **1**, No 1, 138–163.

[8] Skorohod, A.V. (1989). Translated by H. H. McFaden. *Asymptotic Methods in the Theory of Stochastic Differential Equations.* American Mathematical Society, Providence.

[9] Sowers, R. B. Multidimensional reaction-diffusion equations with white noise boundary perturbations. Submitted.

[10] Sowers, R. B. *New Asymptotic Results for Stochastic Partial Differential Equations*, Ph.D. Thesis, Department of Mathematics, University of Maryland, College Park (MD), March 1991.

[11] Walsh, J.B. (1984). An introduction to stochastic partial differential equations. *École d'Éte de Probabilités de Saint-Flour, XIV, Lect. Notes Math. 1180.* Springer, New York.

On the stochastic partial differential equations of Ginzburg-Landau type

T. FUNAKI

Department of Mathematics, School of Science
Nagoya University, Nagoya 464-01, Japan

This paper exposes several recent mathematical results on an interacting system of continuum field distributed over the d-dimensional Euclidean space $\mathbf{R}^d$, which is often referred to as time-dependent Ginzburg-Landau (TDGL) model in physical literatures, e.g. [12]. The evolution law of the system is defined through certain stochastic partial differential equations (SPDE's) on $\mathbf{R}^d$. In Sect. 1, we explain the model from rather heuristic point of view and introduce SPDE's of GL type. The most part of our discussion is devoted to the model for real-valued continuum field; however, the case where the values of the field range over a manifold is also discussed. The existence and uniqueness theorems for these SPDE's are formulated in Sect. 2. Some results on reversible or stationary measures of the SPDE's are summarized briefly. Then mainly from motives of physics two kinds of problems, namely, the hydrodynamic limit and the low temperature limit for the GL model are investigated in Sect.'s 3 and 4, respectively.

1. GL model

The TDGL model describes an interacting system of continuum like spin field distributed over $\mathbf{R}^d$. The configuration is represented by a real-valued continuous function $S = \{S(x), x \in \mathbf{R}^d\}$. The system is governed by the (formal) Hamiltonian or the total energy $\mathcal{H}(S)$, which is called Ginzburg-Landau-Wilson free energy. This energy for configuration S is given by the sum $\mathcal{H}(S) = \mathcal{H}_L(S) + \mathcal{H}_S(S)$ of local-interaction:

$$\mathcal{H}_L(S) = \frac{1}{2}\int_{\mathbf{R}^d} \sum_{|\alpha|,|\beta|\le m} a_{\alpha,\beta}(x) D^\alpha S(x) \cdot D^\beta S(x)\, dx, \tag{1.1}$$

and self-interaction:

$$\mathcal{H}_S(S) = \int_{\mathbf{R}^d} U(x, S(x))\, dx, \tag{1.2}$$

where $|\alpha| = \sum_{i=1}^d \alpha_i$ and $D^\alpha = \left(\frac{\partial}{\partial x_1}\right)^{\alpha_1} \cdots \left(\frac{\partial}{\partial x_d}\right)^{\alpha_d}$ for multi-indices $\alpha = (\alpha_1, \dots, \alpha_d) \in \mathbf{Z}_+^d = \{0, 1, 2, \cdots\}^d$. The coefficients $\{a_{\alpha,\beta}(x)\}$ appearing in $\mathcal{H}_L$ are symmetric smooth functions, i.e., $a_{\alpha,\beta} = a_{\beta,\alpha} \in C_b^\infty(\mathbf{R}^d)$, and satisfy uniform strong ellipticity condition:

$$\sum_{|\alpha|=|\beta|=m} a_{\alpha,\beta}(x)\xi^\alpha\xi^\beta \ge \delta|\xi|^{2m}, \quad x, \xi \in \mathbf{R}^d,\ \delta > 0, \tag{1.3}$$

and non-negativity condition: $\mathcal{H}_L(S) \geq 0$ for all $S \in C_0^\infty(\mathbf{R}^d)$, where $\xi^\alpha = \xi_1^{\alpha_1} \cdots \xi_d^{\alpha_d}$. The self-potential $U(x, s)$ is a real-valued function of space-variable $x \in \mathbf{R}^d$ and spin-variable $s \in \mathbf{R}$ having the form which is a perturbation from a quadratic function of s:

$$U(x,s) = \frac{\gamma}{2}s^2 + V(x,s), \quad \gamma > 0, \tag{1.4}$$

where $V = V(x, s)$ is a *mild* function; the mildness means, in particular, the boundedness of its derivative in s. Notice that the local-interaction $\mathcal{H}_L$ may be rewritten into

$$\mathcal{H}_L(S) = \frac{1}{2}\int_{\mathbf{R}^d} \mathcal{A}S(x) \cdot S(x)\, dx, \tag{1.5}$$

with a symmetric differential operator $\mathcal{A}$ of order $2m$ having the form

$$\mathcal{A}S(x) = \sum_{|\alpha|,|\beta| \leq m} (-1)^{|\alpha|} D^\alpha \{a_{\alpha,\beta} D^\beta S\}(x). \tag{1.6}$$

We assume the condition $m > d/2$.

The statistical mechanical system corresponding to the Hamiltonian $\mathcal{H}$ is described by a Gibbs state which is formally given by

$$\mu(dS) = Z^{-1} e^{-\beta\mathcal{H}(S)}\, dS, \quad dS = \prod_{x \in \mathbf{R}^d} dS(x), \; Z = \text{normalization}. \tag{1.7}$$

Here $\beta > 0$ is a parameter expressing the inverse temperature of the system. Although the Hamiltonian $\mathcal{H}$ itself has no appropriate meaning, the Gibbs state μ can be defined precisely in mathematical terminology. In fact, if $m > d/2$, μ is a probability measure on the space $\mathbf{C} := C(\mathbf{R}^d)$ of all configurations, which satisfies the DLR equation and is given by taking the thermodynamic limit:

$$\mu(dS) = \lim_{n \to \infty} Z_n^{-1} \exp\{-\beta \int_{(-n,n)^d} V(x, S(x))\, dx\}\, \nu(dS), \tag{1.8}$$

where Z_n is a normalizing constant and $\nu = \nu_\beta$ is a centered Gaussian measure on $\mathbf{C}$ with covariance operator $\{\beta(\gamma + \mathcal{A})\}^{-1}$, see [6] for details.

The dynamics which is associated with the Hamiltonian $\mathcal{H}$ is a time evolution having the Gibbs state μ as its equilibrium measure. One can introduce such dynamics $\{S_t(x); t > 0, x \in \mathbf{R}^d\}$ of continuum field by means of the following stochastic differential equation (SDE):

$$dS_t(x) = -\frac{\beta}{2}\Gamma_0 \mathbf{D}\mathcal{H}(x, S_t)dt + \Gamma_0^{\frac{1}{2}} dw_t(x), \quad t > 0,\, x \in \mathbf{R}^d. \tag{1.9}$$

Here Γ_0 is generally a non-negative operator acting on the variable x, $\mathbf{D}\mathcal{H}(x,S)$ is a (formal) functional derivative of $\mathcal{H}$:

$$(1.10) \qquad \mathbf{D}\mathcal{H}(x,S) = \mathcal{A}S(x) + U'(x,S(x)), \quad U'(x,s) = \frac{\partial U}{\partial s}(x,s),$$

and $w_t(x)$ is a cylindrical Brownian motion (c.B.m.) on the space $L^2(\mathbf{R}^d)$, namely, w_t is an $\mathcal{S}'(\mathbf{R}^d)$-valued continuous process such as $\langle w_t, \varphi\rangle/||\varphi||_{L^2(\mathbf{R}^d)}$ is a 1-dimensional standard Brownian motion for every non-zero $\varphi \in \mathcal{S}(\mathbf{R}^d)$, where $\mathcal{S}(\mathbf{R}^d)$ and $\mathcal{S}'(\mathbf{R}^d)$ are the Schwartz's space of rapidly decreasing C^∞-functions and its dual space, respectively. The fluctuation-dissipation theorem permits us to take a non-negative operator Γ_0 freely. In physical literatures, the following two types of SPDE's are discussed particularly by taking $\Gamma_0 = identity$ and $\Gamma_0 = -\Delta(= -\sum_{i=1}^d \frac{\partial^2}{\partial x_i^2})$ in the equation (1.9):

$$(1.11) \qquad dS_t(x) = -\frac{\beta}{2}\mathcal{A}S_t(x)dt - \frac{\beta}{2}U'(x,S_t(x))dt + dw_t(x),$$

and

$$(1.12) \qquad dS_t(x) = \frac{\beta}{2}\Delta\mathcal{A}S_t(x)dt + \frac{\beta}{2}\Delta\{U'(x,S_t(x))\}dt + d\{\mathrm{div}\mathbf{w}_t(x)\},$$

where $\mathrm{div}\mathbf{w}_t(x) = \sum_{i=1}^d \frac{\partial}{\partial x_i} w_t^i(x)$ and $\mathbf{w}_t(x) = \{w_t^i(x)\}_{i=1}^d$ is a system of d independent c.B.m.'s on $L^2(\mathbf{R}^d)$. Notice that $\mathrm{div}\mathbf{w}_t(x)$ and $(-\Delta)^{\frac{1}{2}}w_t(x)$ have the same law.

These two SPDE's exhibit quite different features. In fact, for an arbitrary finite box Λ in $\mathbf{R}^d$, the total charge $\int_\Lambda S_t(x)\,dx$ of the spin field $S_t(x)$ over Λ is preserved as the time passes for the system defined by the latter equation (1.12) by restricting it in Λ and imposing the periodic boundary condition; we discuss such system in Sect. 4 in 1-dimensional case. Especially, because of this fact, the family of stationary measures for the SPDE (1.12) has a rich structure. Indeed, roughly saying, it is parametrized by the set of harmonic functions on $\mathbf{R}^d$, see Sect. 2. In contrast to this property of the equation (1.12), the former equation (1.11) has no such conserved quantity. The equation (1.11) is called the TDGL equation of non-conservative type, while (1.12) is called the TDGL equation of conservative type. The correspondences in the lattice version of the dynamics defined by (1.11) and (1.12) are the stochastic Ising model and the exclusion process, respectively, which appear in the theory of interacting particle systems, [13].

So far, we have been concerned about the real-valued GL model. However, it is natural and interesting to generalize the model for spins taking values on a manifold. In fact, such models are known as stochastic Heisenberg model or nonlinear σ-model in the lattice case. Let M be a complete and compact C^∞-Riemannian manifold with metric $g = \{g_{ij}\}$. For simplicity, we consider the system only on a one-dimensional interval imposing the periodic boundary condition; namely, we take a circle $\mathbf{T} \equiv \mathbf{T}^1 = [0,1)$ in place of the whole Euclidean space $\mathbf{R}^d$. Then, the configuration of M-valued spin field over $\mathbf{T}$ is represented by $S = \{S(x), x \in \mathbf{T}\} \in C(\mathbf{T}, M)$. Consider the energy of S defined by

$$
\begin{aligned}
(1.13) \qquad \mathcal{E}(S) &= \frac{1}{2}\int_{\mathbf{T}} |\frac{dS}{dx}(x)|^2\, dx \\
&= \frac{1}{2}\int_{\mathbf{T}} \sum_{i,j} g_{ij}(S(x))\frac{dS^i}{dx}(x)\frac{dS^j}{dx}(x)\, dx.
\end{aligned}
$$

Note that $\frac{dS}{dx}(x) \in T_{S(x)}M = \{\text{tangent vectors to } M \text{ at } S(x) \in M\}$ and $|v|^2 = g(v,v)$ for $v \in T_s M$. It is well-known that the functional derivative $\mathbf{D}\mathcal{E}(x,S)$ of $\mathcal{E}(S)$ coincides with $-\Delta S(x)$. Here Δ is the nonlinear Laplacian, i.e., $\Delta S(x) = \{(\Delta S(x))^i\}_i \in T_{S(x)}M$, $x \in \mathbf{T}$, is determined by the formula

$$
(1.14) \qquad (\Delta S(x))^i = \frac{d^2 S^i}{dx^2}(x) + \sum_{j,k} \Gamma^i_{jk}(S(x))\frac{dS^j}{dx}(x)\frac{dS^k}{dx}(x), \quad S = \{S^i\}_i,
$$

in terms of local coordinates of M, where Γ^i_{jk} stand for the Christoffel symbols on M, see [3].

It seems quite difficult and even impossible at the first stage to deal with the TDGL equation (1.11) of non-conservative type with $\mathcal{H}$ replaced by $\mathcal{E}$ having the c.B.m. as its random noise because of the nonlinear term appearing in the right hand side of (1.14). We replace the c.B.m. with smooth noise; in other words, we introduce the smooth approximation to the c.B.m. This leads us to the following quasilinear SPDE for $\{S_t(x); t \geq 0, x \in \mathbf{T}\}$:

$$
(1.15) \qquad dS_t(x) = \Delta S_t(x)dt + \mathbf{V}_0(S_t(x))dt + \circ dW_t(x, S_t(x)).
$$

Here $\mathbf{V}_0$ is a C^∞-vector field on M and $W_t(x,s)$ is a $C^\infty(\mathbf{T} \times TM)$-valued Wiener process, namely, $W_t = \sum_{i=1}^\infty w_t^i \mathbf{V}_i$ converging in a proper sense with independent real-valued Brownian motions $\{w_t^i\}$ and $\mathbf{V}_i = \mathbf{V}_i(x,s) \in C^\infty(\mathbf{T} \times TM) = \{C^\infty\text{-sections of the product bundle } \mathbf{T} \times TM \to \mathbf{T} \times M\}$, see [1]. We write $\circ dW_t$ to denote the Stratonovich stochastic differential and $\circ dW_t(x, S_t(x)) = \sum_{i=1}^\infty \mathbf{V}_i(x, S_t(x)) \circ dw_t^i$.

2. Existence-uniqueness results for the SPDE's and their reversible measures

2.1. SPDE's (1.11) and (1.12)

We call $S_t = \{S_t(x), x \in \mathbf{R}^d\}$ a solution (occasionally called a mild solution) of the SPDE's (1.11) or (1.12) if it satisfies the stochastic integral equations (SIE's) obtained by rewriting them formally based on the constant variational formula, namely, S_t is called a solution of (1.11) with initial data S_0 if it satisfies

$$S_t = T_t S_0 + \int_0^t T_{t-s}\, dw_s - \frac{\beta}{2}\int_0^t T_{t-s}\{V'(\cdot, S_s(\cdot))\}\, ds, \tag{2.1}$$

where $\{T_t\}_{t\geq 0}$ is the semigroup generated by the linear operator $-\frac{\beta}{2}(\gamma + \mathcal{A})$. One can introduce a similar SIE for the SPDE (1.12) using the semigroup generated by $\frac{\beta}{2}\Delta(\gamma + \mathcal{A})$. Let us introduce the spaces of configurations; in other words, the spaces on which the solutions of the SPDE's (1.11) and (1.12) live. For $r > 0$, $\mathbf{L}_r^2 = L^2(\mathbf{R}^d, e^{-r\chi(x)}dx)$ denote the Hilbert spaces equipped with natural weighted L^2-inner products, where $\chi(x) = \{1+|x|^2\}^{1/2}$, $x \in \mathbf{R}^d$. The space $\mathbf{C} = C(\mathbf{R}^d)$ is endowed with the usual topology determined from uniform convergence on all bounded sets. Then, the following results are shown by the standard arguments using Eidel'man's estimates on the fundamental solutions of parabolic operators.

THEOREM 2.1. *([7]) We assume $m > d/2$ and $V = V(x,s)$ is a measurable function on $\mathbf{R}^d \times \mathbf{R}$ such that $V(x,\cdot) \in C^1(\mathbf{R})$ for a.e. $x \in \mathbf{R}^d$ and its derivative $V'(x,s)$ in s is bounded and Lipschitz continuous (i.e., $\operatorname{esssup}_{x,s}|V'(x,s)| < \infty$ and $|V'(x,s) - V'(x,\bar{s})| \leq \text{const}\,|s - \bar{s}|$, $x \in \mathbf{R}^d$, $s, \bar{s} \in \mathbf{R}$).*
(i) If the initial data $S_0 \in \mathbf{L}_r^2$ with some $r > 0$, then the SPDE (1.11) has a unique solution S_t satisfying

$$S_t \in C([0,\infty), \mathbf{L}_r^2) \cap C((0,\infty), \mathbf{C}) \text{ a.s.}$$

(ii) If $S_0 \in \mathbf{C}$ in addition to the condition in (i), then $S_t \in C([0,\infty), \mathbf{C})$ a.s. for the solution of the SPDE (1.11).
(iii) Similar results hold also for the SPDE (1.12).

The Hölder continuity of the solutions is also established in [7]. The paper [6] investigate the classes of stationary and reversible probability measures of the SPDE's (1.11) and (1.12); in particular, the following problems are discussed.

(a) The characterization of all reversible measures of (1.11) as the Gibbs states over $\mathbf{R}^d$ by means of the DLR equation — for this purpose, weak C^{m-1}-differentiability of solutions of (1.11) and regularity property of their boundary data are studied by [7].
(b) The construction of reversible measures of (1.11) for a wide class of self-potentials $U(x,s)$.
(c) To show the uniqueness of stationary probability measures of (1.11) under the strict convexity condition (2.2) on $U(x,s)$. Actually, this unique stationary measure μ is given by the formula (1.8). The ergodicity of the process S_t, the solution of (1.11),

and mixing property of μ with respect to the spatial shifts are established. We denote the probability measure μ on the space $\cap_{r>0}\mathbf{L}_r^2 \cap \mathbf{C}$ by $\mu(\mathcal{A}, U)$.

(d) To show, under the strict convexity condition (2.2), that every element of the convex hull of the set $\{\mu(\mathcal{A}, U_\lambda);\ \Delta\lambda = 0, \lambda \in \cap_{r>0}\mathbf{L}_r^2 \cap C^2(\mathbf{R}^d)\}$ is a reversible measure of the SPDE (1.12), where the potential U_λ is defined by $U_\lambda(x,s) = U(x,s) - \lambda(x)s$ for $\lambda = \lambda(x)$.

Here, by the strict convexity condition, we mean

$$U(x,\cdot) \in C^2(\mathbf{R}) \text{ for a.e.} x \in \mathbf{R}^d \text{ and } \operatorname*{essinf}_{x,s} \frac{\partial^2 U}{\partial s^2}(x,s) > 0. \tag{2.2}$$

2.2. SPDE (1.15)

The following theorem is shown for the M-valued SPDE (1.15). The initial data S_0 is taken from the space $C^\infty(\mathbf{T}, M)$, on which we consider the usual topology of uniform C^n-convergence for every $n \geq 0$. The results may be regarded as the probabilistic extensions of those due to Eells and Sampson [3].

THEOREM 2.2. *(i)(Existence of global solutions) Suppose that the manifold M has non-positive sectional curvature K_M. Then there exists a solution S_t, $t \in [0,\infty)$, of the SPDE (1.15).*
(ii)(Regularization of solutions) Every solution $S_t, t \in [0,T]$, of the SPDE (1.15) belongs to the space $C([0,T], C^\infty(\mathbf{T}, M))$ a.s.
(iii)(Pathwise uniqueness of solutions) Let S_t and S'_t, $t \in [0,T]$, be two solutions of the SPDE (1.15) such that $S_0 = S'_0$. Then we have $S_t = S'_t, t \in [0,T]$, a.s.

The meaning of the solutions of (1.15) in this theorem is in the sense of generalized functions on $\mathbf{T}$. The results with outline of the proofs were announced in [4]. The paper containing complete proofs will be published, [9].

3. Hydrodynamic limit

Here we consider the TDGL equation (1.12) of conservative type; take $\beta = 1$ for simplicity. The solution of the SPDE gives the time-evolution of the system at the microscopic level. The problem is to derive the macroscopic kinetic equation by passing from the microscopic to the macroscopic level. For this purpose we introduce the hydrodynamic space-time scaling for the system, namely, the microscopic space-time variable (x,t) and the macroscopic one (y,τ) are connected with each other by the transformation $y = \epsilon x$ and $\tau = \epsilon^2 t$, where $\epsilon > 0$ is a small parameter. Then, the scaled process of the solution $S_t(x)$ of (1.12):

$$\tilde{S}^\epsilon_\tau(y) = S_{\tau/\epsilon^2}(y/\epsilon), \tag{3.1}$$

which gives the time-evolution of the system at the macroscopic level satisfies the following SPDE:

$$(3.2)\qquad d\tilde{S}^{\epsilon}_{\tau}(y)=\frac{1}{2}\Delta A^{\epsilon}\tilde{S}^{\epsilon}_{\tau}(y)d\tau+\frac{1}{2}\Delta\{U'(\frac{y}{\epsilon},\tilde{S}^{\epsilon}_{\tau}(y))\}d\tau+\epsilon^{\frac{d}{2}}d\{\mathrm{div}\mathbf{w}_{\tau}(y)\}.$$

Here A^{ϵ} is an operator determined by

$$(3.3)\qquad A^{\epsilon}f(y)=\sum_{|\alpha|,|\beta|\le m}(-1)^{|\alpha|}\epsilon^{|\alpha|+|\beta|}D^{\alpha}\{a^{\epsilon}_{\alpha,\beta}(y)D^{\beta}f\}(y),\quad a^{\epsilon}_{\alpha,\beta}(y)=a_{\alpha,\beta}(\frac{y}{\epsilon}).$$

Note that $\{\mathrm{div}\mathbf{w}_{\tau/\epsilon^2}\}(y/\epsilon)$ has the same law as $\epsilon^{\frac{d}{2}}\mathrm{div}\mathbf{w}_{\tau}(y)$. The problem is to investigate the asymptotic behavior of $\tilde{S}^{\epsilon}_{\tau}(y)$ as $\epsilon\downarrow 0$.

It is, however, possible to discuss the problem under more general situation where A and U are random. To distinguish the randomness of the c.B.m. and that of (A,U), we denote by $(W,\mathcal{F}_W,\mathbb{P})$ the probability space on which the c.B.m. $\mathbf{w}_t(x)=\mathbf{w}_t(x;\mathrm{w}),\mathrm{w}\in W$, is realized. The solution S_t of (1.12) is sometimes denoted by $S_t(x;\mathrm{w};A,U)$. The random functions $\{a_{\alpha,\beta;\omega}(x)=a_{\alpha,\beta}(x;\omega)\}$ and $U_\omega(y,x,s)=U(y,x,s;\omega)$ of macroscopic space-variable $y\in\mathbf{R}^d$, microscopic space-variable $x\in\mathbf{R}^d$ and spin variable $s\in\mathbf{R}$, $\omega\in\Omega$, are given on another probability space $(\Omega,\mathcal{F}_\Omega,\mathbf{P})$. Denote by A_ω the differential operator A with coefficients $\{a_{\alpha,\beta;\omega}\}$. Then, the time evolution of spin fields we are concerned is the solution $S^{\epsilon}_t(x)=S_t(x;\mathrm{w};A_\omega,U_{\omega,\epsilon})$ of (1.12) with $(A_\omega,U_{\omega,\epsilon})$ in place of (A,U), where $U_{\omega,\epsilon}$ is defined by $U_{\omega,\epsilon}(x,s)=U(\epsilon x,x,s;\omega)$. Denote by $\tilde{S}^{\epsilon}_{\tau}(y)$ the scaled process of $S^{\epsilon}_t(x)$:

$$(3.4)\qquad \tilde{S}^{\epsilon}_{\tau}(y)\equiv\tilde{S}^{\epsilon}_{\tau}(y;\mathrm{w},\omega)=S_{\tau/\epsilon^2}(y/\epsilon;\mathrm{w};A_\omega,U_{\omega,\epsilon}).$$

Assuming the stationarity and ergodicity of $\{a_{\alpha,\beta}(\cdot;\omega)\}$ and $U(y,\cdot,s;\omega)$ with respect to the shift in the variable x, the main result can be formulated as the law of large numbers in the following form. We set $\mathbf{H}^1_r$, $r>0$, the class of all functions $\rho\in\mathbf{L}^2_r$ with derivatives $\nabla\rho\in\{\mathbf{L}^2_r\}^d$ in the sense of generalized functions. We need the strict convexity assumption which is similar to (2.2) on the self potential $U(y,x,s;\omega)$ and some additional technical conditions.

THEOREM 3.1. *([5,8]) Assume that the initial value $\tilde{S}^{\epsilon}_0=S_0\in\cap_{r>0}\mathbf{H}^1_r$ is independent of ϵ. Then, for $\mathbf{P}$-a.s.ω, $\langle\tilde{S}^{\epsilon}_{\tau},\varphi\rangle$ converges to $\langle\rho_\tau,\varphi\rangle$ in probability (w.r.t. $\mathbb{P}$) as $\epsilon\downarrow 0$ for every $\tau>0$ and $\varphi\in C_0^\infty(\mathbf{R}^d)$, where $\langle\tilde{S}^{\epsilon}_{\tau},\varphi\rangle=\int_{\mathbf{R}^d}\tilde{S}^{\epsilon}_{\tau}(x)\varphi(x)\,dx$ etc. The non-random limit $\rho_\tau(y)$ is the unique weak solution of the following nonlinear PDE:*

$$(3.5)\qquad \begin{cases}\frac{\partial}{\partial\tau}\rho_\tau(y)=\frac{1}{2}\Delta\{\bar{\xi}(y,\rho_\tau(y))\}, & y\in\mathbf{R}^d,\\ \rho_0=S_0.\end{cases}$$

The function $\bar{\xi}(y,\rho)$, $y \in \mathbf{R}^d, \rho \in \mathbf{R}$, appearing in (3.5) is defined as follows: The unique Gibbs state associated with $(\mathcal{A}, U)$, i.e., the unique stationary probability measure of the SPDE (1.11), is denoted by $\mu(\mathcal{A}, U)$; see Sect. 2 or [6] for details. Introduce the mean spin function:

$$(3.6) \qquad \bar{\rho}(y,\xi) = E^{\mathbf{P}}[E^{\mu(\mathcal{A}_\omega, U^{\xi}_{y,\omega})}[S(0)]], \quad y \in \mathbf{R}^d, \xi \in \mathbf{R},$$

where $U^{\xi}_{y,\omega}(x,s) = U(y,x,s;\omega) - \xi s$. Then one can prove that $\bar{\rho} \in C^1(\mathbf{R}^d \times \mathbf{R})$ and it is strictly increasing in ξ such that $\lim_{\xi\to\pm\infty} \bar{\rho}(y,\xi) = \pm\infty$. The function $\xi = \bar{\xi}(y,\rho)$ is the inverse function of $\rho = \bar{\rho}(y,\xi)$ in the variable ξ for each $y \in \mathbf{R}^d$.

The method for the proof of Theorem 3.1 is the resolvent approach. See [8] for the heuristic derivation of the limit equation (3.5) based on the hypothesis of the local equilibrium states. The randomness of $(\mathcal{A}, U)$ is averaged out because of the effect of the homogenization caused by the hydrodynamic scaling.

A typical example is the periodic case: Let $\{a_{\alpha,\beta}(x) = a_{\beta,\alpha}(x) \in C_b^\infty(\mathbf{R}^d)\}_{|\alpha|,|\beta|\le m}$ and $U(y,x,s) \in C_b^\infty(\mathbf{R}^d \times \mathbf{R}^d \times \mathbf{R})$ be the given periodic functions with period one in the variable x. Then, this case fits into the above setting. In fact, we may take $\Omega = \mathbf{T}^d =$ the unit d-dimensional torus, $\mathbf{P} =$ the Lebesgue measure on $\mathbf{T}^d$ and define

$$(3.7) \qquad \begin{cases} a_{\alpha,\beta}(x;\omega) = a_{\alpha,\beta}(x+\omega), \\ U(y,x,s;\omega) = U(y,x+\omega,s), \end{cases}$$

for $\omega \in \Omega$.

4. Low temperature limit

Let us consider the SPDE (1.12) on the circle $\mathbf{T}$ instead of the space $\mathbf{R}^d$ taking the simple local-interaction $\mathcal{H}_L(S) = \frac{1}{2}\int_{\mathbf{T}} |\nabla S(x)|^2\, dx$ and the self-potential $U(s)$ which is the function independent of x. Then the SPDE (1.12) has the following form:

$$(4.1) \qquad dS_t(x) = \frac{\beta}{2}\Delta\{-\Delta S_t(x) + U'(S_t(x))\}dt + \nabla dw_t(x), \quad t > 0,\ x \in \mathbf{T},$$

where $\Delta = \frac{d^2}{dx^2}$ and $\nabla = \frac{d}{dx}$. This SPDE has a conservation law, namely, the total spin $\int_{\mathbf{T}} S_t(x)\,dx$ is preserved as the time passes. In particular, $S_0 \in L_0^2$ implies $S_t \in L_0^2$ for all $t > 0$ (a.s.), where L_0^2 is the space of all $S \in L^2 = L^2(\mathbf{T}, dx)$ satisfying $\int_{\mathbf{T}} S(x)\,dx = 0$. The problem is to study the asymptotic behavior of the solution $S_t = S_t^\beta$ of (4.1) as the inverse temperature $\beta \to \infty$, c.f. [2].

The drift term of (4.1) is rewritten into $\frac{\beta}{2}\Delta D\mathcal{H}(x, S_t)\,dt$ at least formally, see the equation (1.9). Therefore, one can expect that the solution S_t^β approach the set of the ground states of $\mathcal{H}$ restricted on the space L_0^2 as $\beta \to \infty$ if $S_0^\beta \in L_0^2$. However, if the self-potential U is of double well type with sufficiently deep bottoms, then the set of all ground states is not a single point but isomorphic to the circle $\mathbf{T}$. More precisely, we suppose the following condition on U :

$$(4.2)\quad \begin{cases} \text{(i)} & U(s) = \frac{\gamma}{2}s^2 + V(s),\ \gamma > 0,\ V \in C_b^\infty(\mathbf{R}) \\ \text{(ii)} & U(s) = -\frac{a}{2}s^2 + \frac{b}{4}s^4,\ a, b > 0,\ \text{for } |s| \le A \\ \text{(iii)} & U \text{ is increasing for } s > A \text{ and decreasing for } s < -A \\ \text{(iv)} & a > (2\pi)^2 = \text{ the least eigenvalue of } -\Delta \text{ on } L_0^2, \end{cases}$$

with sufficiently large A. Two bottoms $e_\pm = \pm\sqrt{\frac{a}{b}}$ of the potential U represents physically different two phases, respectively. It is shown in this special situation that there exists a solution $m = m(x) \in C^\infty(\mathbf{T}) \cap L_0^2$ of an ODE $-\Delta m(x) + U'(m(x)) = 0, x \in \mathbf{T}$, satisfying $\sharp\{x \in \mathbf{T}; m(x) = 0\} = 2$, which is unique except the translation. Fix one such m and set $M = \{m_\xi; \xi \in \mathbf{T}\}$, where m_ξ is defined by translating m: $m_\xi(x) = m(x + \xi)$. Then, M coincides with the set of all ground states of $\mathcal{H}$, i.e., $\min_{S \in L_0^2} \mathcal{H}$ is attained on M ($\mathcal{H}(S) = +\infty$ if $\nabla S \notin L^2(\mathbf{T})$ by convention). For $m \in M$, the eigenspace of the Hessian operator $\mathcal{H}''(m) = P\{-\Delta + U''(m(x))\}\cdot : L_0^2 \to L_0^2$ corresponding to the eigenvalue 0 is a 1-dimensional space spanned by ∇m, where $PS = S - \int_{\mathbf{T}} S(x)\,dx$ is the orthogonal projection of $L^2 \to L_0^2$. Moreover, the Morse index of all other critical points m of $\mathcal{H}$ is larger than 1, i.e., the Hessian operator $\mathcal{H}''(m)$ has at least 1 negative eigenvalue.

The space $H^\alpha = H^\alpha(\mathbf{T}), \alpha \ge 0$, denotes the usual Sobolev space on $\mathbf{T}$, namely, $H^\alpha = \{S \in L_0^2; (-\Delta)^{\frac{\alpha}{2}} S \in L_0^2\}$ is a Hilbert space equipped with inner product $\langle S, S' \rangle_{H^\alpha} = \langle (-\Delta)^\alpha S, S' \rangle$, $S, S' \in H^\alpha$. The following theorem asserts that, in the limit, the phase separation point of S_t^β moves like Brownian motion on $\mathbf{T}$.

THEOREM 4.1. *([10]) Assume the condition (4.2) on the potential U and let S_t^β be the solution of (4.1) such that $S_0^\beta = m_\xi \in M$ with $\xi \in \mathbf{T}$. Then, for every $0 \le t_1 < t_2 < \cdots < t_n < \infty$ and $\delta > 0$, the joint distribution of $(S_{t_1}^\beta, S_{t_2}^\beta, \cdots, S_{t_n}^\beta)$ on the space $\{H^{\frac{1}{2}-\delta}\}^n$ converges weakly to that of $(m_{\xi_{t_1}}, m_{\xi_{t_2}}, \cdots, m_{\xi_{t_n}})$, where $\xi_t = B(\|m\|_{L^2}^{-2} t)$ and $B(t)$ is the Brownian motion on $\mathbf{T}$ starting from ξ.*

The strategy for the proof of this theorem is the following. We first verify that S_t^β stays in the neighborhood of M with high probability as $\beta \to \infty$. For this purpose, we employ $\mathcal{H}$ as the Lyapunov function; however, a smooth approximation of S_t^β is required since $\mathcal{H}(S_t^\beta)$ itself has no meaning. Then the perturbative method is used to identify the limit of S_t^β. The geometry on the state space determined from the inner product $\langle \cdot, \cdot \rangle_{H^{-1}}$ plays an essential role, where H^{-1} is the dual space of H^1. Similar problem is discussed by [11] with the finite-dimensional setting.

References

1. P.H. Baxendale, *Brownian motions in the diffeomorphism group I*, Compositio Math. **53** (1984), 19–50.
2. H.W. Diehl, D.M. Kroll and H. Wagner, *The interface in a Ginsburg-Landau-Wilson model: Derivation of the drumhead model in the low-temperature limit*, Z. Physik **B 36** (1980), 329–333.
3. J. Eells and J.H. Sampson, *Harmonic mappings of Riemannian manifolds*, Amer. J. Math. **86** (1964), 109–160.
4. T. Funaki, *On diffusive motion of closed curves*, in "Probability Theory and Mathematical Statistics (eds. S. Watanabe and Yu.V. Prokhorov) Proceedings, Kyoto 1986," Lecture Notes in Math., **1299**, 86–94, Springer, 1988.
5. ———, *Derivation of the hydrodynamical equation for one-dimensional Ginzburg-Landau model*, Probab. Th. Rel. Fields, **82** (1989), 39–93.
6. ———, *The reversible measures of multi-dimensional Ginzburg-Landau type continuum model*, Osaka J. Math., **28** (1991), 463–494.
7. ———, *Regularity properties for stochastic partial differential equations of parabolic type*, Osaka J. Math., **28** (1991), 495–516.
8. ———, *The hydrodynamic limit for a system with interactions prescribed by Ginzburg-Landau type random Hamiltonian*, preprint (to appear in Probab. Th. Rel. Fields), 1990.
9. ———, *The stochastic partial differential equation with values in a manifold*, preprint (submitted to J. Funct. Anal.), 1991.
10. ———, *The low temperature limit of Ginzburg-Landau model*, in preparation.
11. T.Funaki and H. Nagai, *Degenerative convergence of diffusion process toward a submanifold by strong drift*, preprint (submitted to Stochastics), 1991.
12. P.C. Hohenberg and B.I. Halperin, *Theory of dynamic critical phenomena*, Rev. Mod. Phys., **49** (1977), 435–479.
13. T. Liggett, "Interacting particle systems," Springer, 1985.

STOCHASTIC VARIATIONAL CALCULUS

Takeyuki Hida

Department of Mathematics, Meijo University

Tenpaku, Nagoya, 468 JAPAN

[Summary]

White noise analysis has an aspect of the harmonic analysis that arises from the infinite dimensional rotation group. From this viewpoint we are naturally led to stochastic variational calculus that can be discussed by using certain subgroups of the rotation group.

§0. Introduction

We shall discuss stochastic variational calculus considered as an important development of white noise analysis. For this purpose we first claim that there is an aspect or viewpoint that our white noise analysis can be thought of as a *harmonic analysis* arising from the infinite dimensional rotation group O_∞. For one thing the white noise measure is invariant under the rotation group, and we are able to generalize the finite dimensional harmonic analysis over a sphere that comes from the rotation group $SO(n)$. Indeed, a generalization is obtained successfully in line with the white noise analysis.

It will be shown that there are subgroups of O_∞ with different character and that associated with each subgroup is a part of white noise analysis, where the larger the group the more generalized the white noise analysis. It is further shown that the subgroup consisting of the so-called *whiskers* can lead us to establish the stochastic variational calculus which is in line with the white noise analysis. This situation is illustrated by the 4 story (B ~ III) diagram below.

[Diagram]

- Infinite Dimensional Harmonic Analysis arising from O_∞ -

III. <u>Ultra infinite dimensional case</u>

$\{X(C)\}$ random field living in $(S)^*$,

Applications : Electromagnetic fields, Hadamard equation, and super many time theory $\{\Psi(C)\}$;

$$ih \frac{\delta}{\delta C_P}\Psi(C) = H(P,C)\Psi(C), \quad C \in \mathbf{C}, \quad H : \text{Hamiltonian}$$

diffeomorphism of R^d : $\psi \longrightarrow g \in O_\infty$

whisker $\{\psi_t\} \longrightarrow \{g_t\}$, $g_t g_s = g_{t+s}$

whiskers get innovation; involves conformal group, and deform C to have variation.

II. <u>Infinite dimensional case</u>

$(S) \subset (L^2) \subset (S)^*$

Potthoff-Streit characterization of $(S)^*$

$(S)^* \supset H_n^-$ Hermite polynomials in $x(t)$'s

Lévy group $\mathscr{G} \longrightarrow$ Lévy Laplacian $\Delta_L = \int \partial_t^2 \, (dt)^2$

Gross, Volterra Laplacians Δ_G, Δ_V ; Obata characterization

Gauss kernel $N\exp[-\frac{1}{2c}\int_T x(t)^2 dt]$:

eigenfunctional of Δ_L

Kuo's Fourier transform on $(S)^*$

I. <u>Hyper finite dimensional case</u>

$(L^2) = \bigoplus_n H_n$ and extension like hyper functions on S^1

$\text{supp}(\mu) \simeq S^\infty(\sqrt{\infty})$

Hyper finite group $G_\infty \equiv \underset{d}{<\underline{\lim}} \; G_d$, $G_d \simeq SO(d)$

Laplace-Beltrami operator $\Delta_\infty = -N = \sum_n \{(\frac{\partial}{\partial \xi_n})^2 - \langle x, \xi_n \rangle \frac{\partial}{\partial \xi_n}\}$

H_n : eigenspace of Δ_∞, eigenvalue : $-n$

irreducible representation of G_∞ on H_n

Fourier-Wiener transform

B. Finite dimensional case

$L^2(S^d) = \bigoplus_n H_n$, $S^d \simeq SO(d+1)/SO(d)$, hyper functions on S^d

Rotation group $SO(d+1)$, Δ_d : spherical Laplacian

Fourier series

It is noted that the analysis in each story (except Basement B) corresponds to a subgroup of O_∞.

§1. Background.

Before we come to the main topic, we shall summarize background, from which our approach starts.

Let a Gel'fand triple

$$(1.1) \qquad E \subset L^2(R^1) \subset E^*$$

be given, where E is a nuclear space. A *white noise measure* μ is introduced on E^* by the characteristic functional

$$C(\xi) = \exp[\int |\xi(t)|^2 dt , \qquad \xi \in E,$$

in such a way that

$$(1.2) \qquad C(\xi) = \int_{E^*} \exp[i\langle x, \xi \rangle] \, d\mu(x).$$

The measure space (E^*, μ) is called a *white noise*.

After H. Yoshizawa [9], we define a rotation group $O(E)$, often denoted by O_∞. A continuous linear isomorphism g of E is a *rotation* of E, if it satisfies

(1.3) $\|g\xi\| = \|\xi\|$ for any $\xi \in E$.

The collection $O(E)$ of all the rotations of E forms a group. It is called the *rotation group* of E, or the *infinite dimensional rotation group* in case E is not necessarily specified.

The adjoint of g, denote it by g^*, is determined by the canonical bilinear form $\langle\ ,\ \rangle$ that connects E and E^*:

(1.4) $\langle x, g\xi\rangle = \langle g^*x, \xi\rangle$ for any $x \in E^*$ and any $\xi \in E$.

The collection $O^*(E^*) = \{g^*;\ g \in O(E)\}$ also forms a group and it is isomorphic to $O(E)$ under the mapping

$$g^* \longleftrightarrow g^{-1}.$$

Theorem 1. *The white noise measure* μ *is invariant under* $O^*(E^*)$.

[Remark] With this property one may think that the "support" of μ looks like an infinite dimensional sphere and the μ may be viewed as the uniform probability measure on that sphere.

The *white noise analysis* can therefore be regarded as a harmonic analysis on the Hilbert space $(L^2) = L^2(E^*,\mu)$. However, there is an important remark that μ is $O^*(E^*)$-ergodic, but the entire group $O(E)$ is too big for the ergodicity, unlike finite dimensional case where $SO(n)$ acts. We can find suitable subgroups, with each of which an analysis is associated.

The following theorem shows how large the group $O(E)$ is. Let

(1.5) $(L^2) = \bigoplus_n H_n$

be the Wiener-Itô decomposition of (L^2). Then

Theorem 2. *On each subspace* H_n, $n \geq 1$, *an irreducible unitary reprresentation* $\{U_g,\ g \in O(E)\}$ *of* $O(E)$ *is given by*

(1.6) $(U_g\varphi)(x) = \varphi(g^*x)$.

From the next section onward, various subgroups of $O(E)$ will be found systematically and the associated analysis can be discussed.

§2. Subgroups G_d.

The first simplest example is a subgroup isomorphic to $SO(d)$. Finite dimensional harmonic analysis naturally arises in this case in a similar manner to the case B in the diagram.

Take a complete orthonormal system $\{\xi_k\}$ in $L^2(R^1)$ and let E_d be the d-dimensional subspace of E spanned by the ξ_k's with $1 \le k \le d$. The subgroup G_d is the collection of such g as

i) $g|_{E_d}$ is a rotation of E_d,

ii) $g|_{E_d^\perp}$ is the identity.

Obviously

(2.1) $\quad G_d \simeq SO(d)$.

The ordinary harmonic analysis that arises from G_d will briefly be reviewed as follows. The $(d-1)$-dimensional sphere S^{d-1} is identified with $SO(d)/SO(d-1)$. The $SO(d)$-invariant (or *uniform*) probability measure dP_d on S^{d-1}, is expressed in the form

(2.2) $$dP_d = \frac{1}{2}\,\frac{\Gamma(d/2)}{\pi^{d/2}}\,\{\prod_{i=2}^{d-1} \sin^{i-1}\theta_i\}d\theta_1 d\theta_2 \cdots d\theta_{d-1},$$

where $\theta_1, \theta_2, \cdots, \theta_{d-1}$ are the polar coordinates on S^{d-1} extending over $[0, \pi]$ and $0 \le \theta_{d-1} \le 2\pi$.

The spherical Laplacian (Laplace-Beltrami operator) Δ_d is given by the following formula :

(2.3) $$\Delta_d = \sum_{i=1}^{d-1} \{\prod_{j=1}^{i-1} \sin^2\theta_j\}^{-1} \cdot \{\frac{\partial^2}{\partial\theta_i^2} + (d-1-i)\cot\theta_i \frac{\partial}{\partial\theta_i}\} .$$

The spherical harmonic polynomials, which are the eigenfunctions of Δ_d, of degree p are of the form

(2.4) $\quad Y^{(d)}(p, m_1, m_2, \cdots; \theta_2, \theta_3, \cdots, \theta_{d-1}, \pm\theta_1)$

$$= \exp[\pm i m_1 \theta_1] \prod_{k=1}^{d-2} (\sin \theta_{k+1})^{m_k} C_{m_{k+1}-m_k}^{m_k+k/2} (\cos \theta_{k+1}),$$

where $p = m_{d-1} \geq m_{n-2} \geq \cdots \geq m_1 \geq 0$, p being the degree of $Y^{(d)}$ and where C_k^m is the Gegenbauer polynomial. By using these $Y^{(d)}$, we are able to expand any $L^2(S^{d-1}, P_d)$-function in terms of the $Y^{(d)}$'s. More precisely, denote by H_n be the subspace of $L^2(S^{d-1}, P_d)$ spanned by all the functions of the form (2.4) and of degree n. Then, H_n's are mutually orthogonal and the whole L^2-space admits a decomposition :

(2.5) $$L^2(S^{d-1}, P_d) = \bigoplus_{n=0}^{\infty} H_n.$$

In addition, an *irreducible* unitaty representation of $SO(d)$ is, as is well known, given on each H_n.

Finite, say d, dimensional projection of white noise (E^*, μ) is a superposition of (S^{d-1}, P_d) with different radii, so it is far from G_d-ergodic. However, the limit of G_d is reasonably big for ergodic property.

§3. The space (L^2) of ordinary white noise functionals.

Our harmonic analysis now comes to the hyper finite dimensional case. Start with the direct sum decomposition (1.5) of the Hilbert space (L^2). Since the *support* of μ looks like $S^\infty(\sqrt{\infty})$, we are led to define a subgroup G_∞: with a suitable choice of $n(p)$ we set

(3.1) $$G_\infty \equiv \varprojlim_{p} G_{n(p)} .$$

Keep the complete orthonormal system $\{\xi_K\}$ and introduce the coordinates $x_k = \langle x, \xi_k \rangle$. Define γ_{jk} by

(3.2) $$\gamma_{jk} = x_j \frac{d}{dx_k} - x_k \frac{d}{dx_j}$$

the infinitesimal generator of rotations in the (x_j, x_k)- plane. The system $\{\gamma_{jk}\}$ completely characterizes the so-called infinite dimensional Laplace-Beltrami operator Δ_∞, which is expressed in the form

$$(3.3) \qquad \Delta_\infty = \sum_{k=1}^{\infty} \{(\frac{d}{dx_k})^2 - x_k \frac{d}{dx_k}\}.$$

The operator $N \equiv -\Delta_\infty$ is the number operator, and H_n in (1.5) is the eigenspace for N :

$$(3.4) \qquad N|_{H_n} = n.$$

We can therefore claim that the unitary representation of G_∞ given by

$$(3.5) \qquad (U_g\varphi)(x) = \varphi(g^* x), \qquad g \in G_\infty,$$

is restricted to H_n and the representation of G_∞ on H_n is *irreducible*.

Remark. To give a good interpretation to (3.1), it is convenient to start with $L^2([0,1])$ in place of $L^2(R^1)$ in (1.1). Take a complete orthonormal system $\{w_k\}$, where w_k is a Walsh function formed by the Rademacher functions. The system is *également dense* in the sense of P. Lévy. We take increasing subsystems (members are as many as $n(p)$), the limit of which is the total system $\{w_k\}$. Then we define (3.1). □

So far, a good analogy to §2 is seen, while our general theory enables us to discuss the analysis on (L^2) in line with the *causal calculus*, where the time propagation appears explicitly. First we introduce the differential operators ∂_t, $t \in R^1$, with the help of the $\mathcal{S}$-transform:

$$(3.6) \qquad \partial_t \varphi(x) = \mathcal{S}^{-1} \frac{\delta}{\delta \xi(t)} (\mathcal{S}\varphi)(\xi).$$

Its adjoint ∂_t^* is defined in the usual manner. With these notations the operator Δ_∞ has a coordinate-free expression

$$(3.7) \qquad \Delta_\infty = \int \partial_t^* \partial_t \, dt.$$

It is also possible to have a coordinate free representation of generators of rotations; in place of $\gamma_{j,k}$ in (3.2) we define $\gamma_{s,t}$ by

$$(3.8) \qquad \gamma_{s,t} = \partial_s^* \partial_t - \partial_t^* \partial_s.$$

Like the Laplacian Δ_∞ the above formula for the generator of rotation is fitting for the causal calculus.

Finally we remind that the Fourier-Wiener transform on (L^2) is

(3.9) $$(\mathcal{F}\varphi)(y) = \int_{x\in E^*} \varphi(\sqrt{2}\, x + iy)\,d\mu(x).$$

It satisfies good relationships with O_∞ and Δ_∞ via the formula

(3.10) $$\mathcal{F}\varphi = i^n\varphi \qquad \text{for } \varphi \in H_n.$$

§4. Lévy group and generalized white noise functionals.

Let $\{\xi_k\}$ be a complete orthonormal system in $L^2[0,1]$, and assume that it is *également dense*. The Lévy group is defined as follows : Let π be an automorphism of the set of positive integers. A transformation g_π acting on E is defined in such a way that for $\xi = \sum a_k\xi_k$

(4.1) $$g_\pi\xi = \sum a_k\xi_{\pi(k)}.$$

Let Π be the set of automorphisms π satisfying

(4.2) $$\lim_{N\to\infty} \frac{1}{N}\,\#\{n \le N \,;\, \pi(n) > N\} = 0.$$

Now set

(4.3) $$\mathcal{G} = \{\, g_\pi : \pi \in \Pi,\ g_\pi \in O_\infty \,\}.$$

It forms a group, called the *Lévy group*. The Lévy Laplacian Δ_L is

(4.4) $$\lim_{N\to\infty} \frac{1}{N} \sum_{k=1}^{N} \{\frac{\partial}{\partial\xi_k}\}^2 .$$

It commutes with the Lévy group and its domain is rich in $(S)^*$.

There is a formula, which is formal but suggestive, for Δ_L.

(4.5) $$\Delta_L = \int \partial_t^2\,(dt)^2 \qquad \text{(H.-H.Kuo's formula)}.$$

This expression is useful to carry on actual computations. As is seen from the factor $(dt)^2$, Δ_L annihilates the ordinary white noise functionals, while it acts effectively in the space of generalized white noise functionals which we are going to define.

To fix the idea, we take a self adjoint operator H :

$$H = -(\frac{d}{du})^2 + u^2 + 1.$$

The second quantization technique leads us to introduce

$$(S_p) = D(\Gamma(H)^p) \subset (L^2)$$

and a Gel'fand triple

(4.6) $\quad (S) \subset (L^2) \subset (S)^*,$

where $(S) = \cap (S_p)$.

As for a characterization of $(S)^*$, there ís a powerful criterion. To state the theorem a definition is provided.

Definition Let F be a complex valued functional on the Schwartz space S. F is called a U-*functional* if and only if the following two conditions are satisfied.

(i) F has a ray entire extension on S; i.e. for any $\xi, \eta \in S$ the function $F(\eta + \lambda\xi)$, $\lambda \in R$, has an entire extension $F(\eta + z\xi)$, $z \in \mathbf{C}$.

(ii) There exists $p \geq 0$ and constants K_1, $K_2 > 0$ such that

$$|F(z\xi)| \leq K_1 \exp[K_2|z|^2|\xi|_p^2], \qquad |\ |_p \text{ is the } (S_p)\text{-norm.}$$

Streit-Potthoff's criterion. If $\varphi \in (S)^*$, then its $\mathcal{S}$-transform is a U-functional. Conversely, if F is a U-functional, then there exists φ in $(S)^*$ such that the $\mathcal{S}$-transform of φ is F.

The Kuo's Fourier transform is introduced in such a way that

(4.7) $\quad \hat{\varphi}(y) = \int :\exp[-i\langle x,y\rangle]:_x \varphi(x)\, d\mu(x)$

and we find intimate connections with the notions established above.

§5. Whiskers and variational calculus.

A one-parameter subgroup $\{g_t\}$ of the rotation group O_∞ is called a *whisker* if each g_t comes from a diffeomorphism of the parameter set $\overline{R}^d = R^d \cup \{\infty\}$. If g_t acts in the form

(5.1) $\quad (g_t\xi)(u) = \xi(\psi_t(u))\sqrt{|\psi_t'(u)|}$

with a suitable choice of a family $\{\psi_t;\ t \in R\}$ of functions satisfying

$$\psi_t \circ \psi_s = \psi_{t+s},$$

then $\{g_t\}$ is a whisker.

The conformal group consists of the following whiskers.

1) Shifts. Let S_t^j, being important in causal calculus, be given:

$\xi(u_1,\ldots,u_d) \longrightarrow (S_t^j\xi)(u) = \xi(u_1,\ldots,u_j-t,\ldots,u_d)$, $1 \le j \le d$.

2) Dilation. We introduce the (isotropic) dilations $\{\tau_t\}$:

$$\xi(u) \longrightarrow (\tau_t\xi)(u) = \xi(e^t u)e^{td/2}, \quad u \in R^d.$$

3) Rotations. Let $\tilde{g}$ be a rotation of R^d, i.e. $\tilde{g} \in SO(d)$. Define $g \in O_\infty$ by

$$\xi(u) \longrightarrow (g\xi)(u) = \xi(\tilde{g}u), \quad u \in R^d.$$

4) Special conformal transformation. Let w be the reflection :

$$(w\xi)(u) = \xi(u/|u|^2)|u|^{-d}.$$

Then, a special conformal transformations $\{\kappa_t^j\}$, $1 \le j \le d$, are given by

$$\kappa_t^j = w\, S_t^j\, w.$$

The above subgroups generate $\{d(d+3)/2 + 1\}$-dimensional subgroup (a Lie group) denoted by $C(d)$. It describes significant probabilistic properties, in particular, the *conformal invariance* of white noise. It is our hope that some other whiskers telling us some good meaning would be discovered under a certain principle.

Subgroups arising in this manner can be thought to have entirely different character from the ones in §3, however Remark in §3 would suggests to us to consider it as a limit (in a sense) of members in $G_\infty \vee \mathcal{G}$, since the Rademacher functions separate points of R^1.

We are now ready to use the group $C(d)$ to define $C(d)$-stationary random fields. Take a class $\mathbf{C}$ of $(d-1)$-dimensional manifolds in R^d satisfying certain analytic and geometric properties. Let a C in $\mathbf{C}$ be fixed. Consider $X(C) = X(C,x)$ living in $(S)^*$. Then, we have

$$X(g,x) \equiv X(gC,x) = X(C, g^*x), \quad g \in C(d). \tag{5.2}$$

Such a random field appears in applications, where stochastic variational equation can be expressed in terms of the Lie algebra of $C(d)$ and white noise functionals. More finer analytic treatment of fields of this type as well as generalizations would be good problems to be discussed by using the white noise analysis.

There are other examples of a field with parameter C in applications that can be investigated within our setup. Among others

i) Electromagnetic field and its variation (P. Lévy).

ii) Super many time theory. A relativistic generalization, due to S.Tomonaga, of the Schrödinger equation for the wave function $\Psi(C)$ depending on a surface C in the space-like domain, is given by

$$(5.3) \qquad \left(\frac{h}{2\pi i}\right)\frac{\delta}{\delta C_P} \Psi(C) = - H(P,C)\Psi(C).$$

This equation can be rephrased in our setup expressing C by a smooth function ξ and assuming $\Psi(\xi)$ to be a U-functional, where we can use the Potthoff-Streit criterion. Integrabiulity condition of the equation (5.3) is also discussed in terms of functional analysis (see P. Lévy [1] Part II) having been suggested by Tomonaga's psaper.

Remark. In connection with the use of whiskers, we should like to mention the unitary representation of the group $SL(2,R)$. The infinite dimensional unitary group U_∞ can be introduced in a similar manner to the case of O_∞. Then, we are naturally given the *principal series* of the unitary representation of $SL(2,R)$ by a probabilistic approach. A characterization of this fact is also given.

[References]

[1]. P. Lévy, Problèmes concrets d'analyse fonctionnelle. Gauthier-Villars, 1951.

[2] T. Hida, Brownian motion. Iwanami 1975 (in Japanese) ; english transl. Springer-Verlag 1980.

[3] ——— , Analysis of Brownian functionals. Carleton University Lecture Notes, #13. 1975.

[4] T. Hida, K.-S. Lee and S.-S. Lee, Conformal invariance of white noise. Nagoya Math. J. 98 (1985), 167 - 194.

[5] T. Hida, H.-H. Kuo, J. Potthoff and L. Streit, White noise - An infinite dimensional calculus. (Forthcoming book).

[6] N. Obata. Fock expansion of operators on white noise functionals. to appear in Proceedings of the 3rd Locarno Conference.

[7] Si Si, Variational calculus for Lévy's Brownian motion. Proc.1990 Nagoya Conference on Gaussian Randfom Fields, World Sci. 1991.

[8] J.Potthoff and L.Streit, A characterization of Hida distribution. J. Functional Analysis, 101 (1991), 212 - 229.

[9] H. Yoshizawa, Rotation group of Hilbert space and its application to Brownian motion. Proc. International Conference on Functional Analysis and related topics. (1970) 414 - 423.

A NUCLEAR SPACE-VALUED STOCHASTIC DIFFERENTIAL EQUATION DRIVEN BY POISSON RANDOM MEASURES

G. Kallianpur
J. Xiong
Department of Statistics
University of North Carolina, Chapel Hill

§1 Introduction

Stochastic differential equations (SDE) in infinite dimensional spaces arise in such diverse fields as nonlinear filtering, stochastic partial differential equations, infinite particle systems, chemical kinetics, neurophysiology, etc. In many applications, the driving term comes from "noise" which is generated by a large number of Poisson processes.

The infinite dimensional space which we take as the state space of the processes considered in this paper is the strong topological dual Φ' of a countably Hilbertian nuclear space Φ. Let $(\mathcal{F}_t)_{t\geq 0}$ be a complete, right continuous filtration on a complete probability space $(\Omega,\mathcal{F},P)$. Let $(U,\mathcal{U},\mu)$ be a σ-finite measure space. Let $N(duds)$ be a Poisson random measure on $\mathbb{R}_+\times U$ with intensity measure $\mu(du)ds$ and $\widetilde{N}(duds)$ its compensated random measure.

We consider the stochastic differential equation

$$X_t = X_0 + \int_0^t A(s,X_s)ds + \int_0^t\int_U G(s,X_{s-},u)\widetilde{N}(duds) \tag{1.1}$$

on Φ' where $A: R_+\times\Phi' \to \Phi'$, and $G: R_+\times\Phi'\times U \to \Phi'$.

Most of the applications mentioned above lead to linear stochastic differential equations in a Hilbert, Banach or the dual of a nuclear space and the existence of solution and its properties have been studied by various authors in [5,7,11,12,13, etc]. A Banach space valued SDE with non-linear coefficients and driven by a semimartingale (including a compensated random measure) has been studied by Gyöngy[2]. Both his paper and ours rely on the Galerkin method but there are several differences. The conditions imposed on the coefficients in [2] (especially the coercivity assumption) are not the same as ours and seem to be dictated by the choice of the solution space. In addition, our approach differs from that of [2] in an important respect. Gyöngy, following the method of Krylov and Rozovskii[14], directly aims for a unique strong solution. In the present paper we first obtain a weak solution taking values in some Hilbert space H_{-p_T} to our SDE on $[0,T]$ for any $T>0$ via the solution to a martingale problem and then connect them to obtain an Φ'-valued solution on $[0,\infty)$. Up to this step, the monotonicity condition is not involved. The existence of a unique strong solution is then established by a separate argument that uses a monotonicity condition on the coefficients.

The choice of Φ' as opposed to a Banach or Hilbert space specified in advance is based on two considerations. First of all, the solution considered is for all $t\geq 0$ and not for t restricted to a fixed interval $[0,T]$. It might happen that it is not possible to find a Banach or Hilbert space such that the solution X_t of the SDE lies in this space for all $t\geq 0$. Indeed, there is an example of (1.1), essentially due to G.Kallianpur and S.Ramaswamy[9], which has a unique solution X with the following properties: For $t\in[0,T]$, X_t takes values in a Hilbert space H_{-p_T} and, there is no p such that $X_t \in H_{-p}$ all $t \geq 0$.

Secondly, even if we are only interested in a finite interval, using Φ' still has some technical advantages. Mitoma's paper (1983)[15] about the weak convergence of measures on $D([0.T], \Phi')$ provides a powerful tool for us to establish a solution on Φ'. After we obtain this solution, the regularity of the process is decided by finding the Hilbert space in which its paths lie.

When a large number of Poisson processes are involved as in the case with the fluctuation of neuron potentials, and the amplitude of each Poisson noise is small we can expect it to be approximated by a Wiener process. This leads to a diffusion equation of the form

$$(1.2) \qquad X_t = X_0 + \int_0^t A(s,X_s)ds + \int_0^t B(s,X_s)\, dW_s$$

on Φ' where $A: R_+ \times \Phi' \to \Phi'$, $B: R_+ \times \Phi' \to L(\Phi',\Phi')$, and W is a Φ'-valued Wiener process. Hence the diffusion equation (1.2) may be regarded as a limiting case or a diffusion approximation of the SDE(1.1).

The diffusion equation (1.2) has been studied in [10]. Equation (1.1) has been studied in [4], but the conditions imposed there are very restrictive. Under those conditions, the solution process X_t is of locally bounded variation since the stochastic integral in (1.1) is an ordinary integral with respect to a signed measure. We obtain our basic result under more general conditions that enable us to apply it to a wide class of practical problems.

We state our main theorems in Section2 and give the idea of the proof in Section3. Two results on diffusion approximations are presented in Section4.

§2 Main theorems

We begin by giving some facts about duals of nuclear spaces.

Definition2.1: Φ is called a countably Hilbertian nuclear space, if Φ is a Fréchet space, whose topology is given by an increasing sequence of Hilbertian norms $\|.\|_n$, $n \geq 0$, such that the following is satisfied: If H_n is the completion of Φ with respect to the norm $\|.\|_n$, then for each n there exists $m>n$, such that the canonical injection $H_m \to H_n$ is Hilbert-Schmidt.

Let H_{-n} and Φ' denote the duals of H_n and Φ respectively. Then identifying H_0 with its dual H_0', we have the following sequence of canonical injections:

$$\Phi \to \ldots \to H_2 \to H_1 \to H_0 \equiv H_0' \to H_{-1} \to H_{-2} \to \ldots \to \Phi'.$$

It is well known that

$$\Phi = \bigcap_{n=1}^{\infty} H_n, \quad \Phi' = \bigcup_{n=1}^{\infty} H_{-n}.$$

The following assumption will be made throughout this paper: There exists a sequence (h_m) of elements in Φ, such that (h_m) is a CONS in H_0 and is COS in each space H_n, $n \in \mathbf{Z}$.

Notation:

(1) $h_i^n \equiv h_i \|h_i\|_n^{-1}$, $n\in\mathbf{Z}$, $i\in\mathbf{N}^+$.

(2) For $v\in\Phi'$ and $\phi\in\Phi$ define $v[\phi]\equiv$ the value of the continuous linear functional v at the point ϕ.

(3) $\forall p\in\mathbf{N}^+$, θ_p will denote the surjective linear isometry $H_{-p}\to H_p$ given by

$$\theta_p\Big(\sum_{i=1}^{\infty}\alpha_i\, h_i^{-p}\Big)\equiv\sum_{i=1}^{\infty}\alpha_i\, h_i^{p}\,.$$

where $\{h_i^p\}_{i=1}^{\infty}$ and $\{h_i^{-p}\}_{i=1}^{\infty}$ are CONS of H_p and H_{-p} respectively.

Definition2.2: A probability measure Q on $\mathbf{D}([0,T],\Phi')$ is called a weak solution on [0,T] of SDE(1.1) with initial distribution Q_0 if there exists a stochastic base $(\Omega,\mathcal{F},P,(\mathcal{F}_t))$ and a Poisson random measure N on it with intensity measure μ, a Φ'-valued process X on $(\Omega,\mathcal{F},P,(\mathcal{F}_t))$ such that Q and Q_0 are the distributions of X and X_0 respectively and for any $\phi\in\Phi$, $t\in[0,T]$, we have

$$X_t[\phi]=X_0[\phi]+\int_0^t A(s,X_s)[\phi]ds+\int_0^t\int_U G(s,X_{s-},u)[\phi]\,\tilde{N}(duds)\qquad \text{a.s.} \tag{2.1}$$

If [0,T] can be changed to $[0,\infty)$ and (2.1) hold for any $t\geq 0$, then we call Q on $\mathbf{D}([0,\infty),\Phi')$ a weak solution of SDE(1.1).

We consider the existence of a weak solution of SDE(1.1). Let A: $R_+\times\Phi'\to\Phi'$ and G: $R_+\times\Phi'\times U\to\Phi'$ be measurable functions, where the σ-algebra on Φ' is the Borel σ-algebra with respect to the strong Φ'-topology. To solve the stochastic differential equation (1.1), we make the following ASSUMPTIONS(I):

$\forall T>0$, $\exists\ p_0=p_0(T)\in\mathbf{N}^+$, s.t. $\forall p\geq p_0$, $\exists\ q\geq p$ and a constant $K=K(p,q,T)$ s.t.

(I1) $\forall t\in[0,T]$, we have that $A(t,\cdot)$: $H_{-p}\to H_{-q}$ is continuous; $\forall t\in[0,T]$, $v\in H_{-p}$, we have that $G(t,v,\cdot)\in L^2(U,\mu;H_{-p})$ and, for t fixed, as the $L^2(U,\mu;H_{-p})$-valued function of $v\in H_{-p}$ is continuous.

(I2) (Coercivity) $\forall t\in[0,T]$, $\phi\in\Phi$, we have $2A(t,\phi)[\theta_p(\phi)]\leq K(1+\|\phi\|_{-p}^2)$;

(I3) (Growth) $\forall t\in[0,T]$, $v\in H_{-p}$, we have that $\|A(t,v)\|_{-q}^2\leq K(1+\|v\|_{-p}^2)$, and

$$\int_U \|G(t,v,u)\|_{-p}^2\mu(du)\leq K(1+\|v\|_{-p}^2);$$

Theorem2.1: Let the assumptions (I1)-(I3) hold and, $\forall\phi\in\Phi$, let $E|X_0[\phi]|^2<\infty$. Then (1.1) has a weak solution such that, $\forall T>0$, $\exists p_T$ with

$$E\sup_{0\leq t\leq T}\|X_t\|_{-p_T}^2\leq\tilde{K}(2K,T,E\|X_0\|_{-p_T}^2)<\infty. \tag{2.2}$$

We impose one more condition to ensure that the SDE(1.1) has a unique strong solution.

(I4) (Monotonicity) $\forall t\in[0,T]$, v_1, $v_2\in H_{-p}$

$$2\langle A(t,v_1)-A(t,v_2),v_1-v_2\rangle_{-q}+\int_U \|G(t,v_1,u)-G(t,v_2,u)\|_{-q}^2\mu(du)\leq K\|v_1-v_2\|_{-q}^2.$$

Definition2.3: Let $(\Omega,\mathcal{F},P,\mathbf{F})$ be a stochastic basis and $\tilde{N}(duds)$ a compensated Poisson random measure on $[0,\infty)\times U$. X_0 is a Φ'-valued random variable. Then by a strong solution on Ω to the SDE(1.1) we mean a process X_t defined on Ω . s.t.

(a) X_t is $\mathcal{F}_t$-measurable, Φ'-valued;

(b) $X \in \mathbf{D}([0,\infty), \Phi')$;

(c) There exists a sequence (σ_n) of stopping times on Ω increasing to infinity and independent of ϕ, such that $\forall n$, $\forall \phi \in \Phi$

$$E \int_0^{\sigma_n} \int_U | G(s,X_s,u)[\phi] |^2 \mu(du)ds < \infty$$

$$E \int_0^{\sigma_n} | A(s,X_s)[\phi] |^2 ds < \infty$$

$$E | X_0[\phi] |^2 < \infty;$$

(d) The SDE (1.1) is satisfied for all $t \in [0,\infty)$, P-a.s.

Theorem2.2: Let the assumptions (I1)-(I4) hold, then (1.1) has a unique strong solution if $\forall \phi \in \Phi$, we have $E|X_0[\phi]|^2 < \infty$.

§3 Basic ideas of the proofs.

1. Idea of the proof of Theorem2.1.

Step 1: Using the stochastic Galerkin method, we project SDE(1.1) to the finite dimensional subspace of H_{-p} spanned by $\{h_1^{-p}, \cdots, h_d^{-p}\}$. Let

$$A^d(s,v) = \sum_{k=1}^{d} <A(s, \sum_{l=1}^{d} <v,h_l^{-p}>_{-p} h_l^{-p}),h_k^{-p}>_{-p} h_k^{-p}$$

$$G^d(s,v,u) = \sum_{k=1}^{d} <G(s, \sum_{l=1}^{d} <v,h_l^{-p}>_{-p} h_l^{-p}),h_k^{-p}>_{-p} h_k^{-p}$$

$$X_0^d = \sum_{k=1}^{d} <X_0^d, h_k^{-p}>_{-p} h_k^{-p}$$

and consider the SDE

$$(1.1)' \qquad X_t^d = X_0^d + \int_0^t A^d(s,X_s^d)\, ds + \int_0^t \int_U G^d(s,X_{s-}^d,u)\, \tilde{N}(duds).$$

We note that the conditions (I1)-(I3) hold for A^d and G^d uniformly in d. Furthermore, we have

(A1): If $d \to \infty$, then

(1°) $\forall t \in [0,T]$, $v \in H_{-p}$, $\phi \in \Phi$, we have $A^d(t,v)[\phi] \to A(t,v)[\phi]$;

(2°) $\forall t \in [0,T]$, $v \in H_{-p}$, we have

$$\int_U ||G^d(s,v,u)-G(s,v,u)||_{-p}^2 \mu(du) \to 0.$$

We make the following assumption (to be removed in step5):

(A2): For every d, the SDE(1.1)′ has an H_{-p}-valued weak solution X^d on stochastic bases $(\Omega^d, \mathcal{F}^d, P^d, (\mathcal{F}_t^d))$ such that

$$E^{P^d} \sup_{0 \le t \le T} ||X_t^d||_{-p}^2 \le \tilde{K} < \infty$$

where $\tilde{K}$ is a constant depending only on T,K,p.

Step 2: Let Q^d be the distribution of X^d on $\mathbf{D}([0,T],H_{-p})$, then $\{Q^d\}$ is tight.

To prove this, using Mitoma's argument we show that

(1°) $\{X_0^d\}_{d\in\mathbf{N}}$ is tight in H_{-p};

(2°) For any $\phi\in\Phi$ the sequence $\{ X_t^d[\phi] \}_{d\in\mathbf{N}}$ is tight in $\mathbf{D}([0,T],\mathbf{R})$;

(3°) $\{ Q^d \}_{d\in\mathbf{N}}$ is uniformly p-continuous, i.e. $\forall\epsilon>0$, $\rho>0$, $\exists\ \delta>0$, $\forall d\geq 1$, s.t. if $\|\phi\|_p \leq \delta$,

$$Q^d \left\{ Z\in\mathbf{D}([0,T],\Phi') : \sup_t | <Z_t,\phi> | > \epsilon \right\} \leq \rho.$$

In fact, (1°) is a direct consequence of the projection; (2°) can be obtained by assumption (I3) and the weak convergence theory of real-valued semimartingales; (3°) follows from assumption (A2).

Step 3: To pass to the limit, we need a connecting idea which is the martingale problem formulated below. Let

$$\mathfrak{D}_0^\infty(\Phi')\equiv\left\{ F:\Phi'\rightarrow\mathbf{R} \Big/ \exists h\in C_0^\infty(\mathbf{R}),\phi\in\Phi \text{ s.t. } F(v)=h(v(\phi)) \right\}$$

and, for $F\in\mathfrak{D}_0^\infty(\Phi')$, consider operators $\mathcal{L}_s F$: $\Phi'\rightarrow\mathbf{R}$ defined by

$$\begin{aligned}\mathcal{L}_s F(v)\equiv\ & A(s,v)[\phi]h'(v[\phi]) \\ & + \int_U \{h(v[\phi]+ G(s,v,u)[\phi])-h(v[\phi]) - G(s,v,u)[\phi]h'(v[\phi])\}\mu(du).\end{aligned}$$

For $Z\in\mathbf{D}([0,T],\Phi')$, let

$$M^F(Z)_t\equiv F(Z(t))-F(Z(0))-\int_0^t \mathcal{L}_s F(Z(s))ds.$$

A probability measure Q on $\mathbf{D}([0,T],\Phi')$ is called a solution on [0,T] of the $\mathcal{L}$-martingale problem with initial distribution Q_0 if $\forall F\in \mathfrak{D}_0^\infty(\Phi')$, $\left\{M^F(Z)_t,\ 0\leq t\leq T\right\}$ is a Q-martingale and $Q\circ Z(0)^{-1}= Q_0$.

It is easy to see that Q^d is a solution of the $\mathcal{L}^d$-martingale problem, i.e. $\{M_d^F(Z)_t\}$ is a Q^d martingale. Let Q^* be a cluster point of $\{Q^d\}$ and assume, WLOG, that Q^d converges to Q^* weakly. It is then shown, using Skorohod's theorem, that $M_d^F(Z)_t$ converges in distribution to $M^F(Z)_t$ from which it follows that $\{M^F(Z)_t\}$ is a Q^* martingale. i.e. Q^* is a solution on [0,T] of the $\mathcal{L}$-martingale problem.

Step 4: To show that Q^* is a weak solution is a crucial part of the paper and involves too many details to be gone into here. The main point is to show that

$$M_\phi(t,Z)\equiv Z_t[\phi]-Z_0[\phi] - \int_0^t A(s,Z_s)[\phi]\ ds\ (\forall\phi\in\Phi)$$

is a purely-discontinuous Q^* square integrable martingale which is expressed as a stochastic integral with respect to a Poisson random measure with intensity measure μ.

Step 5: What is left is to prove that the condition (A2) can be removed, namely, Theorem2.1 and the desired estimation holds if $\Phi=\mathbf{R}^d$.

Let j be the Friedrichs mollifier on $\mathbf{R}^d$ and

$$\begin{aligned} A^n(t,x) &= \int A(t,x-n^{-1}z)j(z)dz \qquad \text{for } |x|\leq n, \\ &= A^n(t,nx/|x|) \qquad \text{for } |x|>n, \end{aligned}$$

and let $G^n(t,x,u)$ be defined similarly. We know that (I1)-(I3) hold for A^n and G^n uniformly in n. It is easy to see that, for every n, we have

$$|A^n(t,x)-A^n(t,y)|^2+\int_U |G^n(t,x,u)-G^n(t,y,u)|^2\mu(du)\leq K(1+n)^2|x-y|^2.$$

It follows from [3] that every equation of $(1.1)'$ has a unique solution and, using Burkholder-Davis-Gundy inequality, we get the desired estimation, i.e. (A2) holds for sequences $\{A^n\}$ and $\{G^n\}$. The arguments of step2-4 now show the existence of a weak solution for (1.1) on [0,T].

Step 6: Finally, we construct a weak solution on $[0,\infty)$.

Let $\widetilde{A}(t,v)=A(t+T,v)$ and $\widetilde{G}(t,v,u)=G(t+T,v,u)$, for $0\leq t\leq T$. Then $\widetilde{A}$, $\widetilde{G}$ satisfy the assumptions (I1)-(I3) with $p_0(T)$ and $K(p,q,T)$ changed to $p_0(2T)$ and $K(p,q,2T)$ respectively. Let the initial distribution on Φ' be given by $Q^*\circ Z_T^{-1}$, then we get a measure $\widetilde{Q}^*$ on $\mathbf{D}([0,T],H_{-p_{2T}})$, and hence on $\mathbf{D}([0,T],\Phi')$. Patching up $\widetilde{Q}^*$ with Q^* we get a solution on [0,2T]. Proceeding inductively, we get a weak solution on $[0,\infty)$. This completes the proof of Theorem2.1.

2. Idea of the proof of Theorem2.2.

The idea is to implement Yamada-Watanabe's argument in this setup. The crucial point is the equivalence relation between (1.1) and another kind of SDE driven by an l^2-valued martingale we will establish below.

First, we note that the pathwise uniqueness of (1.1) follows directly from the assumption (I4) and Itô's formula.

Let $(\Omega,\mathfrak{F},P,\mathbf{F})$ be a stochastic basis. An l^2-valued process H_t on $(\Omega,\mathfrak{F},P,\mathbf{F})$ is called a Good process with respect to a CONS (ϕ_n) of $L^2(U,\mathcal{U},\mu)$ if ∃ a Poisson random measure N(duds) with respect to $(\Omega,\mathfrak{F},P,\mathbf{F})$ s.t.

$$H_t = \sum_{n=1}^{\infty} \frac{1}{n} \int_0^t \int_U \phi_n(u)\, \widetilde{N}(du,ds)\, e_n \tag{3.1}$$

where $e_n=(0, \cdots, 0, 1, 0, \cdots) \in l^2$.

We can easily see that the concept of a Good process is well defined and, that two Good processes corresponding to the same CONS (ϕ_n) of $L^2(U,\mathcal{U},\mu)$, but defined on different stochastic bases, have the same distribution on $\left(\mathbf{D}([0,T], l^2), \mathcal{B}\{\mathbf{D}([0,T], l^2)\}\right)$. We denote it by P_G and call it the Good measure with respect to the CONS (ϕ_n). Now, if w is the coordinate process on $\left(\mathbf{D}([0,T], l^2), \mathcal{B}\{\mathbf{D}([0,T], l^2)\}, P_G\right)$, then w is a Good process on an extension of $\left(\mathbf{D}([0,T], l^2), \mathcal{B}\{\mathbf{D}([0,T], l^2)\}, P_G\right)$. Furthermore, the SDE (1.1) is equivalent to the SDE

$$X_t = X_0 + \int_0^t A(s,X_s)ds + \int_0^t \psi(s,X_{s-})dH_s \tag{3.2}$$

where H is defined by (3.1) and $\psi(s, v)$ is a linear operator from l^2 to H_{-p} defined by

$$\psi(s, v)\, e_n \equiv n \int_U G(s,v,u)\phi_n(u)\, \mu(du) \qquad \text{for } s\in[0,T],\ v\in H_{-p}.$$

Using these facts and the standard argument of Yamada-Watanabe[6] (which is shown to be valid in this case), we have Theorem2.2.

§4 Diffusion approximations.

Finally, we consider the diffusion approximation problem mentioned in the introduction. Let A^n: $\mathbf{R}_+\times\Phi'\to\Phi'$ and G^n: $\mathbf{R}_+\times\Phi'\times U\to\Phi'$ be two sequences of measurable mappings on the corresponding spaces. Let N^n be a sequence of stationary Poisson random measures on $\mathbf{R}_+\times U$ with characteristic measures μ^n. Consider a sequence SDE's

$$X_t^n = X_0^n+\int_0^t A^n(s,X_s^n)\,ds + \int_0^t\int_U G^n(s,X_{s-}^n,u)\,\tilde{N}^n(duds) \tag{4.1}$$

with the initial distributions $\{Q_0^n\}$ satisfying $\sup_n \int_{H_{-p}} \|X\|_{-p}^2 Q_0^n(dx) < \infty$.

To ensure the existence and the tightness of the solution sequence, we assume that (I1)-(I3) hold uniformly in n, i.e. make the following assumptions: $\forall T>0$, $\exists\ p_0=p_0(T)\in\mathbf{N}^+$, s.t. $\forall p\geq p_0$, $\exists$ $q\geq p$ and a constant $K=K(p,q,T)$ s.t.

(SI1) $\forall t\in[0,T]$, $A^n(t,\cdot)$: $H_{-p}\to H_{-q}$ are continuous uniformly in n; $\forall t\in[0,T]$, $v\in H_{-p}$, $G^n(t,v,\cdot)\in L^2(U,\mu;H_{-p})$ and, for t fixed, as $L^2(U,\mu;H_{-p})$-valued functions of $v\in H_{-p}$ are continuous uniformly in n.

(SI2) (Coercivity) $\forall t\in[0,T]$, $\phi\in\Phi$, we have $2A^n(t,\phi)[\theta_p(\phi)]\leq K(1+\|\phi\|_{-p}^2)$

(SI3) (Growth) $\forall t\in[0,T]$, $v\in H_{-p}$

$$\|A^n(t,v)\|_{-q}^2\leq K(1+\|v\|_{-p}^2),$$

$$\int_U \|G^n(t,v,u)\|_{-p}^2\mu^n(du)\leq K(1+\|v\|_{-p}^2).$$

Consider the sequence of SDE's(4.1) on [0,T]. For every n, by Theorem2.1, SDE(4.1) has a weak solution X^n on some stochastic basis $(\Omega^n,\mathcal{F}^n,P^n,(\mathcal{F}_t^n))$ and we can prove that if $\{X_0^n\}$ tight in H_{-p}, then $\{Q^n\}$ tight in $\mathbf{D}([0,T],H_{-p})$, where Q^n is the distribution of X^n. To characterize the limit of this sequence, we further make the following assumption.

(A1)'(1°) $X_0^n\overset{\mathcal{D}}{\Rightarrow}X_0$ on H_{-p};
(2°) $\forall t\in[0,T]$, $v\in H_{-p}$, $\phi\in\Phi$, we have $A^n(t,v)[\phi]\to A(t,v)[\phi]$;
(3°) $\forall t\in[0,T]$, $\phi\in\Phi$ and $a>0$ fixed. Uniformly for v in any compact subset of H_{-p}, we have that $|G^n(t,v,u)[\phi]|^2$ integrable with respect to μ^n uniformly in n, $\mu^n\{u: |G^n(t,v,u)[\phi]|>a\}\to 0$ and

$$\int_U |G^n(t,v,u)[\phi]|^2\,\mu^n(du)\to Q(\,B_s(v)^*[\phi],\,B_s(v)^*[\phi]\,).$$

where Q be a continuous quadratic form on Φ and A: $\mathbf{R}_+\times\Phi'\to\Phi'$, B: $\mathbf{R}_+\times\Phi'\to\mathcal{L}(\Phi',\Phi')$ are measurable mappings in the corresponding spaces.

Theorem4.1: Under the assumptions (SI1)-(SI3), (A1)' and the monotonicity condition(D4) below, we have that $Q^n \overset{w}{\Rightarrow} Q^*$ and Q^* is the distribution of the unique strong solution of SDE (1.2).

The next theorem yields a result going in the opposite direction, describing the conditions under which a Φ'-valued diffusion processes can be approximated by processes driven by Poisson random measures. We make the following assumptions:

$\forall T>0$, $\exists\ p_0=p_0(T)\in\mathbf{N}^+$, s.t. $\forall p\geq p_0$, $\exists\ q\geq p$ and a constant $K=K(p,q)$ s.t.

(D1) $\forall t\in[0,T]$, $\phi\in\Phi$, we have that $A(t,\cdot)$: $H_{-p}\to H_{-q}$, $B(t,\cdot)(\cdot)$: $H_{-p}\times H_{-p}\to H_{-p}$ and $Q_{B_t(\cdot)}(\phi,\phi)$: $H_{-p}\to\mathbf{R}$ are continuous;

(D2) (Coercivity) $\forall t\in[0,T]$, $\phi\in\Phi$, we have $2A(t,\phi)[\theta_p(\phi)]\leq K(1+\|\phi\|_{-p}^2)$;

(D3) (Growth) $\forall t\in[0,T]$, $v\in H_{-p}$,

$$\|A(t,v)\|_{-q}^2\leq K(1+\|v\|_{-p}^2),\ \text{and}\quad |\ Q_{B_t(v)}|_{-p,-p}\leq K(1+\|v\|_{-p}^2);$$

(D4) (Monotonicity) $\forall t\in[0,T]$, v_1, $v_2\in H_{-p}$

$$2\langle A(t,v_1)-A(t,v_2),\ v_1-v_2\rangle_{-q}+|\ Q_{B_t(v_1)-B_t(v_2)}|_{-q,-q}\leq K\|v_1-v_2\|_{-q}^2;$$

(D5) $\quad E\ \|X_0\|_{-p}^2<\infty$.

Theorem4.2: Under assumptions (D1)-(D5), the SDE (1.2) has a unique strong solution which can be approximated by a sequence of processes driven by Poisson random measures.

REFERENCES

[1] C.Dellacherie & P.A.Meyer, Probabilities and Potential B, North-Holland, 1982.

[2] I.Gyöngy, On stochastic equation with respect to semimartingales III; Stochastics, Vol 7, pp.231-245, 1982.

[3] I.Gyöngy and N.V.Krylov, On stochastic equation with respect to semimartingales I; Stochastics, Vol 4, pp.1-21, 1980.

[4] G.Hardy, G.Kallianpur and S.Ramasubramanian, A nuclear space-valued stochastic differential equation driven by Poisson random measures; University of North Carolina at Chapel Hill Technical report No. 232.

[5] M.Hitsuda and I.Mitoma, Tightness problem and stochastic evolution equation arising from fluctuation phenomena for interacting diffusions, J. Multivariate Analysis 19, pp.311-328,1986.

[6] N.Ikeda and S.Watanabe, Stochastic differential equations and diffusion processes; North-Holland, 1981.

[7] K.Itô, Foundations of Stochastic Differential Equations in Infinite Dimensional Spaces, Society for Industrial and Applied Mathematics, 1984.

[8] J. Jacod and A.N. Shiryaev, Limit theorem for stochastic processes; Springer-Verlag, 1987.

[9] G.Kallianpur, Stochastic differential equations in duals of nuclear spaces with some applications, University of North Carolina at Chapel Hill Technical Report No. 158.

[10] G.Kallianpur, I.Mitoma and R.L.Wolpert, Diffusion equation in duals of nuclear spaces; Stochastics and stochastic reports, Vol 29, pp.285-329, 1990.

[11] G.Kallianpur and R.L.Wolpert, Infinite dimensional stochastic differential equation models for spatially distributed neuron, Appl. math. optim. 12, pp.125-172, 1984.

[12] G.Kallianpur and R.L.Wolpert, Weak convergence of stochastic neuronal models, Lecture Notes in Biomathematic, Eds. M.Kimura, G.Kallianpur and T.Hida, Stochastic Methods in Biology, pp116-145.

[13] P.Kotelenez, Gaussian approximation to the nonlinear reaction-diffusion equation, Report No. 146, University of Bremen, 1986.

[14] N.Krylov and B.Rozovskii, Stochastic Evolution Equations, Plenum Publishing Corporation, pp.1233-1277, 1981.

[15] I.Mitoma, Tightness of probabilities on $C([0,1],\mathcal{S}')$ and $D([0,1],\mathcal{S}')$; The Annals of Probability, Vol. 11, No 4, pp.989-999, 1983.

Random Vortex Models and Stochastic Partial Differential Equations

Peter Kotelenez*
Department of Mathematics & Statistics
Case Western Reserve University
Cleveland, Ohio
J. Adin Mann Jr.*
Department of Chemical Engineering
Case Western Reserve University
Cleveland, Ohio

Summary: We construct stochastic partial differential equations for the vorticity distribution of a two-dimensional simple fluid. The drift part of these equations is approximately given by the Navier-Stokes Equation and the diffusion part depends on the correlation assumptions for the positions of the centers of the vortices.

1 Introduction

Our aim is to model the time evolution of the vorticity of a two-dimensional incompressible fluid. The restriction to two dimensions is natural in applications like oceanography, where the depth is considered to be negligible in comparison to its planar extension. Although for applications in oceanography one should include the action of the Coriolis force on the vorticity distribution we will neglect its contribution here since we want to be conceptual. Moreover, we believe that it is fairly easy to include the Coriolis contribution into our models, since it acts in the form of an external force on the system. Under the above assumptions we obtain as a macroscopic equation for the distribution of vorticity in a two-dimensional fluid:

$$\left.\begin{aligned} \tfrac{\partial}{\partial t}\mathcal{X}(r,t) &= \nu\Delta\mathcal{X}(r,t) - (U\cdot\nabla)\mathcal{X}(r,t) \\ \mathcal{X}(r,t) &= \ \text{curl}\, U(r,t) = \tfrac{\partial U_2}{\partial r_1} - \tfrac{\partial U_1}{\partial r_2} \\ \nabla\cdot U &\equiv 0. \end{aligned}\right\} \tag{1.1}$$

Here, $U(r,t)$ is the velocity field, $r \in \mathbf{R}^2, \nu \geq 0$ is the kinematic viscosity, Δ the Laplacian, ∇ the gradient, and "$\cdot$" denotes the scalar product in $\mathbf{R}^2$. If $\nu > 0$ we obtain the Navier-Stokes equation for the vorticity. If the fluid is inviscid, i.e., $\nu = 0$, we obtain the Euler equation. Note that by the incompressibility condition $\nabla \cdot U = 0$ we obtain

$$U(r,t) = \int_{\mathbf{R}^2} (\nabla^{\perp} g)(r-q)\mathcal{X}(q,t)dq, \tag{1.2}$$

---

*This research was supported by a grant from ONR. "The physical part, i.e., Remark 1.1, has been written by J. Adin Mann, Jr. alone; the mathematical part, i.e., the rest of the paper without Remark 1.1 has been written by Peter Kotelenez alone."

where $g(r) := -\frac{1}{2\pi}\ell n(\tilde{|r|})$ with $\tilde{|r|}^2 = r_1^2 + r_2^2$ and $\nabla^\perp = (\frac{\partial}{\partial x_2}, \frac{\partial}{\partial x_1})^T$ with "T" denoting the transpose. As a consequence we can obtain the velocity field U, which satisfies the standard Navier-Stokes equation from the vorticity distribution.

Remark 1.1 *This problem is motivated by the more general consideration of the flow fields in liquids. In general velocity fields can be constructed as a sum of contributions from a potential field and the vortex field. That is to say, in three dimensions*

$$U = -\nabla\Phi + K * \mathcal{X}$$

up to a constant where K is the matrix

$$K = \frac{1}{4\pi r^3}\begin{pmatrix} 0 & x_3 & -x_2 \\ -x_3 & 0 & x_1 \\ x_2 & -x_1 & 0 \end{pmatrix}$$

and $*$ is the convolution operation. Many practical situations can be modeled by computing only the vorticity, $\mathcal{X}$, ignoring the potential term or solving for the potential field as a second step. We ignore the potential field in this paper.

It is common practice in the statistical theory of fluids to note formally that the "probability per unit volume" that a molecule k is at the point x is

$$\langle \delta_{x^k}(x) \rangle = \int\int \delta_{x^k}(x) f^{(N)} dx^N dp^N,$$

where $f^{(N)}$ is the N-particle density, dx^N refers to the $3N$ volume element and dp^N the analogous volume element for momentum. It is then possible to use formulas such as that for the momentum per unit volume

$$p(x,t) = \sum_k p_k(t)\delta_{x^k(t)}(x).$$

Here, the momentum p_k is that of the kth particle whose trajectory is $x_k(t)$.

We will use the representation of the vorticity in two dimension as

$$\mathcal{X}(x,t) = \sum_k a_k \delta_{x^k}(x)$$

in the same spirit. However, the "particles" $k = 1, 2, \ldots, N$ are not the N molecules (or atoms) that form the fluid. The coefficient a_k cannot be associated directly with molecular properties as can the density or the velocity (e.g. $\frac{p_k}{\rho} := U_k$). However, a_k can certainly be determined so that $\mathcal{X}$ follows the governing equation (1.1) and therefore a_k can be interpreted physically.

In addition, the fluid is assumed to be simple. That is to say, the pressure tensor is assumed symmetric so that angular momentum conservation is implied directly by the conservation of linear momentum. (This will not be the case, for example, for systems in liquid crystal phases where there are intrinsic torques that come about because of the anisotropic shapes of the molecules.)

In addition it is clear physically that for flow in two dimensions the orientation of rotation (direction of the curl of U) will be preserved as time evolves. This sign will depend on initial conditions.

The above considerations generalize the positivity concept for the density of particles (macroscopically described, e.g., by the reaction-diffusion equation). Let us call (1.1) sign-preserving if its solution $\mathcal{X}(t, \mathcal{X}_0) \geq 0$, provided its initial value satisfies $\mathcal{X}_0 \geq 0$ and $\mathcal{X}(t, \mathcal{X}_0) \leq 0$, provided $\mathcal{X}_0 \leq 0$. Our physical considerations imply that the solution of (1.1) must be sign-preserving; in Section 2 we will give a simple mathematical proof that (1.1) is a sign-preserving partial differential equation (PDE).

Let us briefly describe the content of this paper. In Section 2 we define a system of N point vortices with both positive and negative intensities. Their positions are determined by a $2N$-dimensional (stochastic) ordinary differential equation (ODE), where the stochastic perturbations are i.i.d. $\mathbf{R}^2$-valued standard Wiener processes times a constant $\gamma \geq 0$. Following Marchioro and Pulvirenti [5] we conclude that the empirical process (resp. the expectation of it) associated with the (stochastic) ODE approximates the solution of the Euler equation if $\gamma = 0$ (resp. the solution of the Navier-Stokes equation if $\gamma = \sqrt{2\nu}, \nu > 0$). We derive from this the sign-preserving property for (1.1). Then we show that the empirical process is a signed measure valued Markov process and compute its generator. We derive (formally) a signed measure valued stochastic partial differential equation (SPDE) with the same generator. In Section 3 we define a stochastic (ordinary) differential equation (SDE) for the positions of N vortices where the stochastic perturbations are state-dependent and driven by Brownian sheets. Thus we create correlations between the fluctuations which will be visible if the vortices are close to one another and negligible otherwise. Again we show that the empirical process for this SDE is a signed measure valued Markov process and we compute its generator. Further we derive an SPDE for this generator and show that the empirical process solves the SPDE. Finally we conclude that this SPDE should be solvable on $L_2(\mathbf{R}^2, dr)$, the space of real valued square integrable functions on $\mathbf{R}^2$, where dr is the Lebesque measure.

Let us make precise probabilistic assumptions. $(\Omega, \mathcal{F}, \mathcal{F}_{t\geq 0}, P)$ is a stochastic basis with right continuous filtration. In what follows we will assume that all our stochastic processes are defined on Ω and $\mathcal{F}_t$-adapted. In particular, our stochastic basis should carry two independent Brownian sheets (or set-indexed white noises) $w_i(p,t), p \in \mathbf{R}^2, t \geq 0$, and adaptedness means in this case that $\int_A w_i(dp,t)$ is adapted, where A is an arbitrary Borel set in $\mathbf{R}^2$ with finite Lebesque measure. Naturally, all initial conditions in SDE's, resp. SPDE's have to be adapted.

2 A Random Point Vortex Model without Spatial Correlations

Let $N \in \mathbf{N}, g_N(\tilde{|}x\tilde{|})$ be some smooth approximation of $g(\tilde{|}x\tilde{|}) := -\frac{1}{2\pi}\ell n(\tilde{|}x\tilde{|}), x \in \mathbf{R}^2$, such that $g_n(\tilde{|}x\tilde{|}) \to g(\tilde{|}x\tilde{|})$, as $N \to \infty$. Set

$$K_N(x) := \begin{cases} \nabla^{\perp} g_N(\tilde{|}x\tilde{|}) & \tilde{|}x\tilde{|} \neq 0 \\ 0 & x = 0. \end{cases}$$

Consider N point vortices with intensities $a_i \in \mathbf{R}$ and let x^i be the position of the i-th vortex. We assume that the position satisfies the ordinary (stochastic) differential equation

$$dx^i(t) = \sum_{j=1}^{N} a_j K_N(x^i - x^j)dt + \gamma dm^i(t), \tag{2.1}$$

where $\gamma \geq 0$ and $m^i(t)$ are $\mathbf{R}^2$-valued mean zero square integrable martingales, $i = 1, \ldots, N$. Set

$$\left.\begin{array}{l} \mathcal{X}_N(t) := \sum_{i=1}^N a_i \delta_{x^i(t)}, \\ E\mathcal{X}_N(t) := \mu_N(t), \end{array}\right\} \tag{2.2}$$

where $x^i(t)$ are the solutions of (2.1) with suitable adapted initial conditions. Denote by $\langle \cdot, \cdot \rangle$ the standard scalar product on $L_2(\mathbf{R}^2, dr)$ and also its extension to a duality between distributions and smooth functions. The following facts have been established by Marchioro and Pulvirenti [5].

I. Assume $\gamma = 0$ (i.e., (2.1) is deterministic), and let $\mathcal{X}$ be the solution of the Euler equation (1.1) with $\gamma = 0$. Then there is a sequence $K_N(x) \to K(x) := \nabla^\perp g(|x|)$ as $N \to \infty$ such that $\langle \mathcal{X}_N(0), \varphi\rangle \to \langle \mathcal{X}(0), \varphi\rangle$ as $N \to \infty$ for bounded continuous φ implies for any $t > 0$

$$\langle \mathcal{X}_N(t), \varphi\rangle \to \langle \mathcal{X}(t), \varphi\rangle, \quad \text{as} \quad N \to \infty, \tag{2.3}$$

i.e., $\mathcal{X}_N(t)$ is "approximately" a weak solution of the Euler equation. If one chooses $K(x)$ instead of $K_N(x)$ in (2.1) and assumes that (2.1) has a unique solution for suitable initial values we obtain directly that $\mathcal{X}_N(t)$ is a weak solution of the Euler equation.

II. Choose $\gamma = \sqrt{2\nu}$ and $m^i(t) := \beta^i(t)$, where $\beta^i(t)$ are i.i.d. $\mathbf{R}^2$-valued standard Wiener processes. Further assume that half of the intensities a_j are positive and equal to some $\frac{a^+}{N}; a^+ > 0$, and the other half are negative and equal $\frac{-a^-}{N}, a^- > 0$. Now let $\mathcal{X}$ be the solution of the Navier-Stokes equation (1.1) with $\nu > 0$. Again with the same sequence $K_N(x)$ as in part I and φ being continuous and bounded $\langle \mu_N(0), \varphi\rangle \to \langle \mathcal{X}(0), \varphi\rangle$ as $N \to \infty$ implies for any $t > 0$

$$\langle \mu_N(t), \varphi\rangle \to \langle \mathcal{X}(t), \varphi\rangle, \quad \text{as} \quad N \to \infty. \tag{2.4}$$

Let us draw some conclusions from (2.3), resp. (2.4). First we obtain a mathematical confirmation that the orientation of rotation is preserved as time evolves, see Remark 1.1.

Lemma 2.1 *Both the Euler equation (1.1) with $\nu = 0$ and the Navier-Stokes equation (1.1) with $\nu > 0$ are sign preserving.*

Proof

If we take all $a_j \geq 0$ (resp. all $a_j \leq 0$) we obtain that $\mathcal{X}_N(t)$ and $\mu_N(t)$ are both positive (resp. both negative) measures. Since we can prove (2.4) also for the case of exclusively nonnegative a_j (or exclusively nonpositive a_j) the statement of the lemma follows from (2.3), resp. (2.4).

■

Next we want to show that $\mathcal{X}_N(t)$ defined by (2.2) is a Markov process provided that $\beta^i(t) =: m^i(t)$ are i.i.d. $\mathbf{R}^2$-valued standard Wiener processes. To this end we need more notation. Set

$$\mathcal{M}_N^\pm := \{\mu : \mu = \sum a_i \delta_{x^i}, x^i \in \mathbf{R}^2, a_i \in \mathbf{R}\}.$$

Further, let $\mu = \mu^+ - \mu^-$ be the Jordan decomposition for $\mu \in \mathcal{M}_N^\pm$ and d_p be the Prohorov metric $\mathcal{M}_N^+$, the subset of $\mathcal{M}_N^\pm$, where all $a_j \geq 0$. We define a metric d_p on $\mathcal{M}_N^\pm$ by

$$d_p(\mu_1, \mu_2) := d_p(\mu_1^+, \mu_2^+) + d_p(\mu_1^-, \mu_2^-),$$

with $\mu_1, \mu_2 \in \mathcal{M}_N^\pm$. $(\mathcal{M}_N^\pm, d_p)$ is a compact separable metric space. Let $C_b := C_b(\mathcal{M}_N^\pm, \mathbf{R})$ be the bounded continuous functions from $\mathcal{M}_N^\pm$ into $\mathbf{R}$ and set

$$\begin{aligned} \mathcal{D} := \{L \in C_b : \exists M \in \mathbf{N}, \varphi_i \in C^2(\mathbf{R}^2, \mathbf{R}), i = 1, \ldots, M, f \in C^2(\mathbf{R}^M, \mathbf{R}) \\ \text{such that } L(\mathcal{X}_N) := f(\langle \mathcal{X}_N, \varphi_1\rangle, \ldots \langle \mathcal{X}_N, \varphi_M\rangle)\} \end{aligned} \tag{2.5}$$

where $C^2(\mathbf{R}^k, \mathbf{R})$ are the twice continuously differentiable functions from $\mathbf{R}^k$ into $\mathbf{R}$. Finally we define an approximation of the velocity field through point vortices.

$$U_N(t, x) := \int K_N(x - y) \mathcal{X}_N(t, dy). \tag{2.6}$$

Proposition 2.2 *Assume that $m^i(t) := \beta^i(t)$ $(i = 1,\ldots,N)$ are i.i.d. $\mathbf{R}^2$-valued standard Wiener processes. Then $\mathcal{X}_N(t)$ given by (2.2) is an $\mathcal{M}_N^{\pm}$-valued Markov process and its generator $\mathcal{A}_{N,u}$ is given by*

$$\left.\begin{array}{rl}(\mathcal{A}_{N,u}L)(\mathcal{X}_N) & := \sum_{k=1}^M \frac{\partial}{\partial y_k} f(\langle \mathcal{X}_N, \varphi_1\rangle, \ldots, \langle \mathcal{X}_N, \varphi_M\rangle)\langle \mathcal{X}_N, (U_N\cdot\nabla)\varphi_k + \frac{\gamma^2}{2}\Delta\varphi_k\rangle \\ & +\frac{\gamma^2}{2}\sum_{k,\ell=1}^M \frac{\partial^2}{\partial y_k \partial y_\ell} f(\langle \mathcal{X}_N, \varphi_1\rangle, \ldots, \langle \mathcal{X}_N, \varphi_M\rangle)[\sum_{i=1}^N a_i{}^2 \nabla\varphi_k(x^i)\cdot\nabla\varphi_\ell(x^i)],\end{array}\right\} \quad (2.7)$$

where $L \in \mathcal{D}$.

<u>Proof</u> It follows from Dynkin [2] (Th. 10.13), that $\mathcal{X}_N(t)$ is a Markov process since $(\beta_1(t),\ldots,\beta_N(t)) \sim (\beta_{\pi 1}(t),\ldots,\beta_{\pi N}(t))$, for any permuation $\pi : \{1,\ldots,N\} \to \{1,\ldots,N\}$, where " $\sim$ " means equal in distribution. (2.7) follows from the Itô-formula using (2.6).

■

In what follows we will assume $\gamma^2 = 2\nu (> 0)$, thereby excluding the Euler case. Further we will assume for simplicity that there are positive numbers a^+ and a^- independent of N such that all positive intensities equal $\frac{a^+}{N}$ and all negative intensities equal $\frac{-a^-}{N}$. Note that (2.2) formally satisfies the following "stochastic evolution equation":

$$\left.\begin{array}{rl} d\mathcal{X}_N(t) & = [-(\nabla\cdot U_N)\mathcal{X}_N(t) + \nu\Delta\mathcal{X}_N(t)]dt \\ & + \sum_{i=1}^N \frac{(\pm)a^{\pm}}{N}\delta_{x^i(t)}\sqrt{2\nu}\nabla\cdot d\beta^i(t)\end{array}\right\} \quad (2.8)$$

$(\pm)a^{\pm}$ is a^+ if the i-th intensity is positive and -a^- if the i-th intensity is negative. Thus the stochastic part is a distribution (take the gradient of a test function, then the scalar product with the 2-dimensional Wiener process increments $d\beta^i(t)$ times $\sqrt{2\nu}$ and then apply $\delta_{x^i(t)}$). If we had incompressibility $(\nabla\cdot U_N = 0)$ then (2.8) would be like a stochastically perturbed Navier-Stokes equation. However the $\beta^i(t)$ introduce an artificial tagging for $\mathcal{X}_N(t)$. The Markov generator and other existing models (cf. Dawson [1]) suggest that $d\mathcal{X}_N(t)$ should be driven by an $L_2(\mathbf{R}^2, dr)$-valued cylindrical Brownian motion $W(t)$ rather than by $d\beta^i(t)$. (cf. Kotelenez [3] for the definition). Indeed, let us assume we can give a precise meaning to the following signed measure valued SPDE:

$$d\mathcal{X}_N = [-(\nabla\cdot U_N)\mathcal{X}_N(t) + \nu\Delta\mathcal{X}_N(t)]dt + \sqrt{\frac{2\nu}{N}}F_N(\mathcal{X}_N)\cdot dW, \quad (2.9)$$

where $F_N(\mathcal{X}_N) := \sqrt{N}(\sqrt{-\nabla^T\mathcal{X}_N^+\nabla} - \sqrt{-\nabla^T\mathcal{X}_N^-\nabla})$. For nice functions $\mathcal{X}_N$ $\sqrt{-\nabla^T\mathcal{X}_N^{\pm}\nabla}$ is the positive root of the self-adjoint extension of $-\nabla^T\mathcal{X}_N^{\pm}\nabla$ (via quadratic forms, where $\mathcal{X}_N^{\pm}$ acts as a multiplication operator). Since such a "solution" of (2.9) (if unique) should be a Markov process, we can check by Itô's formula that it should have generator $\mathcal{A}_{N,u}$ given by (2.7) with $\gamma^2 = 2\nu$. Thus (2.9) is the Navier-Stokes equation perturbed by a nonlinear stochastic term which is small of the order $\sqrt{\frac{1}{N}}$. Clearly, (2.9) is a mathematical challenge, and the derivation of the "small" perturbation term gives it some justification. On the other hand (2.9) is highly singular and therefore not easily tractable. Moreover, physically it is unreasonable to be limited by a space-time white noise (the distributional derivative of $W(t)$ with respect to t) as the stochastic driving term in a "stochastic Navier-Stokes equation". The problem seems to be with the choice of independent $\beta^i(t)$ in (2.1) $(m^i(t) := \beta^i(t))$. We will obtain in the next section a different "stochastic Navier-Stokes equation" if we assume suitable spatial correlations in (2.1).

3 A Random Point Vortex Model with Spatial Correlations

Let $w_i(p,t)$ be independent adapted Brownian sheets (or set-indexed white noises) on $\mathbf{R}^2 \times \mathbf{R}_+, i = 1,2$ (cf. Walsh [7] and Kotelenez [3]). Set $w(p,t) := (w_1(p,t), w_2(p,t))^T$ Further, let $\tilde{\Gamma} : \mathbf{R}^4 \to \mathbf{R}_+$ be a Borel measurable function which is symmetric in $x, p \in \mathbf{R}^2$ such that the following two conditions are satisfied:

(i) For any $x \in \mathbf{R}^2$

$$\int_{\mathbf{R}^2} \tilde{\Gamma}^2(x,p)dp = 1; \tag{3.1}$$

(ii) there is a finite constant c such that for any $x, y \in \mathbf{R}^2$

$$\int_{\mathbf{R}^2} [\tilde{\Gamma}(x,p) - \tilde{\Gamma}(y,p)]^2 dp \le c\tilde{|x - y|}^2. \tag{3.2}$$

Set

$$\Gamma(x,p) := \begin{pmatrix} \tilde{\Gamma}(x,p) & 0 \\ 0 & \tilde{\Gamma}(x,p) \end{pmatrix}.$$

With $K_N(x)$ from Section 2 we consider the following system of ordinary stochastic differential equations for the positions (of the centers) of N vortices

$$dx^i(t) = \sum_{j=1}^{N} a_j K_N(x^i - x^j)dt + \sqrt{2\nu}\int_{\mathbf{R}^2} \Gamma(x^i, p)w(dp, dt). \tag{3.3}$$

Remark 3.1 *1. Let $\{\tilde{\phi}_n\}_{n\in\mathrm{N}}$ be a complete orthonormal system (CONS) in $L_2(\mathbf{R}^2, dr)$ and set* $\phi_n := \begin{pmatrix} \tilde{\phi}_n & 0 \\ 0 & \tilde{\phi}_n \end{pmatrix}$. *Then*

$$\int_{\mathbf{R}^2} \Gamma(x,p)w(dp,t) = \sum_{n=1}^{\infty} \sqrt{2}\int_{\mathbf{R}^2} \Gamma(x,q)\phi_n(q)dq\beta^n(t), \tag{3.4}$$

where $\beta^n(t) := \frac{1}{\sqrt{2}}\int \phi_n(q)w(dq,t)$ are $\mathbf{R}^2$-valued i.i.d. standard Wiener processes. Hence (3.3) can be treated as a $2N$-dimensional ordinary Itô-equation which is driven by infinitely many i.i.d. $\mathbf{R}^2$-valued standard Wiener processes. It is well known that an $L_2(\mathbf{R}^2, dr)$-valued standard cylindrical Brownian motion $W(t)$ can be visualized as the (weak) limit of $\sum_{n=1}^{\infty}\beta_n(t)\hat{\phi}_n$, where $\hat{\phi}_n$ is a CONS in $L_2(\mathbf{R}^2, dr)$ (cf. Kotelenez [3]).

2. We now see that (3.2) is a Lipschitz assumption on the stochastic coefficient in (3.3): If $x, y \in \mathbf{R}^2$ we obtain that

$$\left.\begin{array}{l} \sum_{n=1}^{\infty}(\int_{\mathbf{R}^2}(\tilde{\Gamma}(x,q) - \tilde{\Gamma}(y,q))\tilde{\phi}_n(q)dq)^2 = \int_{\mathbf{R}^2}(\tilde{\Gamma}(x,q) - \tilde{\Gamma}(y,q))^2 dq \\ \qquad \le c\tilde{|x-y|}^2, i = 1,2. \end{array}\right\} \tag{3.5}$$

Hence there is a unique solution $X_N(t, X_N(0))$ to (3.3) for each square integrable adapted initial condition $X_N(0)$ and $X_N(t, X_N(0))$ is an $\mathbf{R}^{2N}$-valued Markov process.

3. Assume there are two $\mathbf{R}^2$-valued adapted stochastic processes $y^1(t)$ and $y^2(t)$. Then we easily see that the $\int_0^t\int_{\mathbf{R}^2}\Gamma(y^i(s),p)w(dp,ds)$ are $\mathbf{R}^2$-valued square integrable continuous martingales $(i = 1,2)$ and the mutual quadratic variation of their components is given by

$$\left.\begin{array}{l} \langle\langle \int_0^t\int_{\mathbf{R}^2}\tilde{\Gamma}(y^1(s),p)w_\ell(dp,ds), \int_0^t\int_{\mathbf{R}^2}\tilde{\Gamma}(y^2(s),p)w_\ell(dp,ds)\rangle\rangle \\ \quad = \int_0^t\int_{\mathbf{R}^2}\tilde{\Gamma}(y^1(s),p)\tilde{\Gamma}(y^2(s),p)dpds, \end{array}\right\} \tag{3.6}$$

$\ell = 1,2$. In particular, if $y^1 = x^i$ and $y^2 = x^j$ are the positions of the i-th and j-th vortices, (3.6) gives the correlation of the fluctuation "force" acting on our system of vortices. For $y^1(t) \equiv y^2(t)$ the right hand side of (3.6) equals t by assumption (3.1). The Cauchy-Schwarz inequality and the positivity of $\tilde{\Gamma}$ imply that the time derivative of the mutual quadratic variation in (3.6) is between 0 and 1.

Let us verify (3.1) and (3.2) for a particular example.

Example 3.2 *Set* $\tilde{\Gamma}(x,p) = (\frac{1}{2\pi\varepsilon}e^{-\frac{|x-p|^2}{2\varepsilon}})^{\frac{1}{2}}$ *for some* $\varepsilon > 0$, *and* $x,p \in \mathbf{R}^2$. *Then* $\int_{\mathbf{R}^2}\tilde{\Gamma}(x,p)\tilde{\Gamma}(y,p)dp = e^{-\frac{|x-y|^2}{8\varepsilon}}$ *using the Chapman-Kolmogorov equation. In particular, for* $x = y$ *we obtain (3.1) Moreover,* $\int_{\mathbf{R}^2}(\tilde{\Gamma}(x,p) - \tilde{\Gamma}(y,p))^2 dp = 2(1 - e^{-\frac{|x-y|^2}{8\varepsilon}}) = 2\int_0^{\frac{|x-y|^2}{8\varepsilon}} e^{-s}ds \le \frac{|x-y|^2}{4\varepsilon}$, *which yields (3.2) with* $c = \frac{1}{4\varepsilon}$. *Finally, if we choose* ε *small, i.e., as a function of the magnitude of the intensities which we took* $\sim \frac{1}{N}$ *in Section 2, then we see that the fluctuation "forces" for vortices which are not close to one another are almost uncorrelated, and close vortices generate observable correlations.*

∎

Again we set

$$\mathcal{X}_N(t) := \sum_{i=1}^{N} a_i \delta_{x^i(t)}, \tag{3.7}$$

where now $\{x^i(t)\}_{i=1,\dots,N}$ is the solution of (3.3) with suitable adapted initial conditions. Using the notation and abbreviations from Section 2 we obtain:

Theorem 3.3 $\mathcal{X}_N(t)$ *given by (3.7) is an* $\mathcal{M}_N^{\pm}$*-valued Markov process, and with* $L \in \mathcal{D}$ *its generator* $\mathcal{A}_{N,c}$ *is given by*

$$\left.\begin{array}{l}(\mathcal{A}_{N,c}L(\mathcal{X}_N)) \\ = \sum_{k=1}^{M}\frac{\partial}{\partial y_k}f(\langle\mathcal{X}_N,\varphi_1\rangle,\dots,\langle\mathcal{X}_N,\varphi_M\rangle)\langle\mathcal{X}_n,(U_N\cdot\nabla)\varphi_k + \nu\Delta\varphi_k\rangle \\ +\nu\sum_{k,\ell=1}^{M}\frac{\partial^2}{\partial y_k\partial y_\ell}f(\langle\mathcal{X}_N,\varphi_1\rangle,\dots,\langle\mathcal{X}_N,\varphi_M\rangle)\cdot \\ \cdot\int_{\mathbf{R}^2}[\int_{\mathbf{R}^2}\Gamma(r,p)\nabla\varphi_k(r)\mathcal{X}_N(dr)]\cdot[\int_{\mathbf{R}^2}\Gamma(r,p)\nabla\varphi_\ell(r)\mathcal{X}_N(dr)]dp.\end{array}\right\} \tag{3.8}$$

<u>Proof</u> We can again use Th. 10.13 in Dynkin [2] to show that $\mathcal{X}_N(t)$ is an $\mathcal{M}_N^{\pm}$-valued Markov process. For the verification of (3.8) we use Itô's formula. Let us compute $d\langle\mathcal{X}_N(t),\varphi\rangle$ and $d\langle\langle\mathcal{X}_N(t),\varphi\rangle,\langle\mathcal{X}_N(t),\varphi\rangle\rangle$ (the differential of the mutual quadratic variation of $\langle\mathcal{X}_N(t),\varphi\rangle$ and $\langle\mathcal{X}_N(t),\psi\rangle$), where φ and $\psi \in C^2(\mathbf{R}^2,\mathbf{R})$).

(i)

$$\begin{aligned} d\langle\mathcal{X}_N(t),\varphi\rangle &:= \sum_{i=1}^{N} a_i d\varphi(x^i(t)) = \\ &= \sum_{i=1}^{N} a_i\nabla\varphi(x^i(t))\cdot(\sum_{j=1}^{N} a_j K_N(x^i(t) - x^j(t)))dt \\ &+ \nu\sum_{i=1}^{N} a_i\sum_{\ell=1}^{2}\frac{\partial^2}{(\partial y_\ell)^2}\varphi(x^i(t))\int_{\mathbf{R}^2}\tilde{\Gamma}^2(x^i(t),p)dp\,dt \\ &+ \sqrt{2\nu}\sum_{i=1}^{N} a_i\nabla\varphi(x^i(t))\cdot\int_{\mathbf{R}^2}\Gamma(x^i(t),p)w(dp,dt) \end{aligned}$$

(the mixed second derivatives of φ disappear because $w_1(x,t)$ and $w_2(x,t)$ are assumed to be independent). Hence

$$d\langle\mathcal{X}_N(t),\varphi\rangle = \langle\mathcal{X}_N(t),(U_N\cdot\nabla)\varphi+\nu\Delta\varphi\rangle dt+\sqrt{2\nu}\sum_{i=1}^{N} a_i\nabla\varphi(x^i(t))\cdot\int_{\mathbf{R}^2}\Gamma(x^i(t),p)w(dp,dt) \tag{3.9}$$

(by (3.1)).

(ii)

$$\left.\begin{array}{l} d\langle\langle \mathcal{X}_N(t),\varphi\rangle,\langle \mathcal{X}_N(t),\psi\rangle\rangle = 2\nu\sum_{i,j=1}^N a_i a_j d\langle\langle \varphi(x^i(t)),\psi(x^j(t))\rangle\rangle \\ = 2\nu\sum_{i,j=1}^N a_i a_j \int_{\mathbf{R}^2}[\Gamma(x^i(t),p)\nabla\varphi(x^i(t))]\cdot[\Gamma(x^j(t),p)\nabla\psi(x^j(t))]dp\,dt, \end{array}\right\} \qquad (3.10)$$

where we again used the independence of $w_1(x,t)$ and $w_2(x,t)$. Note that the right hand side of (3.10) can be written as

$$2\nu\int_{\mathbf{R}^2}[\int_{\mathbf{R}^2}\Gamma(r,p)\nabla\varphi(r)\mathcal{X}_N(t,dr)]\cdot[\int_{\mathbf{R}^2}\Gamma(r,p)\nabla\psi(r)\mathcal{X}_N(t,dr)]dp.$$

This finishes the proof. ■

(3.9) implies:

Theorem 3.4 *$\mathcal{X}_N(t)$ given by (3.7) is a weak solution of the SPDE*

$$d\mathcal{X}_N(t) = [\nu\Delta\mathcal{X}_N(t) - (\nabla\cdot U_N)\mathcal{X}_N(t)]dt - \sqrt{2\nu}\nabla\cdot(\mathcal{X}_N(t)\int_{\mathbf{R}^2}\Gamma(\cdot,p)w(dp,dt)). \qquad (3.11)$$

Note that the right hand sides of (3.8) and (3.11) are meaningful both for measures $\mathcal{X}_N(dr)$ and for densities with respect to the Lebesque measure, where instead of $\mathcal{X}_N(dr)$ we write $\mathcal{X}_N(r)dr$. Then the scalar field $\mathcal{X}_N(t,r)$ acts as a multiplication operator on the two-dimensional field $\int_{\mathbf{R}^2}(\Gamma(r,p)w(dp,dt)$ in (3.11). Recall that a mild solution of (3.11) is by definition a solution of the integral equation which we obtain from (3.11) by using the semigroup generated by $\nu\Delta$ on $L_2(\mathbf{R}^2,dr)$ and variation of constants. For more details s. Kotelenez [3]. We expect the following theorem (cf. Kotelenez [4]):

Theorem 3.5 *Assume $\mathcal{X}_N(0)\in L_2(\mathbf{R}^2,dr)$, $\mathcal{X}_N(0)$ is $\mathcal{F}_0$-measurable, and $E\int_{\mathbf{R}^2}\mathcal{X}_N^2(0,r)dr<\infty$. Then there is a unique mild solution $\mathcal{X}_N(t,\mathcal{X}_N(0))$ of (3.11) which is product measurable in (t,r,ω) w.r.t. the product of the Borel σ-algebras and $\mathcal{F}.\mathcal{X}_N(t)=\mathcal{X}_N(t,\mathcal{X}_N(0))$ is adapted and for each $T>0$*

$$E\int_0^T\int_{\mathbf{R}^2}\mathcal{X}_N^2(s,r,x)dr\,ds<\infty. \qquad (3.12)$$

Moreover, the mild solution of (3.11) is an $L_2(\mathbf{R}^2,dr)$-valued Markov process, and its generator is given by (3.8), where $\langle\cdot,\cdot\rangle$ is now the $L_2(\mathbf{R}^2,dr)$ scalar product and $\int_{\mathbf{R}^2}f(r)\mathcal{X}_N(dr)=\int_{\mathbf{R}^2}f(r)\mathcal{X}_N(r)dr$.

Remark 3.6 *(i) The parameter N in (3.11) has to be given a different interpretation than in (2.9) since we do no longer have N distinct point vortices. It is reasonable to link the correlation length of $\tilde{\Gamma}(x,p)$ to N. E.g., in Example 3.2 we could choose $\varepsilon\sim\frac{1}{N}$.*

(ii) The difference between (3.11) and (2.9) comes from the fact that the spatial correlations preserve the double sum in (3.8) (cf. (3.10)), which leads to the quasi-bilinear equation (3.11) ("quasi" because the deterministic part of (3.11) is still nonlinear). On the other hand, the independence assumption on the noise terms in (2.1) implies that only the diagonal terms $\sum a_i^2\nabla\varphi_k(x^i)\cdot\nabla\varphi_\ell(x^i)$ appear in (2.7), which generates the highly singular nonlinear diffusion term in (2.9). To our knowledge Vaillancourt [6] was the first to observe that introducing spatial correlation leads to a phenomenologically different Markov generator. The first named author (P. Kotelenez) would like to thank D. Dawson for having brought Vaillancourt's paper and its potential meaning for SPDE's to his attention. It should be mentioned, however, that the correlations introduced in Vaillancourt's paper are different from ours and apparently do not lead to an SPDE as simple as our equation (3.11).

(iii) By limit arguments we can also show that (3.11) is sign-preserving (cf. Kotelenez [4]).

■

References

[1] Dawson, D.A.: Stochastic evolution equations and related measure processes. J. Multivariate Anal. 5, 1-52 (1975).

[2] Dynkin, E.B.: Markov processes 1. Springer, Berlin-New York (1965).

[3] Kotelenez, P.: Existence, uniqueness and smoothenss for a class of function valued stochastic partial differential equations (preprint #91 – 111 Department of Mathematics and Statistics, Case Western Reserve University).

[4] Kotelenez, P.: Stochastic partial differential equations for particle densities and vortices. (Manuscript in preparation).

[5] Marchioro, C., Pulvirenti, M.: Hydrodynamics in two dimensions and vortex theory. Comm. Math. Phys. 84, 483-503 (1982).

[6] Vaillancourt, J.: On the existence of random McKean-Vlasov limits for triangular arrays of exchangeable diffusions. Stoch. Anal. Appl. (1988).

[7] Walsh, J.B.: An introduction to stochastic partial differential equations. In P.L. Hennequin (ed.) Ecole d'Ete de Probabilites de Saint-Flour XIV-1984. LN in Mathematics 1180. Berlin-Heidelberg-New York, Springer 1986.

On explicit formulas for solutions of evolutionary SPDE's
(a kind of introduction to the theory)

N.V. Krylov
School of Mathematics, University of Minnesota, Minneapolis, MN, 55455

For the author the theory of stochastic partial differential equations started in the fall of 1974 when A.K. Zvonkin and I were preparing, for a conference in Druskininkai, a review [1] on the theory of strong and weak solutions of Ito's stochastic equations. I was necessarily thinking about different representations of solutions. It appeared that one of them can be based on the fact that the solution of Ito's stochastic equation can be viewed as a solution of a SPDE. Actually, at that time the theory of such equations was not developed and in [1] (see also [2]) only a straightforward consequence of such point of view was presented. Now the theory of SPDE's does exist, it is well developed and is well connected with calculations of certain conditional expectations, which give representations of solutions of these SPDE's (see, e. g. [3], [4] and references there). These representations are proved rigorously, but, as it is often the case, the proofs scarcely explain genuine reasons why the representations are true and how in fact they were found. Exactly the latter I want to discuss here in a mere informal way, formal proofs being known for all our formulas but the two last ones (4.13), (4.18), general case of which is relatively new and very cumbersome. The author hopes that this article will serve as a (good) introduction to the theory of SPDE's.

1. Case of a simple backward SPDE with nonrandom coefficients

To begin with let us consider the following one-dimensional Ito's stochastic equation

$$x_t = x + \int_s^t \sigma(r, x_r)dw_r + \int_s^t b(r, x_r)dr, \quad t \geq s, \tag{1.1}$$

where σ and b are smooth (nonrandom) functions of (r, x). Denote the position of solution at the moment t starting at the point x at the moment s by $X(s, x, t)$. For $s \leq r \leq t$ we have $X(s, x, t) = X(r, X(s, x, r), t)$ which simply means that to find the position of solution at time t we can solve equation (1.1) starting from the moment r and from the point reached by the solution at the moment r. If, for t fixed, we introduce the function

$$u(s, x) = X(s, x, t), \quad s \leq t,$$

this evolutional relation will read

$$u(s, x) = u(r, X(s, x, r)), \quad s \leq r \leq t. \tag{1.2}$$

Now take small $\Delta s > 0$ and insert in (1.2) $s - \Delta s$, s instead of s, r. Then we get

$$u(s-\Delta s, x) = u(s, X(s-\Delta s, x, s)) = u(s, x) + u_x(s, x)(X(s-\Delta s, x, s)-x) +$$
$$+ \tfrac{1}{2} u_{xx}(s, x)(X(s-\Delta s, x, s)-x)^2 .$$

From (1.1) it follows that (up to the terms of order not less than Δs)

$$X(s-\Delta s, x, s)-x = \sigma(s, x)(w_s - w_{s-\Delta s}) + b(s, x)\Delta s, \qquad (X(s-\Delta s, x, s)-x)^2 = \sigma^2(s, x)\Delta s.$$

Therefore

$$u(s-\Delta s, x) - u(s, x) = u_x(s, x)\sigma(s, x)(w_s - w_{s-\Delta s}) + [\tfrac{1}{2}\sigma^2(s, x)u_{xx}(s, x) + b(s, x)u_x(s, x)]\Delta s. \qquad (1.3)$$

To write this down in a differential form let us note that the coefficient of $w_s - w_{s-\Delta s}$ is taken at the moment s rather that at the moment $s-\Delta s$, as it should be for the finite difference approximation of ordinary (forward) Ito's stochastic integral. Accordingly, we introduce the backward stochastic integral g_s, $s \le t$, of a random process f_r by the formula

$$g_s = \int_s^t f_r bw_r := \lim \Sigma f_{s_{i+1}}(w_{s_{i+1}} - w_{s_i}), \qquad (1.4)$$

where partitions (s_i) of the interval (s, t) naturally are such that $\max(s_{i+1}-s_i) \to 0$. In (1.4) the notation bw_r is not a misprint but reflects the fact that increments of w are taken backward with respect to values of f. So far bw_r=b(ackward increment of)w_r, whereas dw_r =d(irect increment of)w_r. Naturally, if (1.4) holds for any $s \le t$ (recall that t is fixed) we write

$$bg_s = -f_s bw_s \quad \text{for } s < t.$$

It is worth noting that properties of the backward integral are quite similar to those of direct integral and, for instance, the integral in (1.4) exists if f_r is jointly measurable in ω, t, measurable with respect to the σ-field generated by increments of w on $[r, \infty)$ and locally square integrable with respect to r for any ω. Note that it is exactly the case of the function u(r, x). Note also that in (1.4) we have $f_{s_{i+1}}(w_{s_{i+1}} - w_{s_i}) = f_{s_i}(w_{s_{i+1}} - w_{s_i}) + (f_{s_{i+1}} - f_{s_i})(w_{s_{i+1}} - w_{s_i})$ which implies a useful relation between forward and backward integrals : if f_s has a stochastic (forward) differential then its backward integral g_s _exists_, and

$$bg_s = -f_s bw_s = -f_s dw_s - (df_s)dw_s \qquad \left(g_s = \int_s^t [f_r dw_r + (df_r)dw_r] \right) \qquad (1.5)$$

Returning back to (1.3), taking there s_i, s_{i+1} instead of $s-\Delta s$, s, summing up and passing to the limit we can write now the result as : _for_ $s < t$ _and_ $x \in \mathbb{R}^d$

$$-bu(s, x) = u_x(s, x)\,\sigma(s, x)bw_s + [\tfrac{1}{2}\sigma^2(s, x)u_{xx}(s, x) + b(s, x)u_x(s, x)]ds. \qquad (1.6)$$

Thus we discover that X(s, x, t) as a function of (s, x) satisfies SPDE (1.6). This equation has a property which is very unusual for second order equations. Namely, any smooth function φ of its solution is a solution as well. This can be explained if we note that $v(s, x) := \varphi(u(s, x))$ satisfies the same relation (1.2), the only used in the derivation of (1.6). This feature is common for the first order equations because these equations simply express the fact that the unknown function is constant along characteristics. In our case relation (1.2) is nothing else but an assertion that the function u(r, y) is constant on the curve $y(r) = X(s, x, r)$, $r \in [s, t]$, for s, x, ω being fixed.

Consequently, $(r, X(s, x, r))$, $r \in [s, t]$, is a family of characteristics of equation (1.6) starting from points (s, x).

Our conclusions are true in multidimensional case as well. Namely, let $\sigma^k(r, x)$, $k = 1,\ldots, d_1$, $b(r, x)$ be $\mathbb{R}^d$-valued smooth functions defined for $r \geq 0$, $x \in \mathbb{R}^d$, and let w_r^k, $k = 1,\ldots, d_1$, be independent Wiener processes. Consider the following stochastic equation

$$dx_r = \sigma^k(r, x_r)dw_r^k + b(r, x_r)dr , \quad r > s, \ x_s = x, \tag{1.7}$$

where (and hereafter) we drop the summation sign with respect to $k = 1,\ldots, d_1$ in the first term on the right. Let $X(s, x, r)$ as before be the solution of this equation. To write down the multidimensional counterpart of (1.6) in the most convenient way, for a smooth function $f(x)$ and vectors ξ , $\eta \in \mathbb{R}^d$ we denote

$$f_{(\xi)} = \Sigma f_{x^i} \xi^i , \qquad f_{(\xi)(\eta)} = \Sigma f_{x^i x^j}\xi^i \eta^j. \tag{1.8}$$

Take also a smooth function φ on $\mathbb{R}^d$ and for t fixed define

$$u(s, x) = \varphi(X(s, x, t)). \tag{1.9}$$

By the same argument as before we convince ourselves that $u(s, x)$ _is a solution of the following problem_ :

$$-bu = u_{(\sigma^k)}bw_s^k + [½u_{(\sigma^k)(\sigma^k)} + u_{(b)}]ds \quad \text{for } s < t, \ x \in \mathbb{R}^d \tag{1.10}$$

$$u(t, x) = \varphi(x) \ \text{ on } \mathbb{R}^d . \tag{1.11}$$

2. Generalizations

We now know what at least one solution of a special SPDE looks like. Let us consider more general SPDE's.

1°. _Free terms in equation._ Suppose first that we are interested in the following equation

$$-bu = [u_{(\sigma^k)} + g^k]bw^k_s + [½u_{(\sigma^k)(\sigma^k)} + u_{(b)} + f]ds \quad \text{for } s < t \tag{2.1}$$

with the same boundary data (1.11), same σ^k, b as in (1.10) and some smooth real functions $g^k(s, x)$ and $f(s, x)$. The idea is to transform equation (2.1) to an equation of type (1.10) containing no "right-hand side" terms g^k, f. To this end we introduce a new real variable x^o, consider the space $\mathbb{R}^{d+1} = \{ (x^o, x) : x^o \in \mathbb{R}, x \in \mathbb{R}^d \}$ instead of $\mathbb{R}^d$, and we consider the function

$$v(s, x^o, x) := x^o + u(s, x). \tag{2.2}$$

Note that for vectors $\zeta = (\xi^o, \xi)$, $\gamma = (\eta^o, \eta) \in \mathbb{R}^{d+1}$ in accordance with (1.8) we have

$$v_{(\zeta)} = v_{(\xi^o,\xi)} = v_{x^o}\xi^o + v_{x^1}\xi^1 + \ldots + v_{x^d}\xi^d = \xi^o + u_{(\xi)} , \quad v_{(\zeta)(\gamma)} = v_{(\xi^o,\xi)(\eta^o,\eta)} = u_{(\xi)(\eta)}.$$

Therefore, (2.1) with boundary data (1.11) is equivalent to

$$-bv = v_{(g^k,\ \sigma^k)} bw_s^k + [\tfrac{1}{2} v_{(g^k,\ \sigma^k)(g^k,\ \sigma^k)} + v_{(f,\ b)}]ds\ , \quad s < t, \tag{2.3}$$

$$v(t, x^o, x) = \psi(x^o, x) := x^o + \varphi(x)\ . \tag{2.4}$$

As in the case of equation (1.10), it is natural now, in place of system (1.7), for $r > s$ to consider the following system

$$dx_r^o = g^k(r, x_r)dw_r^k + f(r, x_r)dr, \tag{2.5}$$

$$dx_r = \sigma^k(r, x_r)dw_r^k + b(r, x_r)dr \tag{2.6}$$

with initial data $x_s^o = x^o$, $x_s = x$. If we define the pair $(X^o(s, x^o, x, r), X(s, x^o, x, r))$ as the solution of our system, then, as it is explained above, the function

$$v(s, x^o, x) = \psi(X^o(s, x^o, x, t), X(s, x^o, x, t)) \tag{2.7}$$

will solve the problem (2.3), (2.4). Note that equations (1.7) and (2.6) coincide. Consequently, $X(s, x^o, x, t) = X(s, x, t)$. Furthermore, equation (2.5) does not contain x_r^o on the right, and it means that

$$X^o(s, x^o, x, t) = x^o + \int_s^t f(r, X(s, x, r))dr + \int_s^t g^k(r, X(s, x, r))dw_r^k.$$

All this along with (2.7) and definition (2.2) of the function v allow us to conclude that _the function_

$$u(s, x) := \varphi(X(s, x, t)) + \int_s^t f(r, X(s, x, r))dr + \int_s^t g^k(r, X(s, x, r))dw_r^k \tag{2.8}$$

is a solution of SPDE (2.1) _with boundary data_(1.11). It is useful to compare (2.8) and (1.9) to understand how the presence of "the free terms f, g^k " affects the result.

2°. _Zeroth order terms in equation._ To do the next natural step in generalizations we introduce zeroth order terms in equation (2.1). Namely, we take smooth real functions $\gamma^k(s, x)$, $c(s, x)$, defined for $s \geq 0$ and $x \in \mathbb{R}^d$, and we consider the following equation :

$$-\ bu = [u_{(\sigma^k)} + \gamma^k u + g^k]bw_s^k + [\tfrac{1}{2} u_{(\sigma^k)(\sigma^k)} + u_{(b)} + cu + f]ds, \quad s < t \tag{2.9}$$

with the boundary data (1.11). Now on $\mathbb{R}^{d+1}$ we define a new auxiliary function:

$$v(s, x^o, x) = x^o u(s, x). \tag{2.10}$$

For the same vectors ζ, γ as in 1° we get

$$v_{(\zeta)} = v_{(\xi^o,\ \xi)} = \xi^o u + x^o u_{(\xi)}\ , \quad v_{(\zeta)(\gamma)} = v_{(\xi^o,\xi)(\eta^o,\eta)} = \xi^o u_{(\eta)} + \eta^o u_{(\xi)} + x^o u_{(\xi)(\eta)}\ .$$

After multiplying equation (2.9) by x^o and after simple transformations, using the above formulas, we see that in terms of the function v equation (2.9) looks like

$$-bv = [v_{(x^o\gamma^k,\ \sigma^k)} + x^o g^k] bw_s{}^k + [\tfrac{1}{2} v_{(x^o\gamma^k,\ \sigma^k)(x^o\gamma^k,\ \sigma^k)} + v_{(x^o c,\ b - \sigma^k\gamma^k)} + x^o f] ds \quad \text{for } s < t. \tag{2.11}$$

The boundary condition is $v(t, x) = x^o\varphi(x)$.

Equation (2.11) is quite similar to equation (2.1), the solution of which is given by (2.8) with the process $X(s, x, t)$ defined by system (2.6). That is why we define the pair $(X^o(s, x^o, x, r), X(s, x^o, x, r))$ as the solution of the following system

$$dx_r{}^o = x_r{}^o\gamma^k(r, x_r)dw_r{}^k + x_r{}^o c(r, x_r)dr , \tag{2.12}$$

$$dx_r = \sigma^k(r, x_r)dw_r{}^k + [b(r, x_r) - \sigma^k(r, x_r)\gamma^k(r, x_r)]dr , \ r > s, \tag{2.13}$$

with initial data $x_s{}^o = x^o$, $x_s = x$. In our situation formula (2.8) means that

$$x^o u(s, x) := X^o(s, x^o, x, t)\varphi(X(s, x^o, x, t)) + \tag{2.14}$$

$$+ \int_s^t X^o(s, x^o, x, r)[f(r, X(s, x^o, x, r))dr + g^k(r, X(s, x^o, x, r))dw_r{}^k]$$

is a solution of equation (2.11) with the given boundary data. Let us make some transformations of this formula. Note that equation (2.13) does not contain x^o. Therefore its solution is independent of x^o, and we denote it by

$$\tilde{X}(s, x, r).$$

Moreover, equation (2.12) is linear with respect to $x_r{}^o$ and its solution can easily be found, namely, if we denote

$$\rho(s, x, r) = \exp \left(\int_s^r \gamma^k(p, \tilde{X}(s, x, p))dw_p{}^k + \int_s^r (c - \tfrac{1}{2}|\gamma|^2)(p, \tilde{X}(s, x, p))dp \right),$$

then $X(s, x^o, x, r) = x^o\rho(s, x, r)$, and (2.14) yields

$$u(s, x) = \rho(s, x, t)\varphi(\tilde{X}(s, x, t)) + \int_s^t \rho(s, x, r)[f(r, \tilde{X}(s, x, r)dr + g^k(r, \tilde{X}(s, x, r))dw_r{}^k] .$$

Thus, we received a formula for solution of general completely degenerate equation (2.9).

3°. Case of general SPDE. Let us take an integer $d_o \le d_1$ and let us consider the following shortened version of equation (2.9) :

$$-bu = [u_{(\sigma^\nu)} + \gamma^\nu u + g^\nu]bw_s{}^\nu + [\tfrac{1}{2} u_{(\sigma^k)(\sigma^k)} + u_{(b)} + cu + f]ds \ , \ s < t \ , \tag{2.15}$$

where (and hereafter) in the first term on the right we suppose the summation with respect to the *Greek* index ν running through $1, \dots, d_o$, whereas in the first term in the second square brackets, as usually, the summation is over $k = 1, \dots, d_1$.

Denote by $\mathcal{F}_{s,t}$ the σ-algebra generated by random variables $w_r{}^\nu - w_p{}^\nu$ for $r, p \in [s, t]$, $\nu = 1, \dots, d_o$. It is natural that

$$\mathbb{E}\left(\int_s^t f_r bw_r{}^k \mid \mathcal{F}_{s,t} \right) = \int_s^t \mathbb{E}(f_r \mid \mathcal{F}_{r,t}) bw_r{}^k ,$$

if $k \le d_o$, since then w_r^k is $\mathcal{F}_{s,t}$-measurable, and this expectation is zero if w^k is independent of $\mathcal{F}_{s,t}$, that is if $k > d_o$. With the help of this simple remark, after writing down equation (2.9) in its integral form, we see that if u is a solution of problem (2.9),(1.11), then $\mathbb{E}\{ u(s, x) \mid \mathcal{F}_{s,t} \}$ is a solution of problem (2.15), (1.11). It is true for any γ^k, g^k which coincide with those participating in (2.15) for $k \le d_o$. We can even take $\gamma^k = g^k = 0$ for $k \ge d_o$.

Thus, the solution of this problem (2.15), (1.11) *at a point* (s, x) *is given by the following formula* :

$$u(s, x) = \mathbb{E}\{\rho_t\varphi(x_t) + {}_s\int^t \rho_r[f(r, x_r)dr + g^k(r, x_r)dw_r^k] \mid \mathcal{F}_{s,t} \}, \tag{2.16}$$

where

$$\rho_r = \exp \{ {}_s\int^r \gamma^k(p, x_p)dw_p^k + {}_s\int^r (c - ½|\gamma|^2)(p, x_p)dp \}$$

and x_r *is a solution of the stochastic equation*

$$x_r = x + {}_s\int^r \sigma^k(p, x_p)dw_p^k + {}_s\int^r [b(p, x_p) - \sigma^k(p, x_p)\gamma^k(p, x_p)]dp .$$

Let us discuss how general equation (2.15) is. In the most general linear SPDE's (cf. [3], [4]) the stochastic term on the right is the same as in (2.15) and, instead of the first term in the second square bracket in (2.15), we have the following expression

$$½a^{ij}(s, x)u_{x^ix^j}(s, x),$$

(where we assume the summation convention). It is always supposed that the d×d symmetric matrix $a = (a^{ij})$ is such that $a^{ij}\lambda^i\lambda^j \ge (\sigma^\nu, \lambda)^2$ for any $\lambda \in \mathbb{R}^d$. Therefore, the matrix $\alpha := a - (\sigma^{i\nu}\sigma^{j\nu})$ is symmetric and nonnegative. Let β be the symmetric nonnegative square root of α : $\alpha^{ij} = \beta^{in}\beta^{jn}$. Take $d_1 = d_o + d$ and for $k > d_o$ define σ^k as $(k - d_o)$th column of β. Then $a^{ij} = \alpha^{ij} + \sigma^{i\nu}\sigma^{j\nu} = \sigma^{ik}\sigma^{jk}$ and

$$a^{ij}u_{x^ix^j} = u_{(\sigma^k)(\sigma^k)} .$$

This convinces us that the most general SPDE has, in fact, the form of equation (2.15). Let us stress once again that, however, we don't deal here with rigorous proofs of our conclusions, in particular, we don't touch the very delicate question of smoothness of β. For everything like this we refer the reader to [3], [4].

3. Case of a simple forward SPDE with random coefficients

1°. Comments about backward equation. First, let us go back to equation (1.7) keeping the assumption about deterministic character of σ, b. Recall that the function u(s, x), introduced by (1.9), is constant on the characteristic (r, X(s, x, r)) (see (1.2)), which starts at the nonrandom point (s, x) and "brings back" to this starting point the value of φ at the random ending point corresponding to the moment t. It is simply another interpretation of formula (1.9). Sometimes it is interesting to know how the characteristic ending at a *given* nonrandom point (t, x) goes back . Knowing this, we know where the value $\varphi(x)$ will be spread, hence, we know a level surface of the function u. To obtain an equation for characteristics ending at a given point, we rewrite equation

(1.7) in its integral form :

$$x_t - x_s = {}_s\int^t \sigma^k(r, x_r)dw_r{}^k + {}_s\int^t b(r, x_r)dr\,, \tag{3.1}$$

and we note that x_t and x_s enter this equation quite similarly. Therefore, instead of fixing x_s and solving the equation with respect to x_t , we can, as well, prescribe a fixed value for x_t and solve the equation with respect to x_s . The only inconvenience is that now we will develop the time variable in the backward direction, whereas the stochastic integral in (3.1) is written in the forward direction. In connection with this recall formula (1.5) which expresses the backward stochastic integral (1.4) in terms of ordinary stochastic integral. Then we see that equation

$$x_s = x_t - {}_s\int^t \sigma^k(r, x_r)bw_r{}^k - {}_s\int^t [b(r, x_r) - \sigma^k{}_{(\sigma^k)}(r, x_r)]dr \tag{3.2}$$

coincides with equation (3.1). In differential form we get that

$$bx_s = \sigma^k(s, x_s)bw_s{}^k + [b(s, x_s) - \sigma^k{}_{(\sigma^k)}(s, x_s)]ds \quad \text{for } s < t. \tag{3.3}$$

Define X(s, t, x) as a solution of (3.3) or (3.2) with the condition $x_t = x$ (do not confuse this solution corresponding to the final data (t, x) with the solution X(s, x, t) of (1.7) which correspons to starting data (s, x). In boundry data we always write time variable first.) If our conclusions are true and $\{s, X(s, t, x), s \le t\}$ is the characteristic ending at the point x actually, then for x = X(s, y, t) we must have X(s, t, x) = y , that is we must have

$$X(s, t, X(s, x, t)) = x\,.$$

This relation means that the application $x \to X(s, t, x)$ has an inverse $X^{-1}(s, t, x) = X(s, x, t)$ and, thus, we can rewrite formula (1.9) as

$$u(s, x) = \varphi(X^{-1}(s, t, x))\,. \tag{3.4}$$

2°. <u>Forward version of (3.4).</u> Note that in (3.4) both u and X(s, t, x) are defined with the help of equations (1.10) and (3.3) written in the same time direction. We will invert time in both of them taking the up to now fixed point t as zero on the time axis. Let us consider the following (forward) SPDE

$$du(s, x) = u_{(\sigma^k)}(s, x)dw_s{}^k + [\tfrac{1}{2}u_{(\sigma^k)(\sigma^k)}(s, x) + u_{(b)}(s\, x)]ds,\ s > 0 \tag{3.5}$$

with the initial data

$$u(0, x) = \varphi(x) \quad \text{on } \mathbb{R}^d\,, \tag{3.6}$$

and, in correspondence with (3.3), let us associate with (3.5) the ordinary (forward) stochastic equation

$$dx_s = -\sigma^k(s, x_s)dw_s{}^k - [b - \sigma^k{}_{(\sigma^k)}](s, x_s)ds, \quad s > 0. \tag{3.7}$$

Denote by X(x, s) the solution of (3.7) with the initial data $x_0 = x$. By virtue of (3.4) for the solution of problem (3.5), (3.6) we should have

$$u(s, x) = \varphi(X^{-1}(x, s)). \tag{3.8}$$

Usually one proves (3.8) by proving that $u(s, X(x, s)) = \varphi(x)$, which is, of course, true if $s = 0$, so that it suffices to prove that the stochastic differential of $u(s, X(x, s))$ is zero. The latter one checks using Ventzel's version of Ito's formula, and one immediately sees that the argument does not use the nonrandomness of σ, b, and it is valid for σ, b depending on ω in nonanticipating way as well. *Thus, formula* (3.8) *gives a solution of problem* (3.5), (3.6) *even with random coefficients* σ, b.

4. Generalizations

Here we will go through the same steps as in Section 2 but now we allow to all functions entering our equations to depend on ω in nonanticipating way.

1°. *Free terms in equation.* Instead of (2.1) consider the equation

$$du = [u_{(\sigma^k)} + g^k]dw_s^k + [\tfrac{1}{2}u_{(\sigma^k)(\sigma^k)} + u_{(b)} + f]ds\ , \quad s > 0 \tag{4.1}$$

with the initial data (3.6). As in Sect. 2 for the function v given by (2.2) we get

$$dv = v_{(g^k,\ \sigma^k)}dw_s^k + [\tfrac{1}{2}v_{(g^k,\ \sigma^k)(g^k,\ \sigma^k)} + v_{(f,\ b)}]ds\ . \tag{4.2}$$

Comparing this equation with (3.5) and taking into account that σ, b, g, f are independent of x^o, we see that the following system will play the role of system (3.7)

$$dx_s^o = -g^k(s, x_s)dw_s^k - [f - g^k_{(\sigma^k)}](s, x_s)ds \tag{4.3}$$

$$dx_s = -\sigma^k(s, x_s)dw_s^k - [b - \sigma^k_{(\sigma^k)}](s, x_s)ds\ . \tag{4.4}$$

The solution of (4.4) (that is of (3.7)) is $X(x, s)$, the "solution" of (4.3) is given by

$$X^o(x^o, x, s) = x^o - \int_0^s\{\ g^k(r, X(x, r))dw_r^k - [f - g^k_{(\sigma^k)}](r, X(x, r))dr\ \}.$$

To find the inverse of the application $(x^o, x) \rightarrow (X^o(x^o, x, s), X(x, s))$ we have to solve the system

$$X^o(y^o, y, s) = x^o\ , \qquad X(y, s) = x\ ,$$

which gives

$$y = y(x, s) = X^{-1}(x, s),\ \ y^o = y^o(x^o, x, s) = x^o + \int_0^s \{\ g^k(r, X(y, r))dw_r^k + [f - g^k_{(\sigma^k)}](r, X(y, r))dr\}\ .$$

Now by (3.8), for solution v of equation (4.2) we get $v(s, x) = y^o(x^o, x, s) + \varphi(y(x, s))$, which is equivalent to *the following formula for solution* u *of problem* (4.1), (3.6) :

$$u(s, x) = \varphi(X^{-1}(x, s)) + \int_0^s\{\ g^k(r, X(y, r))dw_r^k + [f - g^k_{(\sigma^k)}](r\ X(y, r))dr\ \}\Big|_{y=X^{-1}(x,\ s)}. \tag{4.5}$$

2°. *Zeroth order terms in equation.* Let us consider the general form of completely degenerate SPDE:

$$du = [u_{(\sigma^k)} + \gamma^k u + g^k]dw_s^k + [\tfrac{1}{2}u_{(\sigma^k)(\sigma^k)} + u_{(b)} + cu + f]ds, \ s > 0 \tag{4.6}$$

with initial data (3.6). For the function v from (2.10), just as (2.11), we find

$$dv = [v_{(x^o\gamma^k,\ \sigma^k)} + x^o g^k]dw_s^k + [\tfrac{1}{2}v_{(x^o\gamma^k,\ \sigma^k)(x^o\gamma^k,\ \sigma^k)} + v_{(x^o c,\ b - \sigma^k\gamma^k)} + x^o f]ds \ . \tag{4.7}$$

This equation is similar to (4.1) and we want to use formula (4.5) to find a solution of (4.7). The role of the process X(x, s), defined by equation (4.4), will be played by solution of the system

$$dx_s^o = -x_s^o\gamma^k dw_s^k - x_s^o[c - |\gamma|^2 - \gamma^k_{(\sigma^k)}]ds \ , \tag{4.8}$$

$$dx_s = -\sigma^k dw_s^k - [b - \sigma^k\gamma^k - \sigma^k_{(\sigma^k)}]ds \ , \tag{4.9}$$

which is easy to write down by applying the rules of construction of the right-hand of (4.4) on the basis of coefficients of (4.1), and by using the following computations (in computations of partial derivatives we note that γ, σ do not depend on the zeroth coordinate) :

$$(x^o\gamma^k,\ \sigma^k)_{(x^o\gamma^k,\ \sigma^k)} = (x^o\gamma^k{}_{(x^o\gamma^k,\ \sigma^k)},\ \sigma^k{}_{(x^o\gamma^k,\ \sigma^k)}) = (x^o\gamma^k\gamma^k + x^o\gamma^k{}_{(\sigma^k)},\ \sigma^k{}_{(\sigma^k)}) \ , \ \gamma^k\gamma^k =: |\gamma|^2 \ .$$

Let us denote $\tilde{X}(x, s)$ the solution of (4.9) with starting point $x_o = x$. Then the solution $\tilde{X}^o(x^o, x, s)$ of equation (4.8) with initial data x^o can be found explicitly, since it is a linear equation : $\tilde{X}_s^o(x^o, x, s) = x^o\rho_s(x)$, where

$$\rho_s(x) := \exp(-{}_0\!\int^s \gamma^k(r, \tilde{X}(x, r))dw_r^k + {}_0\!\int^s(\tfrac{1}{2}|\gamma|^2 - c + \gamma^k_{(\sigma^k)})(r, \tilde{X}(x, r))dr) \ . \tag{4.10}$$

The inverse application to the application $(x^o, x) \rightarrow (\tilde{X}^o(x^o, x, s), \tilde{X}(x, s))$ can easily be found, by formula (4.5) we find v(s, x) and finally by definition (2.10) we conclude that *solution of problem* (4.6), (3.6) *is given by*

$$u(s, x) = F_s(\tilde{X}^{-1}(x, s)) \ ,$$

where

$$F_s(y) := \rho_s^{-1}(y)\varphi(y) + \rho_s^{-1}(y) \ {}_0\!\int^s \rho_r(y) \{ g^k(r, \tilde{X}(y, r))dw_r^k +$$

$$+ [f - g^k\gamma^k - g^k_{(\sigma^k)}](r, \tilde{X}(y, r))dr \} \ . \tag{4.11}$$

3°. Case of general SPDE. As in Sect 2.3 we take an integer $d_o \le d_1$ and consider the equation

$$du = [u_{(\sigma^k)} + \gamma^\nu u + g^\nu]dw_s^\nu + [\tfrac{1}{2}u_{(\sigma^k)(\sigma^k)} + u_{(b)} + cu + f]ds, \ s > 0 \tag{4.12}$$

with the initial condition (3.6). By the same argument as in Sect 2.3 it becomes evident that *for the solution of this equation it holds that*

$$u(s, x) = \mathbb{E}\{ F_s(\tilde{X}^{-1}(x, s)) \mid \mathcal{F}_s \}, \tag{4.13}$$

where $\mathcal{F}_s$ *is the* σ-*field generated by* w_r *for* $r \le s$, $\tilde{X}(x, s)$ *is the solution of* (4.9) *starting at* x, ρ_s *and* F_s *are defined by* (4.10), (4.11), *and* γ^k, g^k *are arbitrary for* $k > d_0$, *for instance, equal to zero.* The comments from Sect 2.3 about generality of equation (4.12) are valid here as well.

4°. Measure-valued solutions. In applications to the theory of filtering one usually gets for the nonnormalized density an equation with operators formally adjoint to the operators entering equation (4.12). As a typical example let us consider the following equation :

$$du = [-(u\tilde{\sigma}^{i\nu})_{x^i} + \tilde{\gamma}^{\nu}u + g^{\nu}]dw_s^{\nu} + [\tfrac{1}{2}(u\tilde{\sigma}^{ik}\tilde{\sigma}^{jk})_{x^ix^j} - (u\tilde{b}^i)_{x^i} + \tilde{c}u + f]ds \tag{4.14}$$

with the initial condition (3.6). One of characteristic features of this equation is that it makes sense even if u is a generalized function, say, a measure, since we can multiply both sides of (4.14) by a test function h and take integrals over $\mathbb{R}^d$ bringing all derivatives from u to h by formal integration by parts. Therefore, it is natural to look for an explicit formula for $\int hu\, dx$.

We will obtain such a formula having multiplied formula (4.13) by h and then by integrating over $\mathbb{R}^d$, but first of all we need to write down equation (4.14) in the form of equation (4.12). Obviously, it will be done if we take arbitrary γ^k, g^k for $k > d_0$ and if in (4.12) we will take

$$\sigma^k = -\tilde{\sigma}^k,\ \gamma^k = \gamma^k - \tilde{\sigma}^{ik}{}_{x^i},\ b^i = -\tilde{b}^i + (\tilde{\sigma}^{ik}\tilde{\sigma}^{jk})_{x^j}, \tag{4.15}$$
$$c = \tilde{c} - \tilde{b}^i{}_{x^i} + \tfrac{1}{2}(\tilde{\sigma}^{ik}\tilde{\sigma}^{jk})_{x^ix^j}.$$

By the way, equation (4.9) in this notations is

$$dx_s = \tilde{\sigma}^k(s, x_s)dw_s{}^k + [\tilde{b} - \tilde{\sigma}^k\gamma^k](s, x_s)ds. \tag{4.16}$$

Its solution (that is solution of (4.9)) starting at the point x is $\tilde{X}(x, s)$. Since equations (4.14), (4.12) coincide, by formula (4.13) for the solution of (4.14) we get

$$\int h(x)u(s, x)dx = \mathbb{E}\left(\int h(x)F_s(\tilde{X}^{-1}(x, s))dx \mid \mathcal{F}_s \right). \tag{4.17}$$

Let us make the change of variables $x = X(y, s)$ in the integral under the sign of expectation. We need, of course, to find the Jacobian of this transformation : $\det(\tilde{X}^i{}_{y^j}) = \det(\tilde{X}_y)$. Recall that if a square matrix Y(s) satisfies the equation $dY(s) = A(s)Y(s)dy_s$, where the process y_s is continuous and has bounded variation, then $\det Y(s) = \det Y(0) \exp \int_0^s \mathrm{tr}A(r)dy_r$. The same is true if y_s is a continuous semimartingale and integrals involved here are understood in Stratonovich's sense . Now let us fix x, $\xi \in \mathbb{R}^d$ and denote $\xi_s = \tilde{X}_{(\xi)}(x, s)$ $(= \tilde{X}_x(x, s)\xi)$. This process satisfies an equation following from equation (4.16), which defines $\tilde{X}(x, s)$, by formal differentiation :

$$d\xi_s = \tilde{\sigma}^k{}_{(\xi_s)}dw_s{}^k + [\tilde{b} - \tilde{\sigma}^k\gamma^k]_{(\xi_s)}ds.$$

Transforming this equation in Stratonovich's form, applying the symbol δ for the corresponding differential (so that $y_s dw_s = y_s\delta w_s - \tfrac{1}{2}dy_s dw_s$, $\tilde{\sigma}^k dw_s{}^k = \tilde{\sigma}^k\delta w_s{}^k - \tfrac{1}{2}\tilde{\sigma}^k{}_{(\tilde{\sigma}^k)}ds$, $\tilde{\sigma}^k{}_{(\xi_s)}dw_s{}^k = \tilde{\sigma}^k{}_{(\xi_s)}\delta w_s{}^k - \tfrac{1}{2}(\tilde{\sigma}^k{}_{(\tilde{\sigma}^k)})_{(\xi_s)}ds$) and using arbitrariness of ξ, we obtain that the matrix $\tilde{X}_x$ satisfies the equation

$$d\tilde{X}^i{}_{xj} = \tilde{\sigma}^k{}_{x^r}\tilde{X}^r{}_{xj}\delta w_s{}^k + [\tilde{b} - \tilde{\sigma}^k\tilde{\gamma}^k - \tfrac{1}{2}(\tilde{\sigma}^k{}_{(\tilde{\sigma}^k)})]_{x^r}\tilde{X}^r{}_{xj}\, ds .$$

Hence,

$$\det(\tilde{X}_y(y, s)) = \exp{}_0\!\int^s\!\big(\operatorname{div}\tilde{\sigma}^k(r, \tilde{X}(y, r))\delta w_r{}^k + \operatorname{div}\,[\tilde{b} - \tilde{\sigma}^k\gamma^k - \tfrac{1}{2}\tilde{\sigma}^k{}_{(\tilde{\sigma}^k)}](r, \tilde{X}(y, r))dr\,\big) =$$

$$= \exp{}_0\!\int^s\!\big(\operatorname{div}\tilde{\sigma}^k dw_r{}^k + [\operatorname{div}[\tilde{b} - \tilde{\sigma}^k\tilde{\gamma}^k - \tfrac{1}{2}\tilde{\sigma}^k{}_{(\tilde{\sigma}^k)}] + \tfrac{1}{2}(\operatorname{div}\tilde{\sigma}^k)_{(\tilde{\sigma}^k)}]dr\,\big).$$

From this and from (4.10) by formulas (4.15) and by easy straightforward computation we get

$$\det(\tilde{X}_y(y, s)) = \rho_s(y)\exp{}_0\!\int^s\!\big(\tilde{\gamma}^k(r, \tilde{X}(y, r))dw_r{}^k + [\tilde{c} - \tfrac{1}{2}|\gamma|^2](r, \tilde{X}(y, r))dr\,\big).$$

Finally, from (4.17) *for the solution* u *of problem* (4.14), (3.6) *and for any test function* h *we have the following formula :*

$$\int h(x)u(s, x)dx = \mathbb{E}\,\big(\int h(\tilde{X}(y, s))G_s(y)dy \mid \mathcal{F}_s \big), \tag{4.18}$$

where

$$G_s(y) := F_s(y)\det(\tilde{X}_y(y, s)) = \Big[\, \varphi(y) + \int_0^s \rho_r(y)\,\big(g^k(r, \tilde{X}(y, r))dw_r{}^k +$$

$$+ [f - g^k\gamma^k - g^k{}_{(\sigma^k)}](r, \tilde{X}(y, r))dr\,\big)\Big] \exp \int_0^s\!\big(\gamma^k(r, \tilde{X}(y, r))dw_r{}^k + [\tilde{c} - \tfrac{1}{2}|\tilde{\gamma}|^2](r, \tilde{X}(y, r))dr\,\big),$$

$\rho_s(x)$, $\tilde{X}(x, s)$ *are defined by* (4.10), (4.9) *(or by* (4.16)*) with* σ, γ, b, c *introduced in* (4.15) *and* $\tilde{\gamma}^k$, g^k *being arbitrary for* $k > d_0$.

Acknowledgment

The author is sincerely grateful to R.B. Sowers for helpful comments and corrections.

References

1. N.V. Krylov, A.K. Zvonkin, *On strong solutions of stochastic differential equations*, Selecta Math. Sov., 1, 1 (1981), 19 - 61.

2. N.V. Krylov, A.Yu. Veretennikov, *On explicit formulas for solutions of stochastic equations*, Math. USSR Sbornik, 29, 2 (1976), 239 - 256.

3. B.L. Rozovskii, *Stochastic Evolution Systems*, Kluwer, Dordrecht, 1990.

4. H. Kunita, *Stochastic flows and stochastic differential equations*, Cambridge Univ. Press, 1990.

CONVOLUTION AND FOURIER TRANSFORM OF HIDA DISTRIBUTIONS

Hui-Hsiung Kuo*

Department of Mathematics
Louisiana State University
Baton Rouge, LA 70803, USA

§1. Introduction

Convolution and the Fourier transform play important roles in the finite dimensional distribution theory. The Fourier transform has been generalized to infinite dimensional distribution theory within the framework of white noise analysis, see e.g. [8, 16, 17, 18, 19]. To generalize convolution to the infinite dimensional case, consider the convolution $f * g$ of two functions f and g defined on $\mathbb{R}^n$:

$$f * g(x) = \int_{\mathbb{R}^n} f(x-y)g(y)\,dy.$$

Obviously, we can not use this formulation to define convolution in the infinite dimensional case since the Lebesgue measure does not exist. On the other hand, the convolution $\nu * \sigma$ of two measures ν and σ on $\mathbb{R}^n$ defined as

$$\nu * \sigma(A) = \int_{\mathbb{R}^n} \nu(A-y)\,d\sigma(y), \quad A \in \mathcal{B}(\mathbb{R}^n)$$

can be generalized to the infinite dimensional case in an obvious way. However, we can not use this formulation to define the convolution of two distributions. To overcome these difficulties, suppose ν, σ and $\nu * \sigma$ are absolutely continuous with respect to the standard Gaussian measure μ with Radon-Nikodym derivatives $\widetilde{\nu}, \widetilde{\sigma}$ and $(\nu * \sigma)^{\sim}$, respectively. Let $\langle\!\langle \cdot,\cdot \rangle\!\rangle$ denote the pairing with respect to the Gaussian measure μ and $\langle \cdot,\cdot \rangle$ the pairing with respect to the Lebesgue measure. Then by informal calculation we have the following:

$$\begin{aligned}
\langle\!\langle (\nu * \sigma)^{\sim}, e^{\langle \cdot,\xi \rangle} \rangle\!\rangle &= \langle \nu * \sigma, e^{\langle \cdot,\xi \rangle} \rangle \\
&= \int_{\mathbb{R}^n} e^{\langle x,\xi \rangle} \int_{\mathbb{R}^n} \nu(dx-y)\,\sigma(dy) \\
&= \int_{\mathbb{R}^n} e^{\langle x,\xi \rangle}\,\nu(dx) \int_{\mathbb{R}^n} e^{\langle y,\xi \rangle}\,\sigma(dy) \\
&= \langle\!\langle \widetilde{\nu}, e^{\langle \cdot,\xi \rangle} \rangle\!\rangle \langle\!\langle \widetilde{\sigma}, e^{\langle \cdot,\xi \rangle} \rangle\!\rangle. \qquad (1.1)
\end{aligned}$$

The last equation gives us a clue as how to define the convolution of two Hida distributions in §3. In this paper we will give some relationships among convolution, the Wick product, the Lévy Laplacian, positivity and the Fourier transform. In particular, we will see that both convolution and the Wick product behave like multiplication with repect to the Fourier transform. However, convolution preserves positivity, while the Wick product does not.

---

* Research supported by NSF Grant DMS-9001859

§2. The space of Hida distributions

In this section we describe briefly the space of Hida distributions. For details, see e.g.[5, 6, 10, 11, 18, 23, 25]. Let $\mathcal{S}(\mathbb{R})$ denote the Schwartz space of rapidly decreasing real-valued functions on $\mathbb{R}$. Let μ be the standard Gaussian measure on the dual space $\mathcal{S}'(\mathbb{R})$ of $\mathcal{S}(\mathbb{R})$, i.e. its characteristic function is given by

$$\int_{\mathcal{S}'(\mathbb{R})} e^{i\langle x,\xi\rangle}\, d\mu(x) = e^{-\frac{1}{2}|\xi|^2}, \quad \xi \in \mathcal{S}(\mathbb{R}),$$

where $|\cdot|$ denotes the $L^2(\mathbb{R})$-norm. For simplicity, let $(L^2) \equiv L^2(\mathcal{S}'(\mathbb{R}), \mu)$. By the Wiener-Itô theorem the space (L^2) has the following decomposition:

$$(L^2) = \bigoplus_{n=0}^{\infty} K_n,$$

where K_n consists of all multiple Wiener integral $I_n(f)$, $f \in \widehat{L}^2(\mathbb{R}^n)$, the symmetric $L^2(\mathbb{R}^n)$ space. Moreover, if $\varphi \in (L^2)$ has the series expansion

$$\varphi = \sum_{n=0}^{\infty} I_n(f_n), \quad f_n \in \widehat{L}^2(\mathbb{R}^n), \tag{2.1}$$

then the (L^2)-norm of φ is given by

$$\|\varphi\| = \left(\sum_{n=0}^{\infty} n! |f_n|^2_{L^2(\mathbb{R}^n)} \right)^{\frac{1}{2}}.$$

Now, let $A = -d^2/du^2 + u^2 + 1$. It is a densely defined self-adjoint operator on $L^2(\mathbb{R})$. Moreover, there exists an orthonormal basis $\{e_n; n \geq 1\}$ for $L^2(\mathbb{R})$ such that $Ae_n = 2ne_n$. Let $\Gamma(A)$ denote the second quantization of A, i.e. for $\varphi = \sum_{n=0}^{\infty} I_n(f_n)$ in (L^2) with

$$\sum_{n=0}^{\infty} n! |A^{\otimes n} f_n|^2_{L^2(\mathbb{R}^n)} < \infty,$$

$\Gamma(A)\varphi$ is defined to be

$$\Gamma(A)\varphi = \sum_{n=0}^{\infty} I_n(A^{\otimes n} f_n).$$

It is easily seen that $\Gamma(A)$ is a densely defined self-adjoint operator on (L^2). For $p \geq 0$, let $(\mathcal{S})_p$ denote the domain of the operator $\Gamma(A)^p$. It is a Hilbert space with the norm $\|\varphi\|_p = \|\Gamma(A)^p \varphi\|$. On the other hand, if $p < 0$, then the norm $\|\cdot\|_p = \|\Gamma(A)^p \cdot\|$ is weaker than $\|\cdot\|$ and we let $(\mathcal{S})_p$ be the completion of (L^2) with respect to $\|\cdot\|_p$. It is easy to see that for $p \geq 0$ the space $(\mathcal{S})_{-p}$ is nothing but the dual space $(\mathcal{S})_p^*$. Let $(\mathcal{S})$ denote the projective limit of $(\mathcal{S})_p$, $p \geq 0$, and let $(\mathcal{S})^*$ be the dual space of $(\mathcal{S})$. Then we have the following sequence of continuous inclusions:

$$(\mathcal{S}) \subset (\mathcal{S})_p \subset (L^2) \subset (\mathcal{S})_p^* \subset (\mathcal{S})^*, \quad p \geq 0.$$

The elements in $(\mathcal{S})$ and $(\mathcal{S})^*$ are called *white noise test functions* and *Hida distributions*, respectively. We mention that each $\Phi \in (\mathcal{S})_p^*$ has the following series expansion

$$\Phi = \sum_{n=0}^{\infty} I_n(f_n), \quad f_n \in \mathcal{S}_p'(\mathbb{R})^{\widehat{\otimes} n},$$

where $I_n(f_n)$ is regarded as a generalized multiple Wiener integral, and

$$\|\Phi\|_{-p} = \left(\sum_{n=0}^{\infty} n! |(A^{-p})^{\otimes n} f_n|_{L^2(\mathbb{R}^n)}^2 \right)^{\frac{1}{2}}.$$

Moreover, we will use the following convention relating the pairing $\langle\langle \cdot, \cdot \rangle\rangle$ of $(\mathcal{S})^*$ and $(\mathcal{S})$ to the inner product of (L^2):

$$\langle\langle \varphi, \psi \rangle\rangle = (\varphi, \overline{\psi})_{(L^2)} = \int_{\mathcal{S}'(\mathbb{R})} \varphi(x)\psi(x)\, d\mu(x), \quad \varphi \in (L^2),\ \psi \in (\mathcal{S}).$$

Define a transformation S on (L^2) [11] by

$$S\varphi(\xi) = \int_{\mathcal{S}'(\mathbb{R})} \varphi(x + \xi)\, d\mu(x), \quad \xi \in \mathcal{S}(\mathbb{R}).$$

By using the translation formula for μ (see e.g. [15]), we can rewrite the above equation as follows:

$$S\varphi(\xi) = \langle\langle \varphi, e^{\langle \cdot, \xi \rangle} \rangle\rangle e^{-\frac{1}{2}|\xi|^2}, \quad \xi \in \mathcal{S}(\mathbb{R}).$$

It is a fact that $e^{\langle \cdot, \xi \rangle}$ is in $(\mathcal{S})$ for any $\xi \in \mathcal{S}(\mathbb{R})$. Therefore the S-transformation can be extended to the space $(\mathcal{S})^*$ of Hida distributions. We will use the same notation to denote the extension, i.e. the S-transformation of $\Phi \in (\mathcal{S})^*$ is given by

$$S\Phi(\xi) = \langle\langle \Phi, e^{\langle \cdot, \xi \rangle} \rangle\rangle e^{-\frac{1}{2}|\xi|^2}, \quad \xi \in \mathcal{S}(\mathbb{R}). \tag{2.2}$$

For $f \in L^2(\mathbb{R})$, we define the renormalization $: e^{\langle \cdot, f \rangle} :$ of $e^{\langle \cdot, f \rangle}$ by

$$: e^{\langle \cdot, f \rangle} : = e^{\langle \cdot, f \rangle - \frac{1}{2}|f|^2}.$$

Then $S\Phi$ can also be written as

$$S\Phi(\xi) = \langle\langle \Phi, : e^{\langle \cdot, \xi \rangle} : \rangle\rangle, \quad \xi \in \mathcal{S}(\mathbb{R}).$$

The infinite dimensional analogue of the finite dimensional Gaussian density function is the following formal expression

$$\exp\left(a\langle x, y \rangle - \frac{1}{2c}|x|^2 \right), \quad x \in \mathcal{S}'(\mathbb{R}),$$

where $a, c \in \mathbb{R}, c > 0$ or $c < -1$ and $y \in \mathcal{S}'(\mathbb{R})$. Note that $|\cdot|$ is the $L^2(\mathbb{R})$-norm so that the above expression is purely formal. Let P_n be a β_n-dimensional orthogonal projection of $L^2(\mathbb{R})$ with range in $\mathcal{S}(\mathbb{R})$ and $P_n \to I$ strongly. Then the function

$$\varphi_n(x) \equiv \exp\left(a\langle P_n x, y \rangle - \frac{1}{2c}|P_n x|^2 \right), \quad x \in \mathcal{S}'(\mathbb{R}),$$

is in (L^2). It is easy to check that its expectation is given by

$$E\varphi_n = \left(\frac{1+c}{c}\right)^{-\beta_n/2} \exp\left(\frac{a^2 c}{2(1+c)}|P_n y|^2\right).$$

Define $\psi_n \equiv \varphi_n / E\varphi_n$. Then $\psi \in (L^2)$ and we can check that

$$S\psi_n = \exp\left(\frac{ac}{1+c}\langle y, P_n\xi\rangle - \frac{1}{2(1+c)}|P_n\xi|^2\right), \quad \xi \in \mathcal{S}(\mathbb{R}).$$

Obviously, for each $\xi \in \mathcal{S}(\mathbb{R})$,

$$\lim_{n\to\infty} S\psi_n(\xi) = \exp\left(\frac{ac}{1+c}\langle y, \xi\rangle - \frac{1}{2(1+c)}|\xi|^2\right), \quad \xi \in \mathcal{S}(\mathbb{R}).$$

Moreover, we can check easily that there exist constants $c > 0$ and $p \geq 0$ such that for all $n \geq 1$ and $\xi \in \mathcal{S}(\mathbb{R})$

$$|S\psi_n(\xi)| \leq \exp(c|A^p\xi|^2).$$

Therefore, by Theorem 2.9 in [23], the sequence $\psi_n, n \geq 1$, converges in $(\mathcal{S})^*$. We will use $G_{ay,c}$ to denote the limit. Thus $G_{ay,c}$ is a Hida distribution with the following S-transformation

$$SG_{ay,c}(\xi) = \exp\left(\frac{ac}{1+c}\langle y, \xi\rangle - \frac{1}{2(1+c)}|\xi|^2\right), \quad \xi \in \mathcal{S}(\mathbb{R}). \tag{2.3}$$

On the other hand, for any $a, c \in \mathbb{C}, c \neq -1$ and $y \in \mathcal{S}'(\mathbb{R})$, we see from Theorem 1.1 in [23] that the right hand side of Eq.(2.3) is the S-transformation of some Hida distribution, which is denoted naturally by $G_{ay,c}$. We will call $G_{ay,c}$ a *Gaussian white noise function.* Note that we allow a to be a complex number and $\mathcal{S}'(\mathbb{R})$ is a real vector space. We will use g_c to denote $G_{0,c}$ so that

$$Sg_c(\xi) = \exp\left(-\frac{1}{2(1+c)}|\xi|^2\right), \quad \xi \in \mathcal{S}(\mathbb{R}). \tag{2.4}$$

It is easy to see that $G_{\frac{1}{c}y,c}$ converges in $(\mathcal{S})^*$ as $c \to 0$. We will use $\widetilde{\delta_y}$ to denote the limit. Then from Eq.(2.3) we have

$$S\widetilde{\delta_y}(\xi) = \exp\left(\langle y, \xi\rangle - \frac{1}{2}|\xi|^2\right), \quad \xi \in \mathcal{S}(\mathbb{R}). \tag{2.5}$$

This is a similar property as in the finite dimensional case. We remark that $\widetilde{\delta_y}$ is the delta function introduced in [14], i.e.

$$\langle\langle \widetilde{\delta_y}, \varphi\rangle\rangle = \widetilde{\varphi}(y), \quad \varphi \in (\mathcal{S}), \tag{2.6}$$

where $\widetilde{\varphi}$ is the continuous version of φ.

Note that from Eqs.(2.3) and (2.5), we have

$$SG_{ay,0}(\xi) = S\widetilde{\delta_0}(\xi) = \exp\left(-\frac{1}{2}|\xi|^2\right), \quad \xi \in \mathcal{S}(\mathbb{R}).$$

Hence $G_{ay,0} = \widetilde{\delta_0}$ for any $a \in \mathbb{C}$ and $y \in \mathcal{S}'(\mathbb{R})$. Note also that by Eq.(2.4), $g_c \to 1$ in $(\mathcal{S})^*$ as $c \to \infty$. Thus we will make the convention $g_\infty = 1$.

§3. Convolution and Wick product

The Wick product was first introduced in [4]. It can be justified by using the characterization theorem in [23]. Let Φ and Ψ be two Hida distributions. It follows from Theorem 1.1 in [23] that the product $(S\Phi)(S\Psi)$ is the S-transformation of a Hida distribution, which is denoted by $\Phi \diamond \Psi$. We call $\Phi \diamond \Psi$ the *Wick product* of Φ and Ψ. It has S-transformation given by

$$S(\Phi \diamond \Psi)(\xi) = S\Phi(\xi)\, S\Psi(\xi), \quad \xi \in \mathcal{S}(\mathbb{R}).$$

The Wick product has also been studied in [20].

Definition 3.1. A signed measure ν on $\mathcal{S}'(\mathbb{R})$ is called a *Hida measure* if the following mapping is continuous

$$\varphi \mapsto \int_{\mathcal{S}'(\mathbb{R})} \widetilde{\varphi}(x)\, d\nu(x), \ \varphi \in (\mathcal{S}),$$

where $\widetilde{\varphi}$ denotes the continuous version of φ.

If ν is a Hida measure, then there exists a unigue Hida distribution, denoted by $\widetilde{\nu}$, such that

$$\langle\!\langle \widetilde{\nu}, \varphi \rangle\!\rangle = \int_{\mathcal{S}'(\mathbb{R})} \widetilde{\varphi}(x)\, d\nu(x), \ \varphi \in (\mathcal{S}).$$

$\widetilde{\nu}$ is also denoted by $d\nu/d\mu$ and called the generalized Radon-Nikodym derivative of ν with respect to the Gaussian measure μ, see [24]. For two signed measures ν and σ on $\mathcal{S}'(\mathbb{R})$ we define their convolution $\nu * \sigma$ by

$$\nu * \sigma(A) = \int_{\mathcal{S}'(\mathbb{R})} \nu(A - x)\, d\sigma(x), \quad A \in \mathcal{B}(\mathcal{S}'(\mathbb{R})).$$

It has been proved in [18] that if ν and σ are two Hida measures, then the convolution $\nu * \sigma$ is also a Hida distribution. Moreover, by using the white noise analogue of Eq.(1.1), we get

$$\langle\!\langle (\nu * \sigma)^{\sim}, e^{\langle \cdot, \xi \rangle} \rangle\!\rangle = \langle\!\langle \widetilde{\nu}, e^{\langle \cdot, \xi \rangle} \rangle\!\rangle \langle\!\langle \widetilde{\sigma}, e^{\langle \cdot, \xi \rangle} \rangle\!\rangle, \quad \xi \in \mathcal{S}(\mathbb{R}).$$

Therefore, from Eq.(2.2), we get

$$S(\nu * \sigma)^{\sim}(\xi) = S\widetilde{\nu}(\xi)\, S\widetilde{\sigma}(\xi)\, e^{\frac{1}{2}|\xi|^2}, \quad \xi \in \mathcal{S}(\mathbb{R}).$$

But from Eq.(2.4), we have $Sg_{-2}(\xi) = e^{\frac{1}{2}|\xi|^2}$. Hence the above equation implies that

$$(\nu * \sigma)^{\sim} = \widetilde{\nu} \diamond \widetilde{\sigma} \diamond g_{-2}. \tag{3.1}$$

Eq.(3.1) suggests us that the convolution of two Hida distributions can be defined as follows.

Definition 3.2. The *convolution* of two Hida distributions Φ and Ψ is the Hida distribution $\Phi * \Psi$ defined by

$$\Phi * \Psi = \Phi \diamond \Psi \diamond g_{-2}.$$

It is obvious from the derivation of Eq.(1.1) that the above convolution reduces to the ordinary convolution of two tempered distributions in the finite dimensional case. Moreover, this convolution can be regarded as multiplication in white noise analysis. It differs from the Wick product by the Gaussian white noise function g_{-2}. The following proposition shows that we can express the Wick product in terms of convolution.

Proposition 3.3. *For any two Hida distributions Φ and Ψ, we have*

$$\Phi \diamond \Psi = \Phi * \Psi * g_{-\frac{1}{2}}.$$

Proof. It is easy to check that for any $a, b \in \mathbb{C} \cup \{\infty\}, a \neq -1, b \neq -1$, we have

$$g_a \diamond g_b = g_{\frac{ab-1}{a+b+2}}, \tag{3.2}$$

where by convention $g_0 = \widetilde{\delta}_0$ and $g_\infty = 1$. It follows that

$$g_{-2} \diamond g_{-\frac{1}{2}} \diamond g_{-2} = g_0 \diamond g_{-2} = g_\infty = 1.$$

Therefore,

$$\Phi * \Psi * g_{-\frac{1}{2}} = \Phi \diamond \Psi \diamond g_{-2} \diamond g_{-\frac{1}{2}} \diamond g_{-2} = \Phi \diamond \Psi \diamond 1 = \Phi \diamond \Psi. \quad \blacksquare$$

In the next theorem we will show how the Lévy Laplacian is related to the Wick product and convolution. Let T be a finite interval of $\mathbb{R}$. Let $\mathcal{D}(\Delta_L)$ consist of all $\Phi \in (S)^*$ satisfying the following conditions:

(1) Φ is of the form

$$\Phi = \sum_{n=0}^{\infty} I_n(f_n),$$

where $\operatorname{supp}(f_n) \subset T^n$ for all n;

(2) $F = S\Phi$ has the first functional derivative $F'(\xi;\cdot) \in L^1(T)$ for all $\xi \in \mathcal{S}(\mathbb{R})$ and the second functional derivative of the form

$$F''(\xi; s,t) = F''_L(\xi;t)\delta(s-t) + f(\xi;s,t), \quad s,t \in T,$$

where $F''_L(\xi;\cdot) \in L^1(T)$ and $f(\xi;\cdot,\cdot) \in L^1(T^2)$ for all $\xi \in \mathcal{S}(\mathbb{R})$, and the function $\int_T F''_L(\cdot;t)\,dt$ is the S-transformation of some Hida distribution.

For $\Phi \in \mathcal{D}(\Delta_L)$, we define the *Lévy Laplacian* $\Delta_L \Phi$ of Φ by

$$\Delta_L \Phi = S^{-1}\left(\frac{1}{|T|}\int_T F''_L(\cdot;t)\,dt\right).$$

It is known that $\Delta_L|_{(L^2)} = 0$, see e.g. [13, 18]. For more information about the Lévy Laplacian, see [2, 3, 5, 7, 18, 21, 22]. The next theorem shows that Δ_L and $\Delta_L + I$ behave like derivations with respect to the Wick product and convolution, respectively.

Theorem 3.4. *If Φ and Ψ are in Δ_L, then $\Phi \diamond \Psi$ and $\Phi * \Psi$ are also in Δ_L. Moreover, we have*

$$\Delta_L(\Phi \diamond \Psi) = (\Delta_L \Phi) \diamond \Psi + \Phi \diamond (\Delta_L \Psi),$$

$$(\Delta_L + 1)(\Phi * \Psi) = ((\Delta_L + 1)\Phi) * \Psi + \Phi * ((\Delta_L + 1)\Psi).$$

Proof. Let $F = S\Phi$ and $G = S\Psi$. By the assumption, we have

$$F''(\xi;s,t) = F''_L(\xi;t)\delta(s-t) + f(\xi;s,t), \quad s,t \in T,$$

$$G''(\xi;s,t) = G''_L(\xi;t)\delta(s-t) + g(\xi;s,t), \quad s,t \in T,$$

where $F''_L(\xi;\cdot), G''_L(\xi;\cdot) \in L^1(T)$ and $f(\xi;\cdot,\cdot), g(\xi;\cdot,\cdot) \in L^1(T^2)$ for all $\xi \in \mathcal{S}(\mathbb{R})$.

Now, let $H = S(\Phi \diamond \Psi)$. Then $H = FG$ and we see immediately that condition (1) of $\mathcal{D}(\Delta_L)$ for $\Phi \diamond \Psi$ is satisfied. Moreover, the second functional derivative of H is easily seen to be

$$\begin{aligned} H''(\xi;s,t) &= F''(\xi;s,t)G(\xi) + F(\xi)G''(\xi;s,t) + F'(\xi;s)G'(\xi;t) + F'(\xi;t)G'(\xi;s) \\ &= H''_L(\xi;t)\delta(s-t) + h(\xi;s,t), \end{aligned}$$

where $H''_L(\xi;t)$ and $h(\xi;s,t)$ are given by

$$H''_L(\xi;t) = F''_L(\xi;t)G(\xi) + F(\xi)G''_L(\xi;t), \tag{3.3}$$

$$h(\xi;s,t) = f(\xi;s,t)G(\xi) + F(\xi)g(\xi;s,t) + F'(\xi;s)G'(\xi;t) + F'(\xi;t)G'(\xi;s).$$

Hence, condition (2) of $\mathcal{D}(\Delta_L)$ for $\Phi \diamond \Psi$ is also satisfied. Therefore, $\Phi \diamond \Psi$ is in $\mathcal{D}(\Delta_L)$ and by Eq.(3.3) we have

$$\Delta_L(\Phi \diamond \Psi) = (\Delta_L \Phi) \diamond \Psi + \Phi \diamond (\Delta_L \Psi).$$

To show the other equality in the theorem, consider the Gaussian white noise function g_c, $c \in \mathbb{C}, c \neq -1$. Let $J = Sg_c$. Then from Eq.(2.4) we get easily that

$$J''(\xi;s,t) = -\frac{1}{1+c}J(\xi)\delta(s-t) + \left(\frac{1}{1+c}\right)^2 \xi(s)\xi(t)J(\xi), \quad s,t \in T.$$

Hence $g_c \in \mathcal{D}(\Delta_L)$ and $\Delta_L g_c = -(1+c)^{-1}g_c$. In particular, we have $\Delta_L g_{-2} = g_{-2}$. Therefore,

$$\begin{aligned} \Delta_L(\Phi * \Psi) &= \Delta_L(\Phi \diamond \Psi \diamond g_{-2}) \\ &= (\Delta_L \Phi) \diamond \Psi \diamond g_{-2} + \Phi \diamond (\Delta_L \Psi) \diamond g_{-2} + \Phi \diamond \Psi \diamond (\Delta_L g_{-2}) \\ &= (\Delta_L \Phi) * \Psi + \Phi * (\Delta_L \Psi) + \Phi * \Psi. \end{aligned}$$

This yields the second equality of the theorem immediately. ■

§4. Positivity

A Hida distribution Φ is called *positive* if $\langle\langle \Phi, \varphi \rangle\rangle \geq 0$ for all $\varphi \in (\mathcal{S}), \varphi \geq 0 \;\; \mu - a.e.$ The next theorem characterizes those positive Gaussian white noise functions $G_{ay,c}$ for $a, c \in \mathbb{C}, c \neq -1, y \in \mathcal{S}'(\mathbb{R})$. First recall that from the end of §2 we have $G_{ay,0} = \widetilde{\delta_0}$. Hence $G_{ay,0}$ is positive. On the other hand, suppose $c \neq 0$ and $G_{ay,c}$ is positive, then a and c must be real numbers. This can be seen as follows. From Eq.(2.3), we have

$$\begin{aligned} SG_{ay,c}(\xi) &= 1 + \frac{ac}{1+c}\langle y, \xi \rangle \\ &+ \frac{1}{2}\left(\frac{a^2c^2}{(1+c)^2}\langle y^{\otimes 2}, \xi^{\otimes 2}\rangle - \frac{1}{1+c}\langle \delta(\cdot - \cdot), \xi^{\otimes 2}\rangle\right) + \cdots. \end{aligned}$$

Therefore,

$$G_{ay,c} = 1 + \frac{ac}{1+c} I_1(y) + \frac{1}{2} I_2 \left(\frac{a^2c^2}{(1+c)^2} y^{\otimes 2} - \frac{1}{1+c} \delta(\cdot - \cdot) \right) + \cdots . \tag{4.1}$$

Here $I_n(f)$ is the generalized multiple Wiener integral of $f \in \mathcal{S}'(\mathbb{R})^{\widehat{\otimes} n}$. Now, consider the following nonnegative test function

$$\varphi = (\langle \cdot, \eta \rangle + r)^2 = 1 + r^2 + 2rI_1(\eta) + I_2(\eta^{\otimes 2}),$$

where $\eta \in \mathcal{S}(\mathbb{R}), |\eta| = 1$ and $r \in \mathbb{R}$. By direct calculation, we get

$$\langle\langle G_{ay,c}, \varphi \rangle\rangle = r^2 + 2\frac{ac}{1+c}\langle y, \eta \rangle r + \frac{c}{1+c} + \frac{a^2c^2}{(1+c)^2}\langle y, \eta \rangle^2 .$$

Since $\langle\langle G_{ay,c}, \varphi \rangle\rangle \geq 0$ for all $\eta \in \mathcal{S}(\mathbb{R}), |\eta| = 1$ and $r \in \mathbb{R}$, we see easily that a and c must be real numbers. Thus we will write $G_{ay,c}$ simply as $G_{x,c}$ for $x \in \mathcal{S}'(\mathbb{R}), c \in \mathbb{R}$ and $c \neq -1$. Note that

$$SG_{x,c}(\xi) = \exp\left(\frac{c}{1+c}\langle x, \xi \rangle - \frac{1}{2(1+c)}|\xi|^2 \right), \quad \xi \in \mathcal{S}(\mathbb{R}). \tag{4.2}$$

Theorem 4.1. *The Hida distribution $G_{x,c}$ is positive if and only if either $c < -1$ or $0 \leq c \leq \infty$.*

Proof. First we prove the sufficiency. When $c = 0$, $G_{x,0} = \widetilde{\delta_0}$ as seen before and so $G_{x,0}$ is positive. For the other cases, consider the measure $p_t(x, \cdot)$ introduced in [1], i.e. $p_t(x, \cdot) = \mu((\cdot - x)/\sqrt{t})$, $x \in \mathcal{S}'(\mathbb{R}), t > 0$. It has been shown in [24] that $p_t(x, \cdot)$ is a Hida measure and

$$\begin{aligned} S(p_t(x,\cdot)^{\sim})(\xi) &= \langle\langle p_t(x,\cdot)^{\sim}, e^{\langle \cdot, \xi \rangle} \rangle\rangle e^{-\frac{1}{2}|\xi|^2} \\ &= e^{-\frac{1}{2}|\xi|^2} \int_{\mathcal{S}'(\mathbb{R})} e^{\langle y, \xi \rangle}\, p_t(x, dy) \\ &= e^{-\frac{1}{2}|\xi|^2} \int_{\mathcal{S}'(\mathbb{R})} e^{\langle x + \sqrt{t}u, \xi \rangle}\, d\mu(u) \\ &= \exp\left(\langle x, \xi \rangle - \frac{1}{2}(1-t)|\xi|^2 \right), \quad \xi \in \mathcal{S}(\mathbb{R}). \end{aligned} \tag{4.3}$$

Suppose $c < -1$ or $0 < c \leq \infty$. Then $\frac{c}{1+c} > 0$ and by comparing Eqs.(4.2) and (4.3), we see easily that

$$G_{x,c} = p_{\frac{c}{1+c}}\left(\frac{c}{1+c}x, \cdot \right)^{\sim}.$$

Here we have used the convention that $\frac{c}{1+c} = 1$ when $c = \infty$. Therefore, $G_{x,c}$ is positive and the sufficiency follows.

Now, we prove the necessity. Suppose $-1 < c < 0$. From Eq.(4.1), we have

$$G_{x,c} = 1 + \frac{c}{1+c} I_1(x) + \frac{1}{2} I_2 \left(\frac{c^2}{(1+c)^2} x^{\otimes 2} - \frac{1}{1+c} \delta(\cdot - \cdot) \right) + \cdots .$$

Consider the the following nonnegative test function as given before

$$\varphi = (\langle \cdot, \eta \rangle + r)^2 = 1 + r^2 + 2rI_1(\eta) + I_2(\eta^{\otimes 2}),$$

where $\eta \in \mathcal{S}(\mathbb{R}), |\eta| = 1$ and $r \in \mathbb{R}$. Then we get

$$\langle\langle G_{x,c}, \varphi\rangle\rangle = \frac{c}{1+c} + \left(r + \frac{c}{1+c}\langle x, \eta\rangle\right)^2.$$

We can choose any such η and then choose $r = -\frac{c}{1+c}\langle x, \eta\rangle$. Thus for such a nonnegative test function φ, we have $\langle\langle G_{x,c}, \varphi\rangle\rangle = \frac{c}{1+c}$. But $\frac{c}{1+c} < 0$ since $-1 < c < 0$. Therefore, $G_{x,c}$ can not be positive and this proves the necessity. ■

We now give a simple example to show that the Wick product does not preserve the positivity. By the above theorem the Hida distribution $g_{\frac{1}{2}}$ is positive. But, from Eq.(3.2), $g_{\frac{1}{2}} \diamond g_{\frac{1}{2}} = g_{-\frac{1}{4}}$ which is not positive. On the other hand, convolution does preserve the positivity.

Theorem 4.2. *If Φ and Ψ are positive Hida distributions, then the convolution $\Phi * \Psi$ is also positive.*

Proof. It has been shown in [9, 26] that a Hida distribution Φ is positive if and only if $\Phi = \widetilde{\nu}$ for some finite Borel measure ν on $\mathcal{S}'(\mathbb{R})$. Therefore, it follows from the assumption that there exist two finite Borel measures ν and σ on $\mathcal{S}'(\mathbb{R})$ such that $\Phi = \widetilde{\nu}$ and $\Psi = \widetilde{\sigma}$. But by Eq.(3.1) and Definition 3.2 we have $(\nu * \sigma)^{\sim} = \widetilde{\nu} * \widetilde{\sigma}$. Hence $\Phi * \Psi = (\nu * \sigma)^{\sim}$ and so $\Phi * \Psi$ is positive. ■

§5. Fourier transform

In this section we will show that the Fourier transform is related to convolution in a similar way as in the finite dimensional case. First we give a brief motivation for the definition of Fourier transform of Hida distributions. For details, see [5, 16, 17, 18]. Consider the finite dimensional Fourier transform

$$\widehat{f}(y) = (2\pi)^{-k/2} \int_{\mathbb{R}^k} e^{-i\langle x, y\rangle} f(x)\, dx. \tag{5.1}$$

Obviously, there are several difficulties to generalize this formulation to infinite dimensional spaces. For instance, the Lebesgue measure does not exist. However, we can rewrite Eq.(5.1) in another expression which can be generalized to the white noise space $(\mathcal{S}'(\mathbb{R}), \mu)$. Define the renormalization of $e^{-i\langle x,y\rangle}$ with respect to the y variable by

$$: e^{-i\langle x,y\rangle} :_y = e^{-i\langle x,y\rangle + \frac{1}{2}|x|^2}.$$

Then $\widehat{f}$ can be rewritten as

$$\widehat{f}(y) = \int_{\mathbb{R}^k} : e^{-i\langle x,y\rangle} :_y f(x)\, d\mu_k(x),$$

where μ_k is the standard Gaussian measure on $\mathbb{R}^k$. Thus it appears that we can define the Fourier transform $\widehat{\Phi}$ of a Hida distribution Φ by

$$\widehat{\Phi}(y) = \int_{\mathcal{S}'(\mathbb{R})} : e^{-i\langle x,y\rangle} :_y \Phi(x)\, d\mu(x). \tag{5.2}$$

But this is just an informal expression. To overcome this difficulty, consider the formal expression for fixed $x \in \mathcal{S}'(\mathbb{R})$,

$$\varphi(y) = e^{-i\langle x,y\rangle}, \quad y \in \mathcal{S}'(\mathbb{R}).$$

Let P_n be the orthogonal projection in §2 and define

$$\varphi_n(y) = e^{-i\langle x, P_n y\rangle}, \quad y \in \mathcal{S}'(\mathbb{R}).$$

Then $\varphi_n \in (L^2)$ and we can check that

$$\begin{aligned} S(\varphi_n/E\varphi_n)(\xi) &= e^{-i\langle x, P_n \xi\rangle} \\ &\to e^{-i\langle x, \xi\rangle}, \quad \xi \in \mathcal{S}(\mathbb{R}). \end{aligned}$$

Therefore, $\varphi_n/E\varphi_n$ converges in $(\mathcal{S})^*$. The limit is denoted by $: e^{-i\langle x,y\rangle} :_y$. It is a renormalization with respect to the y variable. Thus we have

$$S(: e^{-i\langle x,y\rangle} :_y)(\xi) = e^{-i\langle x,\xi\rangle}, \quad \xi \in \mathcal{S}(\mathbb{R}).$$

Now, if we take the S-transformation in Eq.(5.2) informally, then we obtain

$$\begin{aligned} S\widehat{\Phi}(\xi) &= \int_{\mathcal{S}'(\mathbb{R})} e^{-i\langle x,\xi\rangle}\Phi(x)\, d\mu(x) \\ &= \langle\langle \Phi, e^{-i\langle\cdot,\xi\rangle}\rangle\rangle, \quad \xi \in \mathcal{S}(\mathbb{R}). \end{aligned}$$

On the other hand, it has been shown in [23] that for any $\Phi \in (\mathcal{S})^*$, the function $S\Phi(r\xi), r \in \mathbb{R}$, extends to an entire function $S\Phi(z\xi), z \in \mathbb{C}$. Then from Eq.(2.2) we have

$$\langle\langle \Phi, e^{-i\langle\cdot,\xi\rangle}\rangle\rangle = S\Phi(-i\xi)e^{-\frac{1}{2}|\xi|^2}.$$

Therefore,

$$S\widehat{\Phi}(\xi) = S\Phi(-i\xi)e^{-\frac{1}{2}|\xi|^2}, \quad \xi \in \mathcal{S}(\mathbb{R}).$$

Finally, by Theorem 1.1 in [23], $S\Phi(-i\xi)e^{-\frac{1}{2}|\xi|^2}$ is the S-transformation of some Hida distribution. Thus we can define the Fourier transform as follows.

Definition 5.1. Let $\Phi \in (\mathcal{S})^*$. The *Fourier transform* of Φ is defined to be the Hida distribution, denoted by $\widehat{\Phi}$, such that

$$S\widehat{\Phi}(\xi) = \langle\langle \Phi, e^{-i\langle\cdot,\xi\rangle}\rangle\rangle, \quad \xi \in \mathcal{S}(\mathbb{R}),$$

or equivalently,

$$S\widehat{\Phi}(\xi) = S\Phi(-i\xi)e^{-\frac{1}{2}|\xi|^2}, \quad \xi \in \mathcal{S}(\mathbb{R}).$$

Theorem 5.2. *For any Φ and Ψ in $(\mathcal{S})^*$, we have*

$$(\Phi \diamond \Psi)^\frown = \widehat{\Phi} * \widehat{\Psi},$$

$$(\Phi * \Psi)^\frown = \widehat{\Phi} \diamond \widehat{\Psi}.$$

Proof. Direct calculation shows that for any $\xi \in \mathcal{S}(\mathbb{R})$,

$$\begin{aligned}
S(\Phi \diamond \Psi)^{\frown}(\xi) &= S(\Phi \diamond \Psi)(-i\xi)e^{-\frac{1}{2}|\xi|^2} \\
&= S\Phi(-i\xi)S\Psi(-i\xi)e^{-\frac{1}{2}|\xi|^2} \\
&= S\widehat{\Phi}(\xi)e^{\frac{1}{2}|\xi|^2}S\widehat{\Psi}(\xi)e^{\frac{1}{2}|\xi|^2}e^{-\frac{1}{2}|\xi|^2} \\
&= S\widehat{\Phi}(\xi)S\widehat{\Psi}(\xi)e^{\frac{1}{2}|\xi|^2} \\
&= S(\widehat{\Phi} * \widehat{\Psi})(\xi).
\end{aligned}$$

Hence $(\Phi \diamond \Psi)^{\frown} = \widehat{\Phi} * \widehat{\Psi}$. The other equality can be proved by similar calculation as above. Alternatively, observe that $(g_{-2})^{\frown} = g_{-\frac{1}{2}}$ and so

$$\begin{aligned}
(\Phi * \Psi)^{\frown} &= (\Phi \diamond \Psi \diamond g_{-2})^{\frown} \\
&= \widehat{\Phi} * \widehat{\Psi} * (g_{-2})^{\frown} \\
&= \widehat{\Phi} * \widehat{\Psi} * g_{-\frac{1}{2}} \\
&= \widehat{\Phi} \diamond \widehat{\Psi}. \quad \blacksquare
\end{aligned}$$

Note that the above theorem indicates that the convolution of Hida distributions corresponds to the Wick product under the Fourier transform.

At the end of this paper we give an example of stochastic differential equation which can be solved by using the Fourier transform. Let ∂_{t+} and ∂_{t-} denote the forward and backward white noise differentiations, respectively, i.e.

$$S(\partial_{t\pm}\Phi)(\xi) = (S\Phi)'(\xi; t\pm), \quad \xi \in \mathcal{S}(\mathbb{R}),$$

where the prime denotes the functional derivative. Define $\partial_t = \frac{1}{2}(\partial_{t+} + \partial_{t-})$.

Example 5.3. Solve the stochastic differential equation

$$\frac{d}{dt}X(t,x) = \partial_t X(t,x), \quad X(0,x) = c, \tag{5.3}$$

where $t \in \mathbb{R}, x \in \mathcal{S}'(\mathbb{R})$ and c is a constant.

Recall that the Fourier transform takes ∂_t to multiplication by $ix(t)$ [16, 17, 18]. Thus upon taking the Fourier transform of Eq.(5.3) we get the following equation for $u(t,x) = X(t,\cdot)^{\frown}(x)$:

$$\frac{d}{dt}u(t,x) = ix(t)u(t,x), \quad u(0,x) = c.$$

The solution is easily seen to be

$$u(t,x) = c\exp\left(i\int_0^t x(s)\,ds\right) = c\exp(i\langle x, 1_{[0,t]}\rangle).$$

But for any $z \in \mathbb{C}$ and $f \in L^2(\mathbb{R})$, we have

$$:e^{z\langle x,f\rangle}: = e^{z\langle x,f\rangle - \frac{1}{2}z^2|f|^2}.$$

Therefore,

$$u(t,x) = ce^{-\frac{1}{2}t} : \exp(i\langle x, 1_{[0,t]}\rangle) : .$$

Finally, note that $(\widetilde{\delta_y})^\frown = :e^{-i\langle\cdot,y\rangle}:$. Hence the solution $X(t,x)$ of Eq.(5.3) is given by

$$X(t,x) = ce^{-\frac{1}{2}t}(\delta_{-1_{[0,t]}})^\frown.$$

References

[1] Gross, L.: Potential theory on Hilbert space; *J. Func. Anal.* **1** (1967) 123–181

[2] Hida, T.: *Analysis of Brownian Functionals.* Carleton Mathematical Lecture Notes, no.**13** (1975)

[3] Hida, T.: Brownian motion and its functionals; *Ricerche di Matematica* **34** (1985) 183–222

[4] Hida, T. and Ikeda, N.: Analysis on Hilbert space with reproducing kernel arising from multiple Wiener integral; *Proc. Fifth Berkeley Symp. Math Stat. Probab.* **2** (1965) 117–143.

[5] Hida, T., Kuo, H.-H., Potthoff, J. and Streit, L.: *White Noise: An Infinite Dimensional Calculus.* Monograph in preparation

[6] Hida, T. and Potthoff, J.: White noise analysis – an overview; in: *White Noise Analysis – Mathematics and Applications*, T. Hida, H.-H. Kuo, J. Potthoff and L. Streit (eds.), (1990) 140-165, World Scientific

[7] Hida, T. and Saitô, K.: White noise analysis and the Lévy Laplacian; in:*Stochastic Processes in Physics and Engineering*, S. Albeverio etc.(eds.), (1988) 177–184

[8] Ito, Y., Kubo, I. and Takenaka, S.: Calculus on Gaussian white noise and Kuo's Fourier transformation; in: *White Noise Analysis-Mathematics and Applications*, T. Hida, H.-H. Kuo, J. Potthoff and L. Streit (eds.), (1990) 180-207, World Scientific

[9] Kondrat'ev, Yu.G.: Nuclear spaces of entire functions in problems of infinite–dimensional analysis; *Soviet Math. Dokl.* **22** (1980) 588–592

[10] Kubo, I. and Takenaka, S.: Calculus on Gaussian white noise I; *Proc. Japan Acad.* **56A** (1980) 376–380

[11] Kubo, I. and Takenaka, S.: Calculus on Gaussian white noise II; *Proc. Japan Acad.* **56A** (1980) 411–416

[12] Kubo, I. and Takenaka, S.: Calculus on Gaussian white noise III; *Proc. Japan Acad.* **57A** (1981) 433–437

[13] Kubo, I. and Takenaka, S.: Calculus on Gaussian white noise IV; *Proc. Japan Acad.* **58A** (1982) 186–189

[14] Kubo, I. and Yokoi, Y.: A remark on the space of testing random variables in the white noise calculus; *Nagoya Math. J.* **115** (1989) 139–149

[15] Kuo, H.-H.: *Gaussian Measures in Banach Spaces.* Lecture Notes in Math., Vol. 463. Berlin, Heidelberg, New York: Springer-Verlag (1975)

[16] Kuo, H.-H.: On Fourier transform of generalized Brownian functionals; *J. Multivariate Anal.* **12** (1982) 415–431

[17] Kuo, H.-H.: The Fourier transform in white noise calculus; *J. Multivariate Analysis* **31** (1989) 311-327

[18] Kuo, H.-H.: Lectures on white noise analysis;*Preprint* (1990)

[19] Lee, Y.-J.: Analytic version of test functionals, Fourier transform and a characterization of measures in white noise calculus; *Preprint* (1989), to appear in *J. Funct. Anal.*

[20] Meyer, P.A. and Yan, J.A.: A propos des distributions sur l'espace de Wiener; *Séminaire de Probabilités* **XXI**; J. Azéma and M. Yor (ed.s). Berlin, Heidelberg, New York: Springer (1987)

[21] Obata, N.: Analysis of the Lévy Laplacian; *Soochow J. Math.* **14** (1988) 105–109

[22] Obata, N.: A characterization of the Lévy Laplacian in terms of infinite dimensional rotation groups; *Nagoya Math. J.* **118** (1990) 111–132

[23] Potthoff, J and Streit, L.: A characterization of Hida distributions; *Preprint* (1989), to appear in *J. Funct. Anal.*

[24] Potthoff, J. and Streit, L.: Generalized Radon-Nikodym derivatives and Cameron-Martin theory; *Preprint* (1990), to appear in *Proc. Int. Conf. Gaussian Random Fields*, Nagoya, 1990

[25] Potthoff, J. and Yan J.A.: Some results about test and generalized functionals of white noise; to appear in: *Proc. Singapore Probab. Conf.* (1989), L.Y. Chen (ed.)

[26] Yokoi, Y.: Positive generalized white noise functionals; *Hiroshima Math. J.* **20** (1990) 137–157

Splitting–up Approximation for SPDE's and SDE's with Application to Nonlinear Filtering

François LeGland*
INRIA Sophia–Antipolis
2004 Route des Lucioles
F–06565 VALBONNE Cédex

Abstract

A splitting–up approximation is introduced for diffusion processes, based on the successive composition of two stochastic flows of diffeomorphisms over a given partition of the time interval.

In the case where the original diffusion is observed in correlated noise, an equation is derived for the conditional density of the approximating process. This equation is interpreted as a splitting–up approximation of the Zakai equation for the conditional density of the original diffusion process, based on the successive composition of two semigroups, and error estimates are provided using SPDE techniques.

The results presented here are of general interest for the approximation of SDE's and SPDE's, independently of the filtering problem.

In the context of nonlinear filtering, the main interest of splitting–up approximation is that the original Zakai equation is splitted into a second–order deterministic PDE related with the *prediction* step, and a degenerate second–order stochastic PDE related with the *correction* step. This probabilistic interpretation can be used to design further approximation schemes, e.g. in terms of approximating finite–state Markov processes.

1 Nonlinear Filtering Model and SPDE's

Consider the following stochastic differential system, defined on the probability space $(\Omega, \mathcal{F}, P)$

$$\begin{aligned} dX_t &= b(X_t)\,dt + \sigma(X_t)\,dW_t + \rho(X_t)\,dV_t \;, \\ dY_t &= h(X_t)\,dt + dV_t \;, \end{aligned} \tag{1}$$

where the state process $\{X_t\,,\, t \geq 0\}$ and the observation $\{Y_t\,,\, t \geq 0\}$, take values in $\mathbf{R}^m$ and $\mathbf{R}^d$ respectively. Here $\{W_t\,,\, t \geq 0\}$ and $\{V_t\,,\, t \geq 0\}$ are independent standard Wiener processes of appropriate dimension. In addition, the random variable X_0 is independent of the Wiener processes, with probability distribution $p_0(x)\,dx$.

Throughout the paper, it is assumed that the coefficients b, σ, ρ and h are bounded and globally Lipschitz continuous functions defined on $\mathbf{R}^m$, so that the stochastic differential system (1) has a unique strong solution. In addition, the coefficients have bounded derivatives up to some appropriate

*Partially supported by USACCE under Contract DAJA45–90–C–0008.

order. The following definitions are used : $a \triangleq \sigma\sigma^*$ and $c \triangleq \rho\rho^*$. In particular, it is not assumed that either a or c is uniformly elliptic.

Introducing

$$Z_t^s \triangleq \exp\left\{\int_s^t h^*(X_\tau)\, dY_\tau - \tfrac{1}{2}\int_s^t |h(X_\tau)|^2\, d\tau\right\} ,$$

it is standard that, for all $T > 0$ the original probability measure P is equivalent on $[0,T]$ to the *reference probability* measure $P^\dagger$ with Radon–Nikodym derivative $Z_T = Z_T^0$, such that under $P^\dagger$

$$dX_t = b(X_t)\, dt + \sigma(X_t)\, dW_t + \rho(X_t)\, [dY_t - h(X_t)\, dt] \; , \tag{2}$$

where $\{W_t\, , t \geq 0\}$ and $\{Y_t\, , t \geq 0\}$ are independent standard Wiener processes.

The Bayes formula gives

$$\mathbf{E}(\phi(X_t) \mid \mathcal{Y}_t) = \frac{\mathbf{E}^\dagger(\phi(X_t)\, Z_t \mid \mathcal{Y}_t)}{\mathbf{E}^\dagger(Z_t \mid \mathcal{Y}_t)} \; .$$

In addition

$$\mathbf{E}^\dagger(\phi(X_t)\, Z_t \mid \mathcal{Y}_t) = \int \phi(x)\, p_t(x)\, dx \; ,$$

and the unnormalized conditional density $\{p_t\, , t \geq 0\}$ satisfies the Zakai equation [19]

$$dp_t = L^*\, p_t\, dt + \sum_{k=1}^{d} B_k^*\, p_t\, dY_t^k \; , \tag{3}$$

where L and B_k are partial differential operators associated with the stochastic differential system (1)

$$L \triangleq \tfrac{1}{2} \sum_{i,j=1}^{m} [a^{i,j} + c^{i,j}] \frac{\partial^2}{\partial x_i \partial x_j} + \sum_{i=1}^{m} b^i \frac{\partial}{\partial x_i} \; , \qquad \text{and} \qquad B_k \triangleq h_k + \sum_{i=1}^{m} \rho_k^i \frac{\partial}{\partial x_i} \; , \qquad 1 \leq k \leq d \; .$$

The Zakai equation has been extensively studied in [7], [8], [15], [16], [18].

Two possible directions are available for the approximation of the conditional probability distributions

- approximate the Zakai equation (3), using PDE techniques,
- approximate the stochastic differential system (1), e.g. using weak convergence techniques, and derive an equation for the exact conditional probability distribution of the approximate state process.

The first approach has been used in Elliott–Glowinski [4] and Florchinger–LeGland [5], and generally allows to get error estimates. On the other hand, the second approach generally provides a probabilistic interpretation of the approximation, along the ideas of Kushner [13], see DiMasi–Runggaldier [2], DiMasi–Pratelli–Runggaldier [3], Korezlioglu–Mazziotto [6] and Picard [17]. Note that in the second approach, all the above mentionned works deal with the case of independent noise.

The purpose of this paper is to provide, in the case of correlated noise, an approximation based on splitting–up of the SDE (2), using the composition results of Kunita [9]. In addition, the unnormalized conditional density of the approximate state process, is shown to satisfy an approximation of the original Zakai equation, also based on splitting–up and already introduced in [5].

2 Splitting–up Approximations

Consider the following decomposition of the Zakai equation (3)

$$dp_t = L_0^* \, p_t \, dt + \Lambda^* \, p_t \, dt + \sum_{k=1}^{d} B_k^* \, p_t \, dY_t^k \; , \tag{4}$$

where

$$L_0 \triangleq \tfrac{1}{2} \sum_{i,j=1}^{m} a^{i,j} \frac{\partial^2}{\partial x_i \partial x_j} + \sum_{i=1}^{m} b^i \frac{\partial}{\partial x_i} \; , \qquad \text{and} \qquad \Lambda \triangleq L - L_0 = \tfrac{1}{2} \sum_{i,j=1}^{m} c^{i,j} \frac{\partial^2}{\partial x_i \partial x_j} \; .$$

Parallel to this decomposition, there exists a similar decomposition of the state equation (2). Actually the deterministic PDE

$$\dot{p}_t = L_0^* \, p_t \; , \tag{5}$$

is the Fokker–Planck equation associated with the SDE

$$dX_t = b(X_t) \, dt + \sigma(X_t) \, dW_t \; , \tag{6}$$

whereas the degenerate SPDE

$$dp_t = \Lambda^* \, p_t \, dt + \sum_{k=1}^{d} B_k^* \, p_t \, dY_t^k \; , \tag{7}$$

can be connected with the SDE

$$dX_t = \rho(X_t) \, [dY_t - h(X_t) \, dt] \; , \tag{8}$$

in the following way. Define

$$\Xi_{s,t}(x) \triangleq \exp \left\{ \int_s^t h^*(\xi_{s,\tau}(x)) \, dY_\tau - \tfrac{1}{2} \int_s^t |h(\xi_{s,\tau}(x))|^2 \, d\tau \right\} \; , \tag{9}$$

where $\{\xi_{s,t}(\cdot) \, , \, 0 \leq s \leq t\}$ is the stochastic flow of diffeomorphisms associated with (8). For any probability measure $\mu(dx)$ on $\mathbf{R}^m$, define the transformed measure $Q_t^s \, \mu(dx)$ by

$$\langle Q_t^s \, \mu, \phi \rangle = \int \phi(\xi_{s,t}(x)) \, \Xi_{s,t}(x) \, \mu(dx) \; , \tag{10}$$

for any test function ϕ. Using the Itô formula, it can be checked that $\{Q_t^s \, , \, 0 \leq s \leq t\}$ is actually the stochastic semigroup associated with (7), see [8], [10].

Splitting–up approximations are introduced below for both the SDE (2) and the Zakai equation (3), based on a regular partition $0 = t_0 < t_1 < \ldots < t_i < \ldots$ of the time interval $[0, \infty)$, with time step δ.

SDE The approximation α_n to the solution X_{t_n} of the SDE (2), is defined by the following recursion

$$\alpha_{n+1} = \xi_{t_n,t_{n+1}} \circ \zeta_{t_n,t_{n+1}}(\alpha_n) \; , \tag{11}$$

where $\{\zeta_{s,t}(\cdot) \, , \, 0 \leq s \leq t\}$ and $\{\xi_{s,t}(\cdot) \, , \, 0 \leq s \leq t\}$ are the stochastic flows of diffeomorphisms associated with (6) and (8) respectively.

SPDE The approximation $\bar{p}_n$ to the solution p_{t_n} of the Zakai equation (3), is defined by the following recursion

$$\bar{p}_{n+1} = Q^{t_n}_{t_{n+1}} \cdot P^*_\delta \, \bar{p}_n \ , \tag{12}$$

where $\{P_t \, , \, t \geq 0\}$ and $\{Q^s_t \, , \, 0 \leq s \leq t\}$ are the semigroups associated with (5) and (7) respectively.

The connection between the splitting–up approximations (11) and (12) will be investigated in Section 5 below. In the next two sections, convergence results are presented for (11) and (12) respectively.

3 Convergence of Approximation for SDE

The results presented in this section are of general interest for the approximation of SDE's, independently of the nonlinear filtering problem. In addition, the boundedness assumption on the coefficients b, σ, ρ and h, can be replaced by a sub–linear growth assumption.

The approximating sequence $\{\alpha_n \, , \, n \geq 0\}$ introduced in (11) above, can be embedded in the continuous–time process $\{\bar{X}_t \, , \, t \geq 0\}$, defined for $t_n \leq t < t_{n+1}$, by

$$\bar{X}_t = \xi_{t_n,t} \circ \zeta_{t_n,t}(\alpha_n) \ . \tag{13}$$

Define also $\alpha_{n+\frac{1}{2}} = \zeta_{t_n,t_{n+1}}(\alpha_n)$, and the continuous–time processes $\{\bar{\zeta}_t \, , \, t \geq 0\}$ and $\{\bar{\xi}_t \, , \, t \geq 0\}$ by

$$\bar{\zeta}_t = \zeta_{t_n,t}(\alpha_n) \ , \qquad \text{and} \qquad \bar{\xi}_t = \xi_{t_n,t}(\alpha_{n+\frac{1}{2}}) \ , \tag{14}$$

respectively for $t_n \leq t < t_{n+1}$, so that $\bar{X}_t = \xi_{t_n,t}(\bar{\zeta}_t)$. Note that

$$\bar{X}_{t_n} = \alpha_n \ , \qquad \text{and} \qquad \bar{X}_{t_n-} = \lim_{t \uparrow t_n} \bar{X}_t = \alpha_n \ ,$$

so that $\{\bar{X}_t \, , \, t \geq 0\}$ has continuous trajectories, whereas

$$\bar{\xi}_{t_n} = \alpha_{n+\frac{1}{2}} \ , \qquad \text{and} \qquad \bar{\xi}_{t_n-} = \lim_{t \uparrow t_n} \bar{\xi}_t = \alpha_n \ ,$$

so that $\{\bar{\xi}_t \, , \, t \geq 0\}$ is discontinuous at partition points.

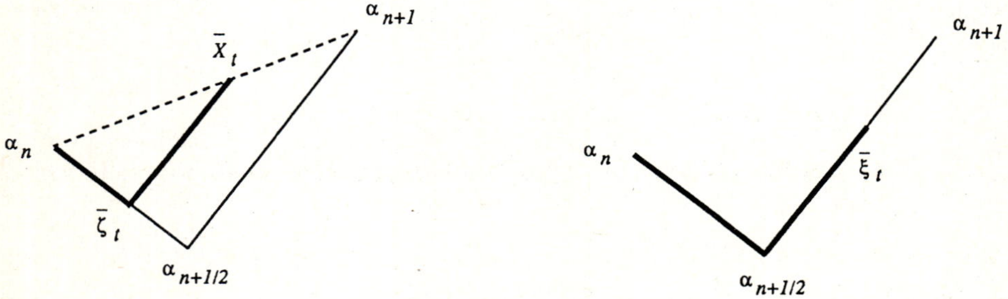

Both approximations have their own interest. Following Kunita [9] it is possible to derive a SDE for $\{\bar{X}_t \, , \, t \geq 0\}$, which makes the convergence analysis easier, see the proof of Proposition 3.1 below. On the other hand, an approximate filtering problem can be defined in terms of the approximation $\{\bar{\xi}_t \, , \, t \geq 0\}$, which provides a probabilistic interpretation of the approximation, see Section 5 below.

Proposition 3.1 *Under suitable regularity assumptions on the coefficients, the approximating process* $\{\bar{X}_t\,,\,t\geq 0\}$ *defined in (13), converges to the solution* $\{X_t\,,\,t\geq 0\}$ *of the original SDE (2), with error rate of order* $O(\sqrt{\delta})$, *i.e.*

$$\mathbf{E}^{\dagger}\left[\sup_{0\leq s\leq t}|X_s-\bar{X}_s|^2\right]\leq C\,\delta\;. \tag{15}$$

In particular for the splitting–up approximation α_n *defined in (11)*

$$\left\{\mathbf{E}^{\dagger}|X_{t_n}-\alpha_n|^2\right\}^{1/2}\leq C\,\sqrt{\delta}\;.$$

PROOF. The basic remark is that the stochastic flows of diffeomorphisms $\{\zeta_{s,t}(\cdot)\,,\,0\leq s\leq t\}$ and $\{\xi_{s,t}(\cdot)\,,\,0\leq s\leq t\}$ are mutually independent under the reference probability measure $P^{\dagger}$. Estimates on stochastic flows of diffeomorphisms can be found in [11], [12]. The proof is actually divided in three steps.

□ *Moment estimates*

The estimates

$$\mathbf{E}^{\dagger}\left[1+|\xi_{s,t}(x)|^2\right]\leq(1+|x|^2)\,e^{\,C\,(t-s)}\;,$$

$$\mathbf{E}^{\dagger}\left[1+|\zeta_{s,t}(x)|^2\right]\leq(1+|x|^2)\,e^{\,C\,(t-s)}\;,$$

and the mutual independence of $\zeta_{t_n,t_{n+1}}(\cdot)$, $\xi_{t_n,t_{n+1}}(\cdot)$ and α_n, give

$$\begin{aligned}\mathbf{E}^{\dagger}\left[1+|\alpha_{n+1}|^2\right] &= \mathbf{E}^{\dagger}\left[1+|\xi_{t_n,t_{n+1}}\circ\zeta_{t_n,t_{n+1}}(\alpha_n)|^2\right]\\ &\leq \mathbf{E}^{\dagger}\left[1+|\zeta_{t_n,t_{n+1}}(\alpha_n)|^2\right]\,e^{\,C\,\delta}\leq\mathbf{E}^{\dagger}\left[1+|\alpha_n|^2\right]\,e^{\,C\,\delta}\;.\end{aligned}$$

By induction

$$\mathbf{E}^{\dagger}\left[1+|\alpha_n|^2\right]\leq\mathbf{E}^{\dagger}\left[1+|X_0|^2\right]\,e^{\,C\,t_n}\;,$$

and therefore

$$\mathbf{E}^{\dagger}\left[|\bar{X}_t|^2+|\bar{\zeta}_t|^2+|\bar{\xi}_t|^2\right]\leq C\,e^{\,C\,t}\;,$$

provided the random variable X_0 is square integrable.

□ *Composition of stochastic flows*

Following [9], the continuous process $\{\bar{X}_t\,,\,t\geq 0\}$ satisfies the following SDE

$$\begin{aligned}d\bar{X}_t &= b_\delta[t,\xi^{-1}_{t_n,t}(\bar{X}_t)]\,dt+\sigma_\delta[t,\xi^{-1}_{t_n,t}(\bar{X}_t)]\,dW_t+\tfrac{1}{2}\chi_\delta[t,\xi^{-1}_{t_n,t}(\bar{X}_t)]\,dt\\ &\quad+\rho(\bar{X}_t)\,[dY_t-h(\bar{X}_t)\,dt]\;,\end{aligned}$$

where the coefficients $b_\delta(t,\cdot)$, $\sigma_\delta(t,\cdot)$ and $\chi_\delta(t,\cdot)$ are defined for $t_n\leq t<t_{n+1}$, by

$$b_\delta(t,x)\triangleq D\xi_{t_n,t}(x)\,b(x)\;,\qquad \sigma_\delta(t,x)\triangleq D\xi_{t_n,t}(x)\,\sigma(x)\;,$$

$$\chi_\delta(t,x)\triangleq\mathrm{Tr}\left[D^2\xi_{t_n,t}(x)\cdot a(x)\right]\;,$$

respectively. Using $\bar{\zeta}_t = \xi^{-1}_{t_n,t}(\bar{X}_t)$ introduced above, gives the simpler expression

$$\begin{aligned} d\bar{X}_t &= b_\delta(t,\bar{\zeta}_t)\,dt + \sigma_\delta(t,\bar{\zeta}_t)\,dW_t + \tfrac{1}{2}\chi_\delta(t,\bar{\zeta}_t)\,dt + \rho(\bar{X}_t)\,[dY_t - h(\bar{X}_t)\,dt] \\ &= b(\bar{X}_t)\,dt + \sigma(\bar{X}_t)\,dW_t + \rho(\bar{X}_t)\,[dY_t - h(\bar{X}_t)\,dt] \\ &\quad + \left[b_\delta(t,\bar{\zeta}_t) - b(\bar{X}_t)\right]\,dt + \left[\sigma_\delta(t,\bar{\zeta}_t) - \sigma(\bar{X}_t)\right]\,dW_t + \tfrac{1}{2}\chi_\delta(t,\bar{\zeta}_t)\,dt\;. \end{aligned}$$

In other words, the approximate process $\{\bar{X}_t\,,\,t \geq 0\}$ satisfies the same equation as the original process $\{X_t\,,\,t \geq 0\}$, except for some additional perturbation terms which need to be estimated.

□ *Estimates on perturbation terms*

Consider for instance the following perturbation term

$$\begin{aligned} \beta_\delta(t) &\triangleq b_\delta(t,\bar{\zeta}_t) - b(\bar{X}_t) = \left[b(\bar{\zeta}_t) - b(\bar{X}_t)\right] + \left[D\xi_{t_n,t}(\bar{\zeta}_t) - I\right]\,b(\bar{\zeta}_t) \\ &= \left[b(\bar{\zeta}_t) - b(\xi_{t_n,t}(\bar{\zeta}_t))\right] + \left[D\xi_{t_n,t}(\bar{\zeta}_t) - I\right]\,b(\bar{\zeta}_t) = \beta^1_\delta(t) + \beta^2_\delta(t)\;. \end{aligned}$$

The estimate

$$\mathbf{E}^\dagger|\xi_{s,t}(x) - x|^2 \leq C\,(1 + |x|^2)\,|t - s|\;,$$

the independence of $\xi_{t_n,t}(\cdot)$ and $\bar{\zeta}_t$, and the Lipschitz property of the drift coefficient, give

$$\mathbf{E}^\dagger|\beta^1_\delta(t)|^2 \leq C\,\mathbf{E}^\dagger|\xi_{t_n,t}(\bar{\zeta}_t) - \bar{\zeta}_t|^2 \leq C\,(t - t_n)\,\mathbf{E}^\dagger\left[1 + |\bar{\zeta}_t|^2\right] \leq C\,\delta\,e^{\,C\,t}\;.$$

Similarily, the estimate

$$\mathbf{E}^\dagger\|D\xi_{s,t}(x) - I\|^2 \leq C\,|t - s|\;,$$

the mutual independence of $D\xi_{t_n,t}(\cdot)$ and $\bar{\zeta}_t$, and the sub–linear growth of the drift coefficient, give

$$\mathbf{E}^\dagger|\beta^2_\delta(t)|^2 \leq \mathbf{E}^\dagger\left[\|D\xi_{t_n,t}(\bar{\zeta}_t) - I\|^2 \cdot |b(\bar{\zeta}_t)|^2\right] \leq C\,(t - t_n)\,\mathbf{E}^\dagger\left[1 + |\bar{\zeta}_t|^2\right] \leq C\,\delta\,e^{\,C\,t}\;.$$

The other perturbation terms can be estimated in the same way, and the result follows from standard estimates on the solution of SDE's. □

Concerning the approximation $\{\bar{\xi}_t\,,\,t \geq 0\}$, the following weaker (non uniform) result can be proved.

Proposition 3.2 *Under suitable regularity assumptions on the coefficients, the approximating process* $\{\bar{\xi}_t\,,\,t \geq 0\}$ *defined in (14) converges to the solution* $\{X_t\,,\,t \geq 0\}$ *of the original SDE (2), with error rate of order* $O(\sqrt{\delta})$*, i.e.*

$$\left\{\mathbf{E}^\dagger|X_t - \bar{\xi}_t|^2\right\}^{1/2} \leq C\,\sqrt{\delta}\;.$$

PROOF. In view of Proposition 3.1 above, it is enough to estimate

$$|\bar{X}_t - \bar{\xi}_t| = |\xi_{t_n,t} \circ \zeta_{t_n,t}(\alpha_n) - \xi_{t_n,t} \circ \zeta_{t_n,t_{n+1}}(\alpha_n)|\;.$$

The estimates

$$\mathbf{E}^\dagger|\xi_{s,t}(x) - \xi_{s,t}(x')|^2 \leq C\,|x - x'|^2\;,$$

$$\mathbf{E}^\dagger|\zeta_{s,t}(x) - \zeta_{s,t'}(x)|^2 \leq C\,(1 + |x|^2)\,|t' - t|\;,$$

and the mutual independence of $\xi_{t_n,t}(\cdot)$, $\{\zeta_{t_n,t}(\cdot)\,,\,t \geq t_n\}$ and α_n, give

$$\mathbf{E}^\dagger|\bar{X}_t - \bar{\xi}_t|^2 \leq C\,\mathbf{E}^\dagger|\zeta_{t_n,t}(\alpha_n) - \zeta_{t_n,t_{n+1}}(\alpha_n)|^2 \leq C\,(t_{n+1} - t)\,\mathbf{E}^\dagger\left[1 + |\alpha_n|^2\right] \leq C\,\delta\;.$$ □

4 Convergence of Approximation for SPDE

Here also, the results presented in this section are of general interest for the approximation of SPDE's, independently of the nonlinear filtering problem.

Consider the splitting–up approximation $\{\bar{p}_n\,,\, n \geq 0\}$ defined in (12) above

$$\bar{p}_{n+1} = Q^{t_n}_{t_{n+1}} \cdot P^*_\delta\, \bar{p}_n\,. \tag{16}$$

For all $r \geq 0$, let H^r denote the space of real-valued Lebesgue–measurable functions on $\mathbf{R}^m$ whose generalized derivatives up to order r are square–integrable, with norm $\|\cdot\|_r$

$$\|u\|_r^2 \triangleq \sum_{|\alpha|\leq r} \|D^\alpha u\|_0^2 < \infty\,.$$

Proposition 4.1 *Under suitable regularity assumptions on the coefficients, the splitting–up approximation $\bar{p}_n$ defined in (16) converges to the solution p_{t_n} of the original Zakai equation (3), with error rate of order $O(\sqrt{\delta})$, i.e.*

$$\left\{\mathbf{E}^\dagger \|p_{t_n} - \bar{p}_n\|_0^2\right\}^{1/2} \leq C\,\sqrt{\delta}\,.$$

PROOF. Just as in the proof of Proposition 3.1, the idea is to get an equation for the continuous–time process $\{v_t\,,\, t \geq 0\}$ defined for $t_n \leq t < t_{n+1}$, by

$$v_t \triangleq Q^{t_n}_t \cdot P^*_{t-t_n}\, \bar{p}_n\,. \tag{17}$$

The proof is actually divided in three steps.

□ *Stability estimates*

Under suitable regularity assumptions on the coefficients and the test function, the following estimate is proved in Krylov–Rozovskii [8] for some appropriate integer r

$$\mathbf{E}^\dagger \|Q^s_t\, \phi\|_r^2 \leq \|\phi\|_r^2\, e^{\,C\,(t-s)}\,,$$

and similarily for the Fokker–Planck equation (5)

$$\|P^*_{t-s}\, \phi\|_r^2 \leq \|\phi\|_r^2\, e^{\,C\,(t-s)}\,.$$

Therefore

$$\mathbf{E}^\dagger \|\bar{p}_{n+1}\|_r^2 = \mathbf{E}^\dagger \|Q^{t_n}_{t_{n+1}} \cdot P^*_\delta\, \bar{p}_n\|_r^2 \leq \mathbf{E}^\dagger \|P^*_\delta\, \bar{p}_n\|_r^2\, e^{\,C\,\delta} \leq \mathbf{E}^\dagger \|\bar{p}_n\|_r^2\, e^{\,C\,\delta}\,.$$

By induction

$$\mathbf{E}^\dagger \|\bar{p}_n\|_r^2 \leq \|p_0\|_r^2\, e^{\,C\,t_n}\,,$$

and therefore

$$\mathbf{E}^\dagger \|v_t\|_r^2 \leq \|p_0\|_r^2\, e^{\,C\,t}\,.$$

□ *Composition of semigroups*

Differentiating (17) with respect to t, gives

$$\begin{aligned} dv_t &= \Lambda^* \, v_t \, dt + \sum_{k=1}^{d} B_k^* \, v_t \, dY_t^k + Q_t^{t_n} \, L_0^* \, P_{t-t_n}^* \, \bar{p}_n \, dt \\ &= L_0^* \, v_t \, dt + \Lambda^* \, v_t \, dt + \sum_{k=1}^{d} B_k^* \, v_t \, dY_t^k + [Q_t^{t_n} \, L_0^* - L_0^* \, Q_t^{t_n}] \, P_{t-t_n}^* \, \bar{p}_n \, dt \\ &= L^* \, v_t \, dt + \sum_{k=1}^{d} B_k^* \, v_t \, dY_t^k + f_t \, dt \; , \end{aligned}$$

where the perturbation term is defined for $t_n \leq t < t_{n+1}$, by

$$f_t \triangleq [Q_t^{t_n} \, L_0^* - L_0^* \, Q_t^{t_n}] \, P_{t-t_n}^* \, \bar{p}_n \; .$$

In other words, the approximate process $\{v_t \, , \, t \geq 0\}$ satisfies the same equation as the original process $\{p_t \, , \, t \geq 0\}$, except for an additional perturbation term which need to be estimated.

□ *Estimate on perturbation term*

It follows from (10) that

$$Q_t^s \, p(\xi_{s,t}(x)) = p(x) \cdot \Xi_{s,t}(x) \cdot [J_{s,t}(x)]^{-1} \; ,$$

where $J_{s,t}(\cdot)$ is the Jacobian (i.e. the determinant of the Jacobian matrix) of the diffeomorphism $\xi_{s,t}(\cdot)$, see [5]. The following estimate can then be obtained

$$\mathbf{E}^\dagger \| \, [Q_t^s \, L_0^* - L_0^* \, Q_t^s] \, \phi \|_0^2 \leq C \, (t-s) \, \|\phi\|_r^2 \; ,$$

for some integer r depending on the regularity of the coefficients and the test function. Therefore

$$\begin{aligned} \mathbf{E}^\dagger \|f_t\|_0^2 &= \mathbf{E}^\dagger \| \, [Q_t^{t_n} \, L_0^* - L_0^* \, Q_t^{t_n}] \, P_{t-t_n}^* \, \bar{p}_n \|_0^2 \\ &\leq C \, (t_n - t) \, \mathbf{E}^\dagger \|P_{t-t_n}^* \, \bar{p}_n\|_r^2 \leq C \, \delta \, \mathbf{E}^\dagger \|\bar{p}_n\|_r^2 \, e^{\, C \delta} \leq C \, \delta \, \|p_0\|_r^2 \, e^{\, C \, t_n} \; . \end{aligned}$$

From this point, the result follows from standard estimates on the solution of SPDE's. □

5 Approximate Nonlinear Filtering Model

The purpose of this section is to investigate the connection between the splitting–up approximations (11) and (12). To this end, define the following approximate model in terms of the approximating sequence $\{\alpha_n \, , \, n \geq 0\}$.

Under the reference probability measure $P^\dagger$, the transition from α_n to α_{n+1} can be decribed as follows. First

$$d\bar{\zeta}_t = b(\bar{\zeta}_t) \, dt + \sigma(\bar{\zeta}_t) \, dW_t \; , \qquad \bar{\zeta}_{t_n} = \alpha_n \; ,$$

which defines the intermediate state

$$\alpha_{n+\frac{1}{2}} = \lim_{t \uparrow t_{n+1}} \bar{\zeta}_t \; .$$

Then

$$d\bar{\xi}_t = \rho(\bar{\xi}_t) \, [dY_t - h(\bar{\xi}_t) \, dt] \; , \qquad \bar{\xi}_{t_n} = \alpha_{n+\frac{1}{2}} \; ,$$

which defines the next state

$$\alpha_{n+1} = \lim_{t \uparrow t_{n+1}} \bar{\xi}_t \ .$$

Here $\{W_t\,,\, t \geq 0\}$ and $\{Y_t\,,\, t \geq 0\}$ are independent standard Wiener processes.

Define next

$$\bar{Z}_t^s \triangleq \exp\left\{\int_s^t h^*(\bar{\xi}_\tau)\, dY_\tau - \tfrac{1}{2}\int_s^t |h(\bar{\xi}_\tau)|^2\, d\tau\right\} ,$$

and let $\bar{P}$ denote the probability measure equivalent on $[0,T]$ to the reference probability measure $P^\dagger$ with Radon–Nikodym derivative $\bar{Z}_T = \bar{Z}_T^0$, such that under $\bar{P}$, the processes $\{\bar{\zeta}_t\,,\, t \geq 0\}$ and $\{\bar{\xi}_t\,,\, t \geq 0\}$ are defined as above, and

$$dY_t = h(\bar{\xi}_t)\, dt + d\bar{V}_t \ ,$$

where $\{W_t\,,\, t \geq 0\}$ and $\{\bar{V}_t\,,\, t \geq 0\}$ are independent standard Wiener processes.

The Bayes formula gives

$$\bar{\mathrm{E}}(\phi(\bar{\xi}_t) \mid \mathcal{Y}_t) = \frac{\mathrm{E}^\dagger(\phi(\bar{\xi}_t)\, \bar{Z}_t \mid \mathcal{Y}_t)}{\mathrm{E}^\dagger(\bar{Z}_t \mid \mathcal{Y}_t)} \ .$$

Let $\{\bar{q}_t\,,\, t \geq 0\}$ denote the unnormalized conditional density of the approximating process $\{\bar{\xi}_{t-}\,,\, t \geq 0\}$, i.e.

$$\mathrm{E}^\dagger(\phi(\bar{\xi}_{t-})\, \bar{Z}_t \mid \mathcal{Y}_t) = \int \phi(x)\, \bar{q}_t(x)\, dx \ . \tag{18}$$

Observe that $\bar{\xi}_{t-} = \xi_{t_n,t}(\alpha_{n+\frac{1}{2}})$ for $t_n < t \leq t_{n+1}$, and $\bar{\xi}_{t_n-} = \alpha_n$. In particular

$$\mathrm{E}^\dagger(\phi(\alpha_n)\, \bar{Z}_{t_n} \mid \mathcal{Y}_{t_n}) = \int \phi(x)\, \bar{q}_{t_n}(x)\, dx \ .$$

The main result of this section is to prove that the splitting–up approximation $\bar{p}_n$ defined in (12), is the unnormalized conditional density of the splitting–up approximation α_n defined in (11), i.e. $\bar{q}_{t_n} = \bar{p}_n$.

Proposition 5.1 *The unnormalized conditional density $\{\bar{q}_t\,,\, t \geq 0\}$ defined in (18), satisfies*

$$\bar{q}_t = Q_t^{t_n} \cdot P_\delta^*\, \bar{q}_{t_n} \ ,$$

for $t_n < t \leq t_{n+1}$. In particular, the approximating sequence $\{\bar{q}_{t_n}\,,\, n \geq 0\}$ satisfies the recursion (12).

PROOF. Observe first that for $t_n \leq t \leq t_{n+1}$

$$\bar{Z}_t^{t_n} = \Xi_{t_n,t}(\alpha_{n+\frac{1}{2}}) \ .$$

Next, if $\phi(\cdot)$ is $\mathcal{Y}_t^{t_n}$–measurable, then the mutual independence of $\zeta_{t_n,t_{n+1}}(\cdot)$ and $\mathcal{F}_{t_n} \vee \mathcal{Y}_t^{t_n}$ gives

$$\mathrm{E}^\dagger(\phi(\alpha_{n+\frac{1}{2}}) \mid \mathcal{F}_{t_n} \vee \mathcal{Y}_t^{t_n}) = \mathrm{E}^\dagger(\phi(\zeta_{t_n,t_{n+1}}(\alpha_n)) \mid \mathcal{F}_{t_n} \vee \mathcal{Y}_t^{t_n}) = P_\delta\, \phi(\alpha_n) \ ,$$

where $\mathcal{F}_{t_n} = \mathcal{Y}_{t_n} \vee \sigma(X_0\,,\, W_t\,,\, 0 \leq t \leq t_n)$.

Combining these two remarks gives, for $t_n < t \leq t_{n+1}$

$$\begin{aligned}
\int \phi(x)\,\bar{q}_t(x)\,dx &= \mathbf{E}^\dagger(\phi(\bar{\xi}_{t-})\,\bar{Z}_t \mid \mathcal{Y}_t) \\
&= \mathbf{E}^\dagger(\bar{Z}_{t_n}\cdot \phi[\xi_{t_n,t}(\alpha_{n+\frac{1}{2}})]\cdot \Xi_{t_n,t}(\alpha_{n+\frac{1}{2}}) \mid \mathcal{Y}_t) \\
&= \mathbf{E}^\dagger(\bar{Z}_{t_n}\cdot \mathbf{E}^\dagger[\,\phi[\xi_{t_n,t}(\alpha_{n+\frac{1}{2}})]\cdot \Xi_{t_n,t}(\alpha_{n+\frac{1}{2}}) \mid \mathcal{F}_{t_n}\vee \mathcal{Y}_t^{t_n}] \mid \mathcal{Y}_t) \\
&= \mathbf{E}^\dagger(\bar{Z}_{t_n}\cdot P_\delta\,[\phi\circ\xi_{t_n,t}\cdot \Xi_{t_n,t}](\alpha_n) \mid \mathcal{Y}_{t_n}\vee \mathcal{Y}_t^{t_n}) \\
&= \int P_\delta\,[\phi\circ\xi_{t_n,t}\cdot \Xi_{t_n,t}](x)\,\bar{q}_{t_n}(x)\,dx \\
&= \int \phi[\xi_{t_n,t}(x)]\cdot \Xi_{t_n,t}(x)\cdot [P_\delta^*\,\bar{q}_{t_n}](x)\,dx \\
&= \int \phi(x)\,[Q_t^{t_n}\cdot P_\delta^*\,\bar{q}_{t_n}](x)\,dx\,. \qquad \square
\end{aligned}$$

The main interest of splitting–up approximation in nonlinear filtering, is that the original equation (3) is splitted into a second–order deterministic PDE (5) related with the *prediction* step, and a degenerate second–order SPDE (7) related with the *correction* step. A similar prediction–correction numerical scheme was obtained by Kushner [13] in the case of independent noises, see also Bensoussan–Glowinski–Rascanu [1] for a related convergence result using PDE technique. This probabilistic interpretation can be used to design further approximation schemes, e.g. in terms of approximating finite–state Markov processes.

References

[1] A. BENSOUSSAN, R. GLOWINSKI, A. RASCANU, Approximation of Zakai equation by the splitting–up method, in: *Stochastic Systems and Optimization (Warsaw–1988)*, (ed. J.Zabczyk) 257–265, Springer–Verlag (LNCIS–136) (1989).

[2] G.B. DI MASI, W.J. RUNGGALDIER, Continuous–time approximations for the nonlinear filtering problem, *Appl.Math.Optim.* **7** (3) 233–245 (1981).

[3] G.B. DI MASI, M. PRATELLI, W.J. RUNGGALDIER, An approximation for the nonlinear filtering problem with error bound, *Stochastics* **14** (4) 247–271 (1985).

[4] R.J. ELLIOTT, R. GLOWINSKI, Approximations to solutions of the Zakai filtering equation, *Stoch.Anal.Appl.* **7** (2) 145–168 (1988).

[5] P. FLORCHINGER, F. LE GLAND, Time–discretization of the Zakai equation for diffusion processes observed in correlated noise, *Stochastics* **35**,233–256 (1991).

[6] H. KOREZLIOGLU, G. MAZZIOTTO, Approximations of the nonlinear filter by periodic sampling and quantization, in: *Analysis and Optimization of Systems, Part 1 (Nice–1984)*, (eds. A.Bensoussan and J.L.Lions) 553–567, Springer–Verlag (LNCIS–62) (1984).

[7] N.V. KRYLOV, B.L. ROZOVSKII, On the Cauchy problem for linear stochastic partial differential equations, *Math.USSR Izvestija* **11** (6) 1267–1284 (1977).

[8] N.V. KRYLOV, B.L. ROZOVSKII, Characteristics of degenerating second-order parabolic Itô equations, *J.Soviet Math.* **32** (4) 336–348 (1982).

[9] H. KUNITA, On the decomposition of solutions of stochastic differential equations, in: *Stochastic Integrals (Durham–1980)* (ed. D.Williams) 213–255, Springer–Verlag (LNM–851) (1981).

[10] H. KUNITA, Stochastic partial differential equations connected with nonlinear filtering, in: *Nonlinear Filtering and Stochastic Control (Cortona–1981)* (eds. S.K.Mitter and A.Moro) 100–169, Springer–Verlag (LNM–972) (1982).

[11] H. KUNITA, Stochastic differential equations and stochastic flows of diffeomorphisms, in: *Ecole d'Eté de Probabilités de Saint-Flour XII (1982)* (ed. P.L.Hennequin) 144–303, Springer–Verlag (LNM–1097) (1984).

[12] H. KUNITA, *Stochastic Flows and Stochastic Differential Equations*, Cambridge University Press (1990).

[13] H.J. KUSHNER, *Probability Methods for Approximations in Stochastic Control and for Elliptic Equations*, Academic Press (1977).

[14] F. LE GLAND, Time discretization of nonlinear filtering equations, in: *28th IEEE CDC (Tampa–1989)* 2601–2606 (1989).

[15] E. PARDOUX, Stochastic partial differential equations and filtering of diffusion processes, *Stochastics* **3** (2) 127–167 (1979).

[16] E. PARDOUX, Filtrage non–linéaire et équations aux dérivées partielles stochastiques associées, in: *Ecole d'Eté de Probabilités de Saint-Flour XIX (1989)* (ed. P.L.Hennequin) 69–163, Springer–Verlag (LNM–1464) (1991).

[17] J. PICARD, Approximation of nonlinear filtering problems and order of convergence, in: *Filtering and Control of Random Processes (ENST/CNET–1983)*, (eds. H.Korezlioglu, G.Mazziotto and J.Szpirglas) 219–236, Springer–Verlag (LNCIS–61) (1984).

[18] B.L. ROZOVSKII, *Stochastic Evolution Systems*, Kluwer (1990).

[19] M. ZAKAI, On the optimal filtering of diffusion processes, *Z.Wahrschein.Verw.Geb.* **11** (3) 230–243 (1969).

REPRESENTATION AND APPROXIMATION OF MARTINGALE MEASURES

S. MELEARD
Laboratoire de Probabilités - Université Paris VI
4 place Jussieu - Tour 56 - 3ème étage - 75252 PARIS cedex 05

INTRODUCTION AND PRINCIPAL RESULTS

Martingale measure Theory was introduced by J.B. Walsh in [W]. The idea was to construct a stochastic calculus for two parameter "space-time" processes having a martingale property in the time variable and a measure property in space. The fundamental example of martingale measure is the "space-time" or two parameter white noise, which appears usually to model some random pertubations depending on space and time in stochastic partial differential equations.

Let us recall the definition of a martingale measure.
$(\Omega,\mathcal{F},\mathcal{F}_t,P)$ is a filtered probability space, E is a Lusin space endowed with its Borel σ-field $\mathcal{E}$.
Let M(.) be a random real-valued function on $\Omega\times\mathbb{R}+\times\mathcal{E}$. M will be called a $(\mathcal{F}_t,P)$-martingale measure if it satisfies the following properties:

. $\forall\ A\in\mathcal{E}$, $M(.,A)$ is a $(\mathcal{F}_t,P)$-square integrable martingale and $M(0,A)=0$.

. $M(t,A\cup B) = M(t,A) + M(t,B)$ P a.s., $\forall$ A,B $\in\mathcal{E}$ such that $A\cap B = \emptyset$, $\forall\ t\in\mathbb{R}+$.

. There exists a non-decreasing sequence (E_n) of Borel sets of E such that:

- $\bigcup_n E_n = E$
- $\forall\ t\in\mathbb{R}+$, $\sup_{A\in\mathcal{E}_n} E(M(t,A)^2)< +\infty$ ($\mathcal{E}_n$ is the Borel σ-field of E_n)
- $\forall\ t\in\mathbb{R}+$, $E(M(t,A_j)^2)$ tends to 0 for all sequence of sets A_j of $\mathcal{E}_n$ decreasing to $\emptyset$.

The martingale measure M is said to be continuous if each M(.,B) is, and orthogonal if $M(.,A)M(.,B)$ is a $(\mathcal{F}_t,P)$-martingale whenever $A\cap B=\emptyset$. If M and N are two $(\mathcal{F}_t,P)$-martingale measures and if $M(.,A)N(.,B)$ is a $(\mathcal{F}_t,P)$-martingale for any A and B in $\mathcal{E}$, then M and N are said to be strongly orthogonal.

When the martingale measure M is assumed to be orthogonal , one can prove [W] the existence of a random positive σ-finite measure $\nu(dx,ds)$ on $E\times\mathbb{R}+$, $\mathcal{F}_t$-predictable (that is: for each Borel set A, the process $(\nu(A\times(0,t]))_t$ is predictable) such that $\forall\ A\in\mathcal{E}$, $\forall\ t\in\mathbb{R}+$, $\nu(A\times(0,t]) = \langle M(A)\rangle_t$ P a.s.
The measure ν is called the covariance measure of M.
The orthogonality of M implies in particular that $\forall\ t\in\mathbb{R}+$, $\forall\ (A,B)\in\mathcal{E}^2$, $\langle M(A),M(B)\rangle_t = \nu(A\cap B,(0,t])$ P a.s. (The covariance kernel is in a certain sense degenerate).

Let us remark that ν can always be disintegrated in the form $\nu(dx,dt) = q_t(dx)dk_t$, where (k_t) is a predictable increasing process and (q_t) a measure-valued predictable process.

Hypothesis: We shall henceforth consider continuous orthogonal martingale measures and use the disintegration of the covariance measure. (The process (k_t) is then continuous).

By an analogous construction to the one of the Itô integral, Walsh [W] defines, for each $\mathcal{P}\otimes\mathcal{E}$ measurable function f defined on $\Omega\times\mathbb{R}+\times E$, ($\mathcal{P}$ is the predictable σ-field on $\Omega\times\mathbb{R}+$), and belonging to $L^2(dP\otimes d\nu)$, the stochastic integral of f with respect to M, denoted by $M(f) = \int_0^t\int_E f(s,x)\, M(dx,ds)$.

We shall present in this paper some recent results about martingale measures and give many applications of them. The aim of this work is to give a comprehensive view of this theory, and to relate apparently different situations thanks to the notion of martingale measure. There are no new results in this survey, but an uniform approach of many different problems. We shall present just the sketches of the proofs and shall give in each case the precise references.

Let us first give some simple examples of martingale measures:

1) if E is a finite space $\{a_1,a_2,\dots,a_n\}$, a martingale measure defined on $E\times\mathbb{R}+$ is uniquely determined by the n square integrable orthogonal martingales $(M_t(a_i))_{i=1}^n$, and conversely if $(m_t^i)_{i=1}^n$ are n square integrable martingales with increasing processes $(C_t^i)_{i=1}^n$, the mapping $M_t(A) = \sum_{i=1}^n m_t^i\, \delta_{\{a_i\}}(A)$ defines an orthogonal martingale measure on E with covariance measure $\sum_{i=1}^n dC_t^i\, \delta_{\{a_i\}}(da)$.

2) More generally, consider a Lusin space E, an E-valued predictable process (u_t) and a square integrable martingale (m_t) with increasing process C_t. Then the mapping $M_t(A) = \int_0^t 1_A(u_s)\, dm_s$ defines an orthogonal martingale measure with covariance measure $\delta_{u_s}(da)\, dC_s$.

If the martingale m is continuous, then M is continuous.

3) "Space-time" white noise: Let $\tilde{W}$ be a centered Gaussian measure on $(E\times\mathbb{R}+,\mathcal{E}\otimes\mathcal{B}(\mathbb{R}+),\mu)$, where μ is a σ-finite measure on $E\times\mathbb{R}+$. One knows that for each Borel set A of E, $W_t(A) = \tilde{W}(A\times(0,t])$ is a Gaussian process with independent increments and cadlag paths. It is easy to show that W defines an orthogonal martingale measure with deterministic covariance measure μ, with respect to its natural filtration.

When the measure μ is continuous (in the sense that $\mu(A\times\{t\}) = 0\ \forall\ A,\ \forall\ t)$, the martingale measure W is continuous.

Definition: we call white noise a martingale measure defined from a Gaussian non-atomic measure (in the above sense).

White noises are completely characterized by the deterministic nature of their covariance measure:

Proposition: ([W], [Ek-Me]) If M is a continuous orthogonal martingale measure, M is a white noise if and only if its covariance measure is deterministic.

The white noises are thus the reference processes in the theory of continuous orthogonal martingale measures, as the Brownian motion is in the theory of square integrable continuous martingales. Indeed, we have even proved that each continuous orthogonal martingale measure is in fact the image measure of a time changed white noise, in the following sense:

Theorem: ([Ek-Me]) Let M be a continuous orthogonal martingale measure on $E\times\mathbb{R}_+$, with covariance measure $q_t(dx)dk_t$. Let λ be a deterministic σ-finite diffuse measure (on a Lusin space U) such that : $\forall\ t\in\mathbb{R}_+,\quad q_t(E) \le \lambda(U)$. Then there exists a $\mathcal{P}\otimes\mathcal{U}$ measurable process φ, with values in $E\cup\{\text{cemetery point } \delta\}$ and a white noise W on $U\times\mathbb{R}_+$, with covariance measure $\lambda(du)dt$, defined on an extension of the probability space Ω, such that:

$$\forall f \in L^2(q_t(dx)dk_t),\ \forall\ t\in\mathbb{R}_+,\quad M_t(f) = \int_0^{k_t}\int_U f(\varphi(s,u))\ W(du,ds).$$

For the proof of this result, two different ideas have been developpped: the first is a generalization of the Skorohod representation theorem ([Ek-Le]) which proves that the random measure q_t can be represented as image measure of the measure λ by a predictable process φ with values in $E\cup\{\delta\}$:

$$\forall\ t\in\mathbb{R}_+,\quad \forall\ A\in\mathcal{E},\quad q_t(A) = \int_U 1_A(\varphi(t,u))\ \lambda(du)$$

and conversely, there exists a predictable kernel $Q_t(x,du)$ from E to U which satisfies: $\forall\ t\in\mathbb{R}_+,\quad \forall\ B\in\mathcal{U},\quad \int_U 1_B(u)\ f(\varphi_t(u))\ \lambda(du) = \int_E f(x)\ Q_t(x,B)\ q_t(dx).$

The second idea is an extension theorem, proved in [Ek-Me] with techniques similar to those in classical stochastic calculus:

Extension theorem: Let E and F be two Lusin spaces, M a continuous orthogonal martingale measure on $E\times\mathbb{R}_+$, with covariance measure $q_t(dx)dk_t$, and $r_t(x,dy)$ a predictable transition kernel from E to F.

Then there exists on an extension of Ω a continuous orthogonal martingale measure $\hat{M}$ on $E\times F\times\mathbb{R}_+$, with covariance measure $q_t(dx)r_t(x,dy)dk_t$, with first marginal M. (that is: $\forall\ A\in\mathcal{E},\ \forall\ t\in\mathbb{R}_+,\quad \hat{M}_t(A\times F) = M_t(A)\quad P$ a.s.).

This extension result will be used later to give a useful representation of a class of vector square integrable martingales as stochastic integrals of martingale measures. But let us first give an application of the theory of martingale measures to branching processes. The use of this notion has allowed to answer a question formulated by Dawson in 1975 [Da] (and partially resolved in one dimension by Konno-Shiga [Ko-Sh] in 1988):

Are measure-valued branching processes solutions to stochastic differential equations, and in which space ?

If X is a continuous measure-valued branching process, it is proved in [Ro] that X satisfies the following martingale property:
if A is the generator of the underlying diffusion process,

$\forall f\in\mathcal{D}(A)$, $\forall t\in\mathbb{R}+$, $\langle X_t,f\rangle - \langle X_o,f\rangle - \int_o^t \langle X_s,(A+b)f\rangle\, ds = M_t^f$, where M^f

is a continuous square integrable martingale with increasing process

$\int_o^t \langle X_s, cf^2\rangle\, ds = \int_o^t \int_{\mathbb{R}^d} f^2(x)\, c(x)\, X_s(dx)ds$, where b and c are bounded continuous functions, c being positive. (The function b is related to the mean of the reproduction law, the function c to the variance of this law).

In [Me-Ro2], we interpret then the measure $c(x)X_s(dx)ds$ as the covariance measure of a continuous orthogonal martingale measure and we prove that there exists effectively a continuous orthogonal martingale measure M on $\mathbb{R}^d\times\mathbb{R}+$, with covariance measure $c(x)X_s(dx)ds$ such that: $\forall f\in\mathcal{D}(A)$, $M_t(f) = M_t^f$.
We obtain thus the process X as solution of a stochastic differential equation in the space of $L^2(\Omega)$-valued vector measures: $dX_t = (A + b)^* X_t\, dt + dM_t$.

To obtain a more explicit relation between X and M, we write M as image measure of a white noise (following the above described result):
$X_t = \lambda o\varphi_t^{-1}$, $M = Wo\varphi^{-1}$ and then $dX_t = (A + b)^* X_t\, dt + \sqrt{c}\, dWo\varphi_t^{-1}$.

Unfortunately, we don't have uniqueness for the solutions of this equation, since we have in particular no information about the regularity of φ_t as function of X_t. We get also an evolution equation.

We can generalize these results to discontinuous branching processes [Me-Ro3] and obtain a similar equation with an additional Poisson part.

We shall now present some results concerning a certain class of vector martingales which we shall see come from an interesting class of random martingale problems. We shall give some applications of these results.

Our study concerns a family of d continuous square integrable martingales $(m_t^i)_{i=1}^d$ saytisfying the following hypotheses (H):
(H) : $\forall i\in\{1,\dots,d\}$, $m_o^i = 0$

$$\forall i,j \in \{1,\dots,d\},\ \forall t\in\mathbb{R}+,\ \langle m^i,m^j\rangle_t = \int_o^t\int_E a_{ij}(s,x)\, q_s(dx)\, dk_s$$

where: E is a Lusin space, (q_t) is a predictable finite measure (on E) valued process, (k_t) is a continuous increasing process, $a(s,x) = \sigma(s,x)\sigma^*(s,x)$ is a

$\mathcal{P}\otimes\mathcal{E}$ measurable matrix with $a_{ij}(s,x) \in L^1(q_s(dx)dk_s)$ $\forall$ i,j $\in$ {1,...,n},a.s..
For this class, we obtain a very interesting representation theorem.

Representation Theorem: ([Ek-Me])
Let us consider a family of d continuous square integrable martingales $(m_t^i)_{i=1}^d$ saytisfying the hypotheses (H).
Then there exist on an extension of the probability space d continuous orthogonal martingale measures $(M^k)_{k=1}^d$, strongly orthogonal, with covariance measure $q_s(dx)dk_s$ such that:

$$\forall\ i\in\{1,\dots,d\},\qquad m_t^i = \sum_{k=1}^{d} \int_0^t \int_E \sigma_{ik}(s,x)\, M^k(dx,ds).$$

This theorem generalizes the particular case where the matrix a doesn't depend on x and where the martingales are represented as stochastic integrals with respect to d orthogonal martingales whith increasing process k_t (up to a multiplicative constant). (cf. for example [Ik-Wa]). As in this particular case, the proof of the theorem uses techniques of construction and extension of martingale measures, in particular the extension theorem above.

This theorem allows us to give a pathwise representation of the solutions of the following martingale problem:

We consider on a given probability space $(\Omega,\mathcal{F},\mathcal{F}_t,P)$ a predictable process (q_t) whith values in the set of finite measures on a Lusin space E.
We introduce the random generator $\mathcal{L}^q$ defined as follows:
$\forall\ f \in C_b^2(\mathbb{R}^d)$, $\forall\ t \in \mathbb{R}+$, $\forall\ y \in \mathbb{R}^d$,
$\mathcal{L}^q f(t,y) = \int_E \mathcal{L}f(t,y,x)\, q_t(dx)$, where $\mathcal{L}$ is the second order differential operator $\mathcal{L}f(t,y,x) = 1/2 \sum_{i,j} a_{ij}(s,y,x) \frac{\partial^2 f}{\partial y_i \partial y_j} + \sum_i b_i(s,y,x) \frac{\partial f}{\partial y_i}$
with bounded coefficients a and b and $a(s,y,x) = \sigma(s,y,x)\sigma^*(s,y,x)$.
σ can be degenerate.
Following [Ja-Me2], we define a solution of this random martingale problem: let C be the space of continuous functions from $\mathbb{R}+$ into $\mathbb{R}^d$ and $(\mathcal{C}_t)$ the associated filtration. We introduce the extension $\tilde{\Omega} = C\otimes\Omega$, $\tilde{\mathcal{F}}_t = \bigcap_{s>t} (\mathcal{C}_s\otimes\mathcal{F}_s)$.

If (X_t) designs the canonical process on C, we shall call solution of the martingale problem associated with $\mathcal{L}^q$ a probability measure $\tilde{P}$ defined on $\tilde{\Omega}$, extension of P, such that $\forall\ f \in C_b^2(\mathbb{R}^d)$,

$$(\mathcal{P})\quad f(X_t) - f(X_0) - \int_0^t \mathcal{L}^q f(s,X_s)\, ds = f(X_t) - f(X_0) - \int_0^t\int_E \mathcal{L}f(s,X_s,x)\, q_s(dx)ds$$

is a $\tilde{P}$ martingale.

For example, if the space E is the finite set of points $\{a_1,\dots,a_N\}$, then $q_t(da) = \sum_{j=1}^N q_t^j\, \delta_{\{a_j\}}$, where for each t, $1 \geq q_t^j \geq 0$ and $\sum_{j=1}^N q_t^j = 1$.

One can then easily prove that $\tilde{P}$ is the law of a process x solution of the stochastic differential equation:

$$dx_t = \sum_{j=1}^{N} b(t,x_t,a_j)\, q_t^j\, dt + \sum_{j=1}^{N} \sigma(t,x_t,a_j)\, (q_t^j)^{1/2}\, dW_t^j \,, \quad \text{where} \quad (W^j)_{j=1}^{N} \text{ is a}$$

N-dimensional Brownian motion defined on an extension of the space $\tilde{\Omega}$. The process M defined by $M(A\times(0,t]) = \sum_{j=1}^{N} \int_0^t (q_s^j)^{1/2}\, 1_{\{a_j\in A\}}\, dW_s^j$ is a martingale measure.

By applying the representation theorem above and usual ideas of stochastic calculus, we generalize this result to any Lusin space E and then give a pathwise interpretation of the probability measure $\tilde{P}$:

Theorem: ([Ek-Me])

1) - $\tilde{P}$ is the law of a continuous d-dimensional process X solution of the stochastic differential equation:

$$(E) \quad \forall\, i\in\{1,..,d\}, \quad dX_t^i = \sum_{k=1}^{d} \int_E \sigma_{ik}(t,X_t,x)\, M^k(dx,dt) + \int_E b_i(t,X_t,x)\, q_t(dx)dt$$

where (M^k) is a family of continuous orthogonal martingale measures, strongly orthogonal, with covariance measure $q_t(dx)dt$.

2) - If the functions σ and b are Lipschitz continuous in the $\mathbb{R}^d$-variable, uniformly on (t,x), there is strong existence and uniqueness of solutions for (E), in the sense that existence and pathwise uniqueness are obtained for given d-dimensional martingale measure M and process q.

The choice of M is of course not unique, except in the case where q is deterministic since M is then the white noise with intensity $q_t(dx)dt$.

Let us give now some applications of this result.

A - An interacting particle system.

We give here the principal ideas of a work developed in [Me-Ro1].
The interacting particle system we consider moves in $\mathbb{R}^{dn}$ following a Markov process whose generator $\mathcal{L}^n$ is given by :

$$\mathcal{L}^n f(x^1,\dots,x^n) = \sum_{k=1}^{n} L_{(k)}^{\mu_x^n} f(x^1,\dots,x^n) \quad \text{where} \quad L_{(k)}^{\mu_x^n} f(x^1,\dots,x^n) = L^{\mu_x^n} f_{(k)}(x^k).$$

$f_{(k)}$ is equal to the function f acting only on the kth variable , the others variables being frozen, μ_x^n is the empirical measure of x ($\mu_x^n = 1/n \sum_{k=1}^{n} \delta_{x^k}$) and for a probability measure m on $\mathbb{R}^d$, L^m is defined on each function f of $C_b^2(\mathbb{R}^d)$ by $L^m f(y) = 1/2 \sum_{i,j} a_{ij}(y,m) \dfrac{\partial^2 f}{\partial y_i \partial y_j} + \sum_i b_i(y,m) \dfrac{\partial f}{\partial y_i}$.

We suppose moreover that

$$a(x,m) = \int \sigma(x,y)\sigma^*(x,y)\; m(dy) \quad \text{and} \quad b(x,m) = \int b(x,y)\; m(dy).$$

One can easily prove that there is a unique solution P^n to the martingale problem associated with $\mathcal{L}^n$.

By applying the above results, we obtain that P^n is the law of the solution of the following system of stochastic differential equations:

$$X_t^i = X_o^i + \sum_{k=1}^{n} \int_o^t \int_{\mathbb{R}^d} \sigma_{ik}(X_s^i,x)\, M^k(dx,ds) + \int_o^t b(X_s^i,\mu_s^n)\, ds ,$$

where $\mu^n = 1/n \sum_{k=1}^{n} \delta_{X^k}$, M^k being n strongly orthogonal d-dimensional continuous martingale measures.

The process X^1 is thus a semimartingale and we have its decomposition in martingale part and bounded variation part. We can then apply well known criteria to prove the tightness of the sequence of the laws of (X^1) (under P^n). That implies also the tightness of the laws of the empirical measures μ^n (first step to obtain the propagation of chaos for the sequence P^n). To obtain the uniqueness of limit values, we necessary need the pathwise representation. The limit is solution of a non linear martingale problem, obtained as fixed point of a contraction. We will not here detail more precisely this model, and refer to the paper mentioned above [Me-Ro1].

B - Relaxed control problem

We study a model of controlled diffusion which is solution of the equation

$(\mathcal{E})$ $\quad dx_t^u = b(t,x_t^u,u_t)\, dt + \sigma(t,x_t^u,u_t)\, dB_t, \qquad x_o^u = z$

where B is a $\mathcal{F}_t$ Brownian motion, b and σ are continuous bounded functions, respectively from $\mathbb{R}+\times\mathbb{R}^d\times E$ into $\mathbb{R}^d$ and from $\mathbb{R}+\times\mathbb{R}^d\times E$ into the space of (d×d) matrices, b and σ being uniformly Lipschitz continuous in the $\mathbb{R}^d$ variable (we don't need a non-degenerate assumption on σ), $z\in\mathbb{R}^d$, and the control process u is $\mathcal{F}_t$-predictable with values in a metric compact set E. (We denote by $\mathcal{U}$ the set of such processes).

The cost function J on the time interval [0,1] is given by:

$$J(z,u) = E\ [\ \int_o^1 h(s,x_s^u,u_s)\, ds + g(x_1^u)\ ]$$

where the functions g and h are assumed continuous and bounded, respectively from $\mathbb{R}^d$ and $\mathbb{R}+\times\mathbb{R}^d\times E$ into $\mathbb{R}$.

The aim of the control theory is to optimize this cost function over the processes u in $\mathcal{U}$. We shall call value function the function $V(z) = \inf_{u\in\mathcal{U}} J(z,u)$ and optimal control a control on which this infimum is attained if there exists one. But an optimal control does not necessarily exists in $\mathcal{U}$, the set $\mathcal{U}$ of E-valued predictable processes not being endowed with a compact topology. The usual idea in control theory ([FN1],[FN2],[ENP1],[ENP2]) is to embed this set in the set $\mathcal{R}$ of probability measures on E×[0,1] which disintegrate in the form $q_t(dx)dt$ where q_t is a transition probability measure from $\mathbb{R}+$ into $\mathbb{R}$, the embedding being given by the application: $\psi((u_t)) = \delta_{u_t}(dx)dt$.

The set $\mathcal{R}$ is called set of relaxed controls and is compact for the weak topology. One can also prove ([Ja-Me1]) that the weak convergence is in $\mathcal{R}$

equivalent to the stable convergence. (The test functions can be just taken measurable in time and continuous in space). In fact, thanks to a lemma known under the name of chattering lemma, one can approximate each element of $\mathcal{R}$ in the following way:

<u>Chattering lemma:</u> ([ENP1]) Let (q_t) be a predictable process with values in the space of probability measures on E. There exists a sequence of E-valued predictable processes (u_t^k) such that the sequence of random measures $(\delta_{u_t^k}(dx)dt)$ converges weakly on $E\times[0,1]$ to $q_t(dx)dt$, P a.s., when k tends to ∞

One generalizes the notion of a diffusion process x^u controlled by a process u of $\mathcal{U}$ to the one of a diffusion process X^q controlled by a relaxed control $q_t(dx)dt$. We relax the model by making the dependence on the dynamics linear in the control (in a certain sense, the model becomes simpler): the operator of the controlled diffusion (controlled by $q_t(da)dt \in \mathcal{R}$) is given with respect to the generators $\mathcal{L}^a$ of the processes x^a, $a \in E$:

$$\mathcal{L}_t^q f(y) = \int_E \mathcal{L}_t^x f(y)\, q_t(dx)$$

$$= \int_E \left(\sum_{i=1}^d f'_{y_i}(y)\, b_i(t,y,x) + 1/2 \sum_{i,j=1}^d f''_{y_i y_j}(y)\, a_{ij}(t,y,x) \right) q_t(dx)$$

where $a(t,y,x) = \sigma(t,y,x)\sigma^*(t,y,x)$.

The representation theorem gives a stochastic representation of these controlled diffusions:

(E) $\quad dX_t^q = \int_E b(t,X_t^q,x)\, q_t(dx)dt + \int_E \sigma(t,X_t^q,x)\, M(dx,dt)$ where M is a d-dimensional continuous martingale measure with covariance measure $q_t(dx)dt$.

The cost function of this relaxed model will be

$$J(z,q) = E\left(\int_0^1\int_E h(s,X_t^q,x)\, q_s(dx)\, ds + g(X_1^q) \right).$$

The problem of relaxed control is then easier to solve, thanks to the compactness of the set $\mathcal{R}$, and the existence of an optimal control in this class has been proved. ([ENP2]). One has then to compare the relaxed problem to the initial problem and in particular the cost functions of each problem.

That was the first motivation to prove the following stability theorem, which is in fact interesting by itself.

<u>Stability theorem:</u> ([Me])

Let M be a continuous orthogonal martingale measure on $E\times[0,T]$ with finite covariance measure ν.

Let us assume that there exists a sequence of random predictable measures (ν^n) converging weakly to ν on $E\times[0,T]$, P almost surely and such that for each n, ν^n and ν have the same second marginals ($\nu^n(E\times.) = \nu(E\times.)$ P a.s.).

Then, there exists on an extension of the probability space a sequence of orthogonal continuous martingale measures M^n defined on $E\times[0,T]$, with covariance measure ν^n, such that :

. For each predictable bounded function φ from $\Omega\times E\times[0,T]$ to $\mathbb{R}$, continuous in the E-variable,

$$\lim_{n\to+\infty} E[(M_t^n(\varphi) - M_t(\varphi))^2\ ] = 0.$$

Ideas of the proof:

1) - For almost all $\omega \in \Omega$, we construct a sequence of random probability measures (m^n) on $E\times E\times[0,T]$ satisfying :

$m^n(\omega,E,da',dt) = \nu(\omega,da',dt)$; $m^n(\omega,da,E,dt) = \nu^n(\omega,da,dt)$

and converging weakly to a random probability measure m defined on $E\times E\times[0,T]$ by $m(\omega,da,da',dt) = \nu(\omega,da,dt)\ \delta_a(da')$.

We use for this construction a generalization of Skorohod's representation theorem ([S]) to product spaces.

2) - We construct thanks to the extension theorem a sequence of continuous orthogonal martingale measures $(\hat{M}^n)$ defined on $E\times E\times[0,T]$ having for each n the dual predictable projection of m^n as covariance measure and a second marginal equal to M.

3) - Let us consider now for each n the first marginal M^n of the martingale measure $\hat{M}^n$. The sequence (M^n) is the one we were looking for, as it is easy to verify : For each bounded continuous function φ from E to $\mathbb{R}$, $\forall\ n\in\mathbb{N}$, $\forall\ t\in[0,T]$

$$M_t^n(\varphi) = \int_0^t\int_{E\times E} \varphi(a)\ \hat{M}^n(da,da',ds) \quad \text{and} \quad M_t(\varphi) = \int_0^t\int_{E\times E} \varphi(a')\ \hat{M}^n(da,da',ds).$$

Thus

$$E[(M_t^n(\varphi) - M_t(\varphi))^2] = E\Big[\Big(\int_0^t\int_{E\times E} (\varphi(a)-\varphi(a'))\ \hat{M}^n(da,da',ds)\Big)^2\Big]$$
$$= E\Big[\int_0^t\int_{E\times E} (\varphi(a)-\varphi(a'))^2\ m^n(da,da',ds)\Big]$$

which converges to

$$E\Big[\int_0^t\int_{E\times E} (\varphi(a)-\varphi(a'))^2\ \nu(da,ds)\ \delta_a(da')\Big] = 0\ .$$

The generalization to the stochastic integrals is obtained thanks to the equivalence in $\mathcal{R}$ between weak convergence and stable convergence.

An amusing application to this theorem is to obtain each continuous orthogonal martingale measure defined on a compact set as limit of a sequence of stochastic integrals with respect to a Brownian motion.

Corollary : Let E be a metric compact space and M a continuous orthogonal martingale measure with covariance measure $q_t(dx)dt$ on $E\times[0,1]$, where q is a probability measure-valued process.

Then there exists a sequence of predictable E-valued processes (u^k) and a

Brownian motion B defined on an extension of the probability space such that : for each continuous bounded function ϕ from E to $\mathbb{R}$, $\forall\, t \in [0,1]$

$$\lim_{k\to+\infty} E\left[\left(M_t(\phi) - \int_0^t \phi(u_s^k)\, dB_s \right)^2 \right] = 0.$$

A similar result for a more general covariance measure can be deduced by time change.

Remark : For each function ϕ, $M(\phi)$ is a continuous martingale with increasing process $\int_0^t \left(\int_E \phi^2(x)\, q_s(dx) \right) ds$ and can be represented as a stochastic integral with respect to a Brownian motion W^ϕ :

$$M_t(\phi) = \int_0^t V_s^\phi\, dW_s^\phi \quad \text{where} \quad V_s^\phi = \left(\int_E \phi^2(x)\, q_s(dx) \right)^{1/2}.$$

It is clear that V^ϕ is not linear in ϕ and thus the Brownian motion W^ϕ depends on ϕ.

The interest of the above corollary is to give an approximation of $M_t(\phi)$ in $L^2(\Omega)$ by stochastic integrals of a Brownian motion independent of ϕ.

Ideas of the corollary proof :

The chattering lemma allows to approximate the random probability measure $q_t(dx)dt$ by a sequence of atomic measures of the form $\delta_{u_t^k}(dx)dt$ for the weak topology on the space of measures on $E\times[0,1]$, this being true for almost all ω of Ω. The u^k's are predictable E-valued processes.

By the stability theorem, we construct a sequence of continuous orthogonal martingale measures M^k with covariance measure $\delta_{u_t^k}(dx)dt$, which converges to M

As seen in example 2, M^k can then be written as : $M_t^k(A) = \int_0^t 1_A(u_s^k)\, dm_s^k$ where m^k is the continuous martingale defined by $m_t^k = M_t^k(E)$.

But the construction of M^k ensures in fact that for each k of $\mathbb{N}$, $M_t^k(E) = M_t(E)$ M(E) is a continuous martingale with increasing process t, and is thus a Brownian motion (independent of k) that we denote by B.

Let us now come back to the control problem and let us apply the above approximation results. To a relaxed control $q_t(dx)dt$ and to the associated controlled diffusion X^q, we adjoin a continuous orthogonal martingale measure M such that X^q is solution of the following stochastic differental equation :

$$\text{(E)} \qquad dX_t^q = \int_E b(t,X_t^q,x)\, q_t(dx)dt + \int_E \sigma(t,X_t^q,x)\, M(dx,dt), \qquad X_0^q = z.$$

To the pair $(q_t(dx)dt,\ M(dx,dt))$, we associate the sequence $(\delta_{u_t^k}(dx)dt,\ M^k(dx,dt))$ as described in the corollary. We also introduce for each k the process X^k solution of the same equation as (E) where the couple $(q_t(dx)dt,\ M(dx,dt))$ is replaced by $(\delta_{u_t^k}(dx)dt,\ M^k(dx,dt))$. Since there exists a Brownian motion B such that for each k, $M_t^k(A) = \int_0^t 1_A(u_s^k)\, dB_s$, it is easy to see that

X^k is in fact solution of the initial stochastic differential equation :

$$(\mathcal{E}) \qquad dx_t^u = b(t,x_t^u,u_t)\, dt + \sigma(t,x_t^u,u_t)\, dB_t, \qquad x_o^u = z$$

where u^k is substituted for u.

We can then state the main following theorem : the relaxed control problem is approximated in L^2 by the initial control problem :

Theorem : ([Me])

1) $\lim_{k\to+\infty} E\,[\sup_{t\in[0,1]} (X_t^k - X_t^q)^2] = 0.$

2) Let J_k be the sequence of cost functions associated with X^k. There exists a sub-sequence (J_{k_l}) of the sequence (J_k) which converges to J, where J is the cost function associated with the relaxed diffusion X^q.

Ideas of the proof :

The first step is obtained by using the convergence of the covariance measures $\delta_{u_t^k}(dx)dt$ and of the martingale measures M^k, and thanks to the hypothesis of continuity and Lipschitz continuity on the $\mathbb{R}^d$-variable of the coefficients σ and b.

The convergence of a sub-sequence of (J_k) is then a direct consequence of the convergence of the sequence (X^k) in L^2, since the functions h and g which appear in the cost function are bounded and continuous.

REFERENCES

[Da] D.A. Dawson, *Stochastic evolution equations and related measure processes*, J. Multivariate Anal. 5, 1975.

[Ek-Me] N. El Karoui, S. Méléard, *Martingale measures and stochastic calculus*, Probat ity Theory and Related Fields 84, (1990), pp. 83-101.

[Ek-Le] N. El Karoui, J.P. Lepeltier, *Représentation des processus ponctuels multivariés à l'aide d'un processus de Poisson*, Z. W. Verw. Geb. 39, 111-133 (1977).

[ENP1] N. El Karoui, D. Huu Nguyen, M. Jeanblanc-Picqué, *Existence of an optimal markovian filter for the control under partial observations*, SIAM J. of Control and Optimisation, 26 n°4 (1988), pp. 1025-1061

[ENP2] N. El Karoui, D. Huu Nguyen, M. Jeanblanc-Picqué, *Compactifications methods in the control of degenerate diffusions: existence of an optimal control.*, Stochastics, 20, (1987), pp. 169-219.

[FN1] W.H. Fleming, M. Nisio, *On the existence of optimal stochastic controls*, J. Math. Mech. 15 (1966), pp. 777-794.

[FN2] W.H. Fleming, M. Nisio, *On stochastic relaxed control for partially observed diffusions*, Nagoya Math. Journal 93,(1984), pp. 71-108.

[Ik-Wa] N. Ikeda, S. Watanabe, Stochastic Differential Equations and Diffusion Processes, Amsterdam/Kodansha: North Holland Mathematical Library 1981.

[Ja-Me1] J. Jacod, J. Memin, *Sur un type de convergence intermédiaire entre la convergence en loi et la convergence en probabilité*, Séminaire de Probabilités XV,L.N. n° 850,(1980), Springer.

[Ja-Me2] J. Jacod , J. Memin, *Weak and strong solutions of stochastic differential equations: existence and stability*, Proc. Durham Symp.1980, L.N. n° 851,(1981), Springer.

[Ko-Sh] N. Konno, T. Shiga, *Stochastic differential equations for some measure-valued diffusions*, Prob. Theory and Rel. Fields (1979), 201-226.

[Me] S. Méléard, *Martingale measure approximations. Application to the control of diffusions,* Prépublication n°69 du Laboratoire de Probabilités de l'Université Paris 6.

[MeRo1] S. Méléard, S. Roelly, *Systèmes de particules et mesures martingales: un théorème de propagation du chaos*, Séminaire de Probabilités n°22, L.N. n°1321, (1988), pp. 438-448, Springer.

[MeRo2] S. Méléard, S. Roelly, *A generalized equation for a continuous measure branching process*, Proceedings Trento 1988 "Stochastic Partial Differential Equations II", L.N. n°1390, (1990),pp. 171-185, Springer.

[Me-Ro3] S. Méléard, S. Roelly, *Discontinuous measure-valued branching processes and generalized stochastic equations*, to appear in Math. Nachrichten.

[Ro] S. Roelly-Coppoletta, *A criterion of convergence of measure-valued processes: application to measure branching processes*, Stochastics, Vol. 17, (1986), 43-65.

[S] A.V. Skorohod, Studies in the Theory of Random Processes, Reading, Mass, Addison Wesley 1965.

[W] J.B. Walsh, *An introduction to stochastic partial differential equations*, Ecole d'été de Probabilités de Saint-Flour, n°XIV, 1984, Springer.

Backward Stochastic Differential Equations and Quasilinear Parabolic Partial Differential Equations

E. Pardoux
Université de Provence
UFR MIM
F–13331 Marseille Cedex 3
and INRIA

S. Peng
Institute of Mathematics
Shandong University
Jinan, 250100
China

Introduction

A new class of backward stochastic differential equations has been studied by the authors in [3], and it has been used by the second author in [4], in order to give a probabilistic formula for the given solution of a system of parabolic partial differential equation.

The aim of the present paper is to study the regularity properties of the solution of the backward SDE (in short BSDE), and to deduce a converse of the results of [4], namely to show that a given function expressed in terms of the solution of the BSDE solves a certain system of parabolic PDEs. Our result generalizes the well–known Feynman–Kac formula (see Remark 3.3 below). It gives an existence and uniqueness result for a *system* of quasilinear (and *possibly degenerate*) parabolic equations. We also obtain an existence result for the viscosity solution of a quasilinear parabolic equation.

We shall extend our approach in a forthcoming publication, to the case of systems of quasilinear parabolic stochastic partial differential equations. Our approach may also prove useful for solving certain equations on manifolds.

The paper is organised as follows. In section 1, we shall state our assumptions, and recall some results from previous work. In section 2, we establish some estimates and regularity results for the solution of the BSDE, in section 3 we shall relate it to a system of quasilinear parabolic PDEs. Finally, in 4, we relate the solution of the one–dimensional BSDE to the viscosity solution of a quasilinear parabolic PDE, under much weaker assumptions.

1 Preliminaries

In all what follows, we shall work on a fixed finite time interval $[0,T]$. We suppose given on a probability space $(\Omega,\mathcal{F},P)$ a d–dimensional standard Wiener process $\{W_t;\ t\in[0,T]\}$. For $0\le t\le r\le T$, we define $\overset{o}{\mathcal{F}}{}^t_r=\sigma\{W_s-W_t;\ t\le s\le r\}$ and $\mathcal{F}^t_r$ denotes the completion of $\overset{o}{\mathcal{F}}{}^t_r$ with the P–null sets of $\mathcal{F}$. We shall write $\mathcal{F}_r$ for $\mathcal{F}^0_r$ and $\mathcal{F}^t$ for $\mathcal{F}^t_T$.

For any $0\le t\le r\le T$, $p\in\mathbb{N}$, we denote by $M^2(t,r;\mathbb{R}^p)$ the subset of $L^2(\Omega\times(t,r),\ dP\times ds,\ \mathbb{R}^p)$ consisting of $\mathcal{F}^t_s$–progresively measurable processes.

$C^k(\mathbb{R}^p,\mathbb{R}^q)$, $C^k_{l,b}(\mathbb{R}^p,\mathbb{R}^q)$, $C^k_p(\mathbb{R}^p,\mathbb{R}^q)$ will denote respectively the set of functions of class C^k from $\mathbb{R}^p$ into $\mathbb{R}^q$, the set of those functions of class C^k whose partial derivatives of order less than or equal to k are bounded (and hence the function itself growths at most like a linear function of the variable x at infinity), and the set of those functions of class C^k which, together with all their partial derivatives of order less than or equal to k, grow at most like a polynomial function of the variable x at infinity.

We are given $b \in C^3_{l,b}(\mathbb{R}^d;\mathbb{R}^d)$ and $\sigma \in C^3_{l,b}(\mathbb{R}^d;\mathbb{R}^{d\times d})$, and for each $t \in [0,T)$, $x \in \mathbb{R}^d$, we denote by $\{X^{t,x}_s,\ t \le s \le T\}$ the unique strong solution of the following SDE :

$$\begin{cases} dX^{t,x}_s &= b(X^{t,x}_s)ds + \sigma(X^{t,x}_s)dW_s,\ t \le s \le T \\ X^{t,x}_t &= x \end{cases} \tag{1}$$

It is well–known (see e.g. Stroock [8]) that the random field $\{X^{t,x}_s;\ 0 \le t \le s \le T,\ x \in \mathbb{R}^d\}$ has a version which is a.s. jointly continuous in (t,s,x), together which its x partial derivatives of order one and two.

Moreover, $\sup_{t\le s\le T}(|X^{t,x}_s| + |\nabla X^{t,x}_s| + |D^2X^{t,x}_s|) \in \cap_{p\ge 1}L^p(\Omega)$, for each t and x, where $\nabla X^{t,x}_s$ and $D^2X^{t,x}_s$ denote respectively the first and second order partial derivative of $X^{t,x}_s$ with respect to x.

Let us now recall the notion of derivation on Wiener space. We denote by $\mathbf{S}$ the set of random variables ξ of the form :

$$\xi = \varphi(W(h_1),\dots,W(h_n))$$

where $f \in C^\infty_b(\mathbb{R}^n)$, $h_1,\dots,h_n \in L^2(0,T;\mathbb{R}^d)$, and $W(h_1) = \int_0^T(h_i(s), dW_s)$ is the Wiener integral of h_i with respect to $\{W_s;\ 0 \le s \le T\}$ [$(\cdot,\cdot)$ denotes the scalar product in $\mathbb{R}^d$]. To such a random variable ξ, we associate a "derivated process" $\{D_r\xi\,;\ r \in [0,T]\}$ defined as :

$$D_r\xi = \sum_{i=1}^n \frac{\partial\varphi}{\partial x_i}(W(h_1),\dots,W(h_n))h_i(r)$$

For $\xi \in \mathbf{S}$, we define its 1,2–norm by :

$$||\xi||^2_{1,2} = E(\xi^2) + E\int_0^T |D_r\xi|^2 dr$$

One can show (see e.g. Nualart–Pardoux [2]) that $D.\ :\mathbf{S} \to L^2(\Omega\times(0,T);\ \mathbb{R}^d)$ is closable, as an unbounded operator from $L^2(\Omega)$ into $L^2(\Omega\times(0,T);\ \mathbb{R}^d)$, hence $D.$ can be extended as an operator from its domain which coincides with $\mathbb{D}^{1,2} \triangleq \overline{\mathbf{S}}^{||\cdot||_{1,2}}$ into $L^2(\Omega\times(0,T);\ \mathbb{R}^d)$. Note that if $\xi \in \mathbb{D}^{1,2}$ is $\mathcal{F}^t_s$ measurable, $D_r\xi = 0$ for $r \in [0,T]\setminus[t,s]$. We shall denote by $D^i_r\xi$, $1 \le i \le d$, the i–th component of $D_r\xi$.

It follows from the closedness property of $D.$ that the following results can be proved by approximation.

Lemma 1.1 *For any $0 \le t < s \le T$, $x \in \mathbb{R}^d$, $X^{t,x}_s \in (\mathbb{D}^{1,2})^d$, and a version of $\{D_rX^{t,x}_s;\ s,r \in [0,T]\}$ is given by :*

(i) $D_rX^{t,x}_s = 0,\ r \in [0,T]\setminus(t,s]$

(ii) For any $t < r \le T$, $\{D_rX^{t,x}_s;\ r \le s \le T\}$ is the unique solution of the linear SDE :

$$\begin{aligned} D_rX^{t,x}_s &= \sigma(X^{t,x}_s) + \int_r^s b'(X^{t,x}_\alpha)D_rX^{t,x}_\alpha d\alpha \\ &\quad + \int_r^s \sigma'_i(X^{t,x}_\alpha)D_rX^{t,x}_\alpha dW^i_\alpha \end{aligned}$$

where we use the convention of summation over the repeated index i, from $i = 1$ to $i = d$, and σ_i denotes the i–th column of the matrix σ.

We now introduce the BSDE. Let $f : [0,T] \times \mathbb{R}^d \times \mathbb{R}^k \times \mathbb{R}^{k\times d} \to \mathbb{R}^k$ be such that for any $s \in [0,T]$, $(x,y,z) \to f(s,x,y,z)$ is of class C^3 and moreover :

(i) $f(s,\cdot,0,0) \in C_p^3(\mathbb{R}^d;\ \mathbb{R}^k)$,

(ii) the first order partial derivatives in y and z are bounded on $[0,T] \times \mathbb{R}^d \times \mathbb{R}^k \times \mathbb{R}^{k\times d}$, as well as their derivatives of order one and two with respect to x, y, z.

Let $g \in C_p^3(\mathbb{R}^d)$. For any $t \in [0,T)$ and $x \in \mathbb{R}^d$, let $\{(Y_s^{t,x}, Z_s^{t,x});\ t \le s \le T\}$ denote the unique element of $M^2(t,T;\ \mathbb{R}^k) \times M^2(t,T;\ \mathbb{R}^{k\times d})$ which solves the following BSDE (see [3]) :

$$Y_s^{t,x} = g(X_T^{t,x}) + \int_s^T f(X_r^{t,x}, Y_r^{t,x}, Z_r^{t,x})dr - \int_s^T Z_r^{t,x} dW_r,\ t \le s \le T\,. \tag{2}$$

For further reference, let us indicate the method of construction of the solution $(Y_\cdot^{t,x}, Z_\cdot^{t,x})$. Dropping the superscript t,x for convenience, we construct the solution in three steps.

First, given arbitrary $\overline{Y} \in M^2(t,T;\ \mathbb{R}^k)$ and $\overline{Z} \in M^2(t,T;\ \mathbb{R}^{k\times d})$, we solve the equation

$$Y_s = g(X_T) + \int_s^T f(X_r, \overline{Y}_r, \overline{Z}_r)dr - \int_s^T Z_r dW_r,\ t \le s \le T\,, \tag{3}$$

whose unique solution is given explicitly by :

$$Y_s = E^{\mathcal{F}_s^t}[g(X_T) + \int_s^T f(X_r, \overline{Y}_r, \overline{Z}_r)dr]\,,$$

$\{Z_s,\ t \le s \le T\}$ is the unique element of $M^2(t,T;\ \mathbb{R}^{k\times d})$ which is given by Itô's representation theorem of Brownian martingales (see e.g. Karatzas–Shreve [1]), such that :

$$\int_t^T Z_s dW_s = g(X_T) + \int_t^T f(X_r, \overline{Y}_r, \overline{Z}_r)dr - E\left[g(X_T) + \int_t^T f(X_r, \overline{Y}_r, \overline{Z}_r)dr\right]$$

Next, given an arbitrary element $\overline{Y} \in M^2(t,T;\ \mathbb{R}^k)$, we solve the equation

$$Y_s = g(X_T) + \int_s^T f(X_r, \overline{Y}_r, Z_r)dr - \int_s^T Z_r dW_r,\ t \le s \le T\,, \tag{4}$$

by the iterative procedure :

$$\begin{aligned} Z_s^0 &\equiv 0\,, \\ Y_s^{n+1} &= g(X_T) + \int_s^T f(X_r, \overline{Y}_r, Z_r^n)dr - \int_s^T Z_r^{n+1} dW_r,\ t \le s \le T,\ n \in \mathbb{N}\,, \end{aligned}$$

where the n–th equation is solved with the help of the first step.

We finally solve equation (2) by the iterative procedure :

$$\begin{aligned} Y_s^0 &\equiv 0\,, \text{and for } n \in \mathbb{N}\,, \\ Y_s^{n+1} &= g(X_T) + \int_s^T f(X_r, Y_r^n, Z_r^{n+1})dr - \int_s^T Z_r^{n+1} dW_r,\ t \le s \le T, \end{aligned}$$

where the n–th equation is solved with the help of the second step.

The convergence of the two approximating sequences is proved respectively on page 57–58 and 59 of [3].

2 Regularity of the solution of the backward SDE

We first estimate higher order moments of (Y). We again suppress the superscript t, x for notational convenience. For $z \in \mathbb{R}^{k\times d}$, we denote $||z|| = \sqrt{Tr(zz^*)}$.

Lemma 2.1 *For any $p \geq 1$,*

$$E\left(\sup_{t\leq s\leq T} |Y_s|^p\right) < \infty \tag{5}$$

$$E\left[\left(\int_t^T |Z_s|^2\, ds\right)^{p/2}\right] < \infty \tag{6}$$

Proof : From Itô's formula,

$$|g(X_T)|^2 = |Y_s|^2 - 2\int_s^T (f(X_r, Y_r, Z_r), Y_r)dr + 2\int_s^T (Y_r, Z_r dW_r) + \int_s^T ||Z_r||^2 dr \;,$$

and for any $p \geq 2$,

$$|g(X_T)|^{2p} \geq |Y_s|^{2p} - 2p\int_s^T |Y_r|^{2(p-1)}(f(X_r, Y_r, Z_r), Y_r)dr$$

$$+2p\int_s^T |Y_r|^{2(p-1)}(Y_r, Z_r dW_r) + p\int_s^T |Y_r|^{2(p-1)}||Z_r||^2 dr \;,$$

hence for some constant C,

$$|Y_s|^{2p} + p\int_s^T |Y_r|^{2(p-1)}||Z_r||^2 dr \leq |g(X_T)|^{2p}$$

$$+2Cp\int_s^T |Y_r|^{2(p-1)}(|Y_r| + |Y_r|^2 + |Y_r|\,||Z_r||)dr - 2p\int_s^T |Y_r|^{2(p-1)}(Y_r, Z_r dW_r)$$

where we have used the assumption on f. Had we done the same calculation with the function $u \to u^p$ remplaced by

$$\varphi_{n,p}(u) = (u \wedge n)^p + pn^{p-1}(u-n)^+, \; n \in \mathbb{N} \;,$$

we would have obtained :

$$\varphi_{n,p}(|Y_s|^2) + \int_s^T \varphi'_{n,p}(|Y_r|^2)||Z_r||^2 dr$$

$$\leq \varphi_{n,p}(|g(X_T)|^2) + 2C\int_s^T \varphi'_{n,p}(|Y_r|^2)(|Y_r| + |Y_r|^2 + |Y_r|\,||Z_r||)dr$$

$$-2\int_s^T \varphi'_{n,p}(|Y_r|^2)(Y_r, Z_r dW_r)$$

We can take the expectation in the last equation, and let $n \to +\infty$, in order to deduce by monotone convergence :

$$E(|Y_s|^{2p}) + pE\int_s^T |Y_r|^{2(p-1)}||Z_r||^2 dr \leq E(|g(X_T)|^{2p})$$

$$+2CpE\int_s^T |Y_r|^{2(p-1)}(|Y_r| + |Y_r|^2 + |Y_r|\,||Z_r||)dr$$

It then follows by Hölder's inequality that there exists $C(p)$ such that :

$$E(|Y_s|^{2p}) + \frac{p}{2}E\int_s^T |Y_r|^{2(p-1)}||Z_r||^2 dr \leq E(|g(X_T)|^{2p})$$

$$+C(p)E\int_s^T (1 + |Y_r|^{2p})dr\,, \quad t \leq s \leq T\,.$$

It follows from Gronwall's Lemma that

$$\sup_{t\leq s\leq T} E(|Y_s|^{2p}) < \infty \tag{7}$$

and hence

$$E\int_t^T |Y_r|^{2(p-1)}||Z_r||^2 dr < \infty \tag{8}$$

for an arbitrarily large p.

Now

$$|Y_s|^{2p} \leq |g(X_T)|^{2p} + C(p)\int_t^T (1 + |Y_r|^{2p})dr$$

$$-2p\int_s^T |Y_r|^{2(p-1)}(Y_r, Z_r dW_r)$$

$$E\left(\sup_{t\leq s\leq T} |Y_s|^{2p}\right) \leq E\left[|g(X_T)|^{2p} + C(p)\int_t^T (1 + |Y_r|^{2p})dr\right.$$

$$\left.-2p\int_t^T |Y_r|^{2(p-1)}(Y_r, Z_r dW_r)\right]$$

$$+2pE\left[\sup_{t\leq s\leq T}\int_t^s |Y_r|^{2(p-1)}(Y_r, Z_r dW_r)\right]$$

Hence (5) follows from the Burkholder–Davis–Gundy inequality, (7) and (8) (which again holds with an arbitrarily large p).

Finally, for any $t \leq a \leq s \leq b \leq T$,

$$\int_a^s Z_r dW_r = Y_s - Y_a + \int_a^s f(r, X_r, Y_r, Z_r)dr$$

$$\sup_{a\leq s\leq b}\left|\int_a^s Z_r dW_r\right| \leq 2\sup_{t\leq s\leq T}|Y_s| + \int_a^b |f(r, X_r, Y_r, Z_r)|dr$$

Hence, from Burkholder–Davis–Gundy's inequality, for any $p \geq 2$, $\exists\, c_p$ s.t.

$$\frac{1}{c_p}E\left[\left(\int_a^b \|Z_r\|^2 dr\right)^{p/2}\right] \leq E\left(\sup_{a\leq s\leq b}\left|\int_0^s Z_r dW_r\right|^p\right)$$
$$\leq c_p\left(1+E\left[\left(\int_a^b |Z_r| dr\right)^p\right]\right)$$
$$\leq c_p\left(1+(b-a)^{p/2}E\left[\left(\int_a^b |Z_r|^2 dr\right)^{p/2}\right]\right)$$

Hence, provided $b-a \leq c_p^{-4/p}$,

$$E\left[\left(\int_a^b \|Z_r\|^2 dr\right)^{p/2}\right] < \infty ,$$

and (6) follows. □

Let us now express Z in terms of the Wiener space derivative of Y.

Proposition 2.2 *$Y, Z, \in L^2(t,T;\ \mathbb{D}^{1,2})$, and a version of $\{D_\theta Y_s,\ D_\theta Z_s;\ t \leq \theta \leq T,\ t \leq s \leq T\}$ is given by :*

(i) $D_\theta Y_s = 0,\ D_\theta Z_s = 0;\ t \leq s < \theta \leq T$

(ii) For any fixed $\theta \in [t,T]$ and $1 \leq i \leq d$, $\{D_\theta^i Y_s,\ D_\theta^i Z_s;\ \theta \leq s \leq T\}$ is the unique solution of the BSDE :

$$D_\theta^i Y_s = g'(X_T)D_\theta^i X_T + \int_s^T F_i(r,\ D_\theta^i Y_r,\ D_\theta^i Z_r)dr - \int_s^T D_\theta^i Z_r dW_r\ , \tag{9}$$

where

$$F_i(r,u,v) = f_x'(X_r,Y_r,Z_r)D_\theta^i X_r + f_y'(X_r,Y_r,Z_r)u + f_z'(X_r,Y_r,Z_r)v\ .$$

Moreover, for any $1 \leq i \leq d$, $\{D_s^i Y_s,\ t \leq s \leq T\}$ is a version of $\{(Z_s)_i,\ t \leq s \leq T\}$ (where $(Z_s)_i$ denote the i-th column ot he matrix Z_s).

Before proving the Proposition, let us establish the following simple but very useful Lemma, which is a particular case of a much more general result in Ustunel [9]. We nevertheless include a complete proof for the convenience of the reader.

Lemma 2.3 *Let $Z \in M^2(t,T;\ \mathbb{R}^d)$ be such that $\xi \triangleq \int_t^T (Z_s, dW_s)$ satisfies $\xi \in \mathbb{D}^{1,2}$. Then*

$$Z_i \in L^2(t,T;\ \mathbb{D}^{1,2}),\ 1 \leq i \leq d\ , \tag{10}$$

and

$$D_s^i \xi = (Z_s)_i + \int_s^T D_s^i Z_r dW_r,\ ds \times dP\ a.e. \tag{11}$$

Proof : The fact that (10) implies $\xi \in \mathbb{D}^{1,2}$ and (11) is well known (see e.g. Nualart–Pardoux [2], Proposition 3.4). Hence we only need to prove (10). Let us assume that $d = 1$ for notational convenience.

Note that if (10) holds, then

$$||\xi||_{1,2}^2 = 2E\int_t^T Z_s^2 ds + E\int_t^T\int_t^T ||D_s Z_r||^2 ds dr \,.$$

Hence the result follows if we show that the set

$$\{\xi = \int_t^T Z_s dW_s\,;\; Z \in L^2(t,T\,;\; \mathbb{D}^{1,2})\}$$

is dense in $\mathbb{D}^{1,2} \cap L^2(\Omega, \mathcal{F}_T^t, P)$. But that follows from the fact that the above set contains $\{\xi \in \mathbb{S} \cap L^2(\Omega, \mathcal{F}_T^t, P)\,;\; E\xi = 0\}$, which can be seen from Ocone's formula (see e.g. [2] Corollary A2)

$$\xi = \int_t^T E(D_s\xi/\mathcal{F}_s)dW_s \,,$$

which applies to such ξ's. □

We can now proceed to the :

Proof of Proposition 2.2 We restrict ourselves to the case $d = 1$. We first remark that the fact that equation (9) has a unique solution follows easily from the results of [3], since $f_x'(X_r, Y_r, Z_r)D_\theta X_r$ is bounded in $L^p(\Omega)$, $p \geq 1$, and $f_y'(X_r, Y_r, Z_r)$, $f_z'(X_r, Y_r, Z_r)$ are bounded.

We first consider equation (3) with

$$\overline{Y}, \overline{Z} \in M^2(t,T\,;\; \mathbb{R}^k) \cap L^2(t,T\,;\; (\mathbb{D}^{1,2})^k).$$

Hence $g(X_T) + \int_s^T f(X_s, \overline{Y}_s, \overline{Z}_s)ds \in \mathbb{D}^{1,2}$, and it follows from Lemma 2.3 that $Z \in L^2(t,T\,;\; (\mathbb{D}^{1,2})^k)$, and consequently $Y_s \in \mathbb{D}^{1,2}$, $t \leq s \leq T$, and for $t \leq \theta \leq s$, $1 \leq i \leq d$

$$\begin{aligned} D_\theta^i Y_s &= g'(X_T)D_\theta^i X_T + \\ &\int_s^T [f_x'(X_r, \overline{Y}_r, \overline{Z}_r)D_\theta^i X_r + f_y'(X_r, \overline{Y}_r, \overline{Z}_r)D_\theta^i \overline{Y}_r + f_z'(X_r, \overline{Y}_r, \overline{Z}_r)D_\theta^i \overline{Z}_r]dr \\ &- \int_s^T D_\theta^i Z_r dW_r \end{aligned}$$

We next consider equation (4) with

$$\overline{Y} \in M^2(t,T\,;\; \mathbb{R}^k) \cap L^2(t,T\,;\; (\mathbb{D}^{1,2})^k).$$

From the last result, the corresponding approximating sequence satisfies $Y^n, Z^n \in L^2(t,T;\; (\mathbb{D}^{1,2})^k)$, $n \in \mathbb{N}$, and for $t \leq \theta \leq s$, $1 \leq i \leq d$, $n \in \mathbb{N}$,

$$\begin{aligned} D_\theta^i Y^{n+1} &= g'(X_T)D_\theta^i X_T \\ &+ \int_s^T [f_x'(X_r, \overline{Y}_r, Z_r^n)D_\theta^i X_r + f_y'(X_r, \overline{Y}_r, Z_r^n)D_\theta^i \overline{Y}_r + f_z'(X_r, \overline{Y}_r, Z_r^n)D_\theta^i Z_r^n]dr \\ &- \int_s^T D_\theta^i Z_r^{n+1} dW_r,\; \theta \leq s \leq T \end{aligned}$$

Using estimates very similar to those on pages 57–58 of [3], we first show that $E\int_\theta^T |D_\theta^i Z_r^n|^2 dr \le C$ and $E\int_\theta^T |D_\theta^i Y_r^n|^4 dr \le C$. Hence

$$E\int_s^T (D_\theta^i Y_r^{n+2} - D_\theta^i Y_r^{n+1})[(f_x'(X_r,\overline{Y}_r,Z_r^{n+1}) - (f_x'(X_r,\overline{Y}_r,Z_r^n))D_\theta^i X_r$$

$$+(f_y'(X_r,\overline{Y}_r,Z_r^{n+1}) - (f_y'(X_r,\overline{Y}_r,Z_r^n))D_\theta^i \overline{Y}_r$$

$$+(f_z'(X_r,\overline{Y}_r,Z_r^{n+1}) - (f_y'(X_r,\overline{Y}_r,Z_r^n))D_\theta^i Z_r^n]dr$$

tends to zero, as $n \to \infty$. One can then show that $D_\theta^i Y^n$, $D_\theta^i Z^n$ are Cauchy in $L^2(\theta,T;\ (\mathbb{D}^{1,2})^k)$, and the limit satisfies, for $t \le \theta \le s \le T$, $1 \le i \le d$,

$$D_\theta^i Y_s = g'(X_T)D_\theta^i X_T$$

$$\int_s^T [f_x'(X_r,\overline{Y}_r,Z_r)D_\theta^i X_r + f_y'(X_r,\overline{Y}_r,Z_r)D_\theta^i \overline{Y}_r + f_z'(X_r,\overline{Y}_r,Z_r)D_\theta^i Z_r]dr$$

$$-\int_s^T D_\theta^i Z_r dW_r$$

The same kind of procedure applies to the second approximating sequence, which proves all but the last statement of the Proposition.

Finally, since for $t < \theta \le s \le T$,

$$\begin{aligned} Y_s &= Y_t - \int_t^s f(X_r,Y_r,Z_r)dr + \int_t^s Z_r dW_r\ , \\ D_\theta^i Y_s &= (Z_\theta)_i - \int_\theta^s [f'(X_r,Y_r,Z_r)D_\theta^i X_r + f_y'(X_r,Y_r,Z_r)D_\theta^i Y_r \\ &\qquad + f_z'(X_r,Y_r,Z_r)D_\theta^i Z_r]dr \\ &\quad - \int_\theta^s D_\theta^i Z_r dW_r\ , \end{aligned}$$

for a.e. s, the jump of $D_\theta Y_s$ at $\theta = s$ equals Z_s. With the version of $D_\theta Y_s$ that we have choosen above, that means exactly that

$$D_s Y_s = Z_s\ ,\ s \text{ a.e.}$$

□

We next want to show that $\{D_s Y_s\,;\ t \le s \le T\}$ processes an a.s. continuous version. For that sake, we first recall that the matrix valued process $\{\nabla X_s = \left(\frac{\partial X_s^i}{\partial x^j}\right)_{1\le i,\,j\le d};\ t \le s \le T\}$ solves the following SDE :

$$\nabla X_s = I + \int_s^t b'(X_r)\nabla X_r dr + \int_s^t \sigma_l'(X_r)\nabla X_r dW_r^l$$

The next formula is an immediate consequence of the uniqueness of the solution of the SDE satisfied by $D_\theta X_s$:

$$D_\theta X_s = \nabla X_s (\nabla X_\theta)^{-1} \sigma(X_\theta)\,,\ t \le \theta \le s \le T\,. \tag{12}$$

Let $\{\nabla Y_s, \nabla Z_s\,;\ t \le s \le T\}$ be the unique element of $M^2(t,T;\ \mathbb{R}^{k\times d}) \times M^2(t,T;\ \mathbb{R}^{k\times d\times d})$ which solves :

$$\nabla Y_s = g'(X_T)\nabla X_T \tag{13}$$

$$+\int_s^T [f'_x(X_r,Y_r,Z_r)\nabla X_r + f'_y(X_r,Y_r,Z_r)\nabla Y_r + f'_z(X_r,Y_r,Z_r)\nabla Z_r]dr$$

$$-\int_s^T \nabla Z_r dW_r,\ t \le s \le T\,.$$

We shall later interpret ∇Y_s (resp. ∇Z_s) as the matrix of first order partial derivatives of Y_s (resp. Z_s) with respect to x (x denoting again the initial condition for X_t). For the time being, let us establish the :

Lemma 2.4 *For $t \le \theta \le s \le T$,*

$$D_\theta Y_s = \nabla Y_s (\nabla X_\theta)^{-1} \sigma(X_\theta)\ ,$$

and the process $\{D_s Y_s\,;\ t \le s \le T\}$ as defined by Proposition 2.2 is a.s. continuous.

Proof : We deduce from the uniqueness of the solution of equation (9) and formula (12) that :

$$D^i_\theta Y_s = \nabla Y_s (\nabla X_\theta)^{-1} \sigma_i(X_\theta)\,,\ t \le \theta \le T\,.$$

The first of the statements follows, hence

$$D_s Y_s = \nabla Y_s (\nabla X_s)^{-1} \sigma(X_s)$$

and the continuity of $D_s Y_s$ follows from that of $\nabla Y_s, \nabla X_s$ and X_s □

From Proposition 2.2 and Lemma 2.4, we deduce that $\{Z_s\,;\ t \le s \le T\}$ has an a.s. continuous version, and we shall from now on identify $\{Z_s\}$ with its continuous version. An immediate consequence of Proposition 2.2 and Lemma 2.4 is now :

Lemma 2.5 *For any $0 \le t \le s \le T$, $x \in \mathbb{R}^d$,*

$$Z^{t,x}_s = \nabla Y^{t,x}_s (\nabla X^{t,x}_s)^{-1} \sigma(X^{t,x}_s)$$

and in particular

$$Z^{t,x}_t = \nabla Y^{t,x}_t \sigma(x)\,.$$

□

Since one can establish $L^p(\Omega)$ estimates for $\sup_s |\nabla Y_s|$ as was done for $\sup_s |Y_s|$ in Lemma 2.1, we deduce from the last Lemma :

Lemma 2.6 *For any $p \ge 1$,*

$$E\left(\sup_{t\le s\le T} ||Z^{t,x}_s||^p\right) < \infty$$

□

We now study the regularity with respect to x of $Y_t^{t,x}$. Our proof is an adaptation of that of Stroock [8] for the usual SDEs, but we include the continuity with respect to t of $Y_s^{t,x}$ and its $x-$derivatives. Let us first introduce some notations. If g is a function of $x \in \mathbb{R}^d$, for $h \in \mathbb{R} \setminus \{0\}$, let $\Delta_h^i g(x) \triangleq h^{-1}[g(x + he_i) - g(x)]$, $1 \leq i \leq d$, where e_i denotes the i–th vector of an arbitrary orthonormal basis of $\mathbb{R}^d$.

Let us state the main technical steps for the process $X^{t,x}$ for further reference. We omit the proofs which are adaptations of those in [8]. Note that we define $X_s^{t,x}, Y_s^{t,x}, Z_s^{t,x}$ for any $(s,t) \in [0,T]^2$, $x \in \mathbb{R}^d$ by letting $X_s^{t,x} = X_{s \vee t}^{t,x}$ and similarly for $Y_s^{t,x}$, while $Z_s^{t,x} = 0$ for $s < t$.

Lemma 2.7 *For any $p \geq 2$, there exists a constant c_p such that for any $t, t' \in [0,T]$, $x, x' \in \mathbb{R}^d$, $i \in \{1, \ldots, d\}$, $h, h' \in \mathbb{R} \setminus \{0\}$,*

$$E\left(\sup_{0 \leq s \leq T} |X_s^{t,x}|^p\right) \leq c_p(1 + |x|^p) \tag{14}$$

$$E\left(\sup_{0 \leq s \leq T} |X_s^{t,x} - X_s^{t',x'}|^p\right) \leq c_p(1 + |x|^p)\left(|x - x'|^p + |t - t'|^{\frac{p}{2}}\right) \tag{15}$$

$$E\left(\sup_{0 \leq s \leq T} |\Delta_h^i X_h^{t,x}|^p\right) \leq c_p \tag{16}$$

$$E\left(\sup_{0 \leq s \leq T} |\Delta_h^i X_s^{t,x} - \Delta_{h'}^i X_s^{t',x'}|^p\right) \tag{17}$$

$$\leq c_p(1 + |x|^p)\left(|x - x'|^p + |h - h'|^p + |t - t'|^{\frac{p}{2}}\right)$$

□

It then follows immediately, using Kolmogorov's Lemma :

Corollary 2.8 *For any $t \in [0,T]$, $x \in \mathbb{R}^d$, the mapping $x \to X_s^{t,x}$ is a.s. differentiable, and the matrix of partial derivatives $\nabla X_s^{t,x}$ possesses a version which is a.s. continuous in (s,t,x).*

□

Iterating the argument, we obtain the existence of jointly continuous second derivatives.

We now follow the same procedure to establish :

Theorem 2.9 $\{Y_s^{t,x};\ (s,t) \in [0,T]^2,\ x \in \mathbb{R}^d\}$ *has a version whose trajectories belong to $C^{0,0,2}([0,T]^2 \times \mathbb{R}^d)$.*

Before proceeding to the proof, let us state the Corollary which we shall need in the next section :

Corollary 2.10 *For any $t \in [0,T]$, the mapping $x \to Y_t^{t,x}$ is of class C^2, the function and its partial derivatives of order one and two being continuous in (t,x)*

□

Proof of Theorem 2.9 We shall only prove the analog of Lemma 2.7. Going back to the proof of Lemma 2.1, and using (14) we deduce that for any $p \geq 2$, there exist c_p and q such that :

$$E\left(\sup_{0\leq s\leq T}|Y_s^{t,x}|^p\right) \leq c_p(1+|x|^q) \tag{18}$$

and moreover

$$E\left[\left(\int_t^T ||Z_s^{t,x}||^2 ds\right)^{p/2}\right] \leq c_p(1+|x|^q) \,. \tag{19}$$

Next for $t \vee t' \leq s \leq T$,

$$\begin{aligned}
Y_s^{t,x} - Y_s^{t',x'} &= \left(\int_0^1 g'\left(X_T^{t',x'} + \lambda(X_T^{t,x} - X_T^{t',x'})\right) d\lambda\right) [X_T^{t,x} - X_T^{t',x'}] \\
&+ \int_s^T \left(\varphi_r(t,x\,;\,t',x')[X_r^{t,x} - X_r^{t',x'}] + \psi_r(t,x\,;\,t',x')[Y_r^{t,x} - Y_r^{t',x'}]\right. \\
&\left. +\chi_r(t,x\,;\,t',x')[Z_r^{t,x} - Z_r^{t',x'}]\right) dr \\
&- \int_s^T (Z_r^{t,x} - Z_r^{t',x'}) dW_r
\end{aligned}$$

where

$$\begin{aligned}
\varphi_r(t,x\,;\,t',x') &= \int_0^1 f'_x(\Sigma_{r,\lambda}^{t,x\,;\,t',x'})d\lambda \\
\psi_r(t,x\,;\,t',x') &= \int_0^1 f'_y(\Sigma_{r,\lambda}^{t,x\,;\,t',x'})d\lambda \\
\chi_r(t,x\,;\,t',x') &= \int_0^1 f'_z(\Sigma_{r,\lambda}^{t,x\,;\,t',x'})d\lambda
\end{aligned}$$

and

$$\begin{aligned}
\Sigma_{r,\lambda}^{t,x\,;\,t',x'} &= \left(r, X_r^{t',x'} + \lambda(X_r^{t,x} - X_r^{t',x'}),\ Y_r^{t',x'} + \lambda(Y_r^{t,x} - Y_r^{t',x'}),\right. \\
&\left. Z_r^{t',x'} + \lambda(Z_r^{t,x} - Z_r^{t',x'})\right) .
\end{aligned}$$

Combining the argument of Lemma 2.1 with (15), we obtain :

$$E\left(\sup_{0\leq s\leq T}\left|Y_s^{t,x} - Y_s^{t',x'}\right|^p\right) \leq c_p(1+|x|^q) \times \left(|x-x'|^p + |t-t'|^{p/2}\right) \tag{20}$$

In fact we should restrict the sup to $t \vee t' \leq s \leq T$, but (20) follows then easily from that restricted result. We have moreover :

$$E\left[\left(\int_{t\wedge t'}^T ||Z_s^{t,x} - Z_s^{t',x'}||^2 ds\right)^{p/2}\right] \leq c_q(1+|x|^q)\left(|x-x'|^p + |t-t'|^{p/2}\right) \tag{21}$$

We next have :

$$\Delta_h^i Y_s^{t,x} = \int_0^1 g'(X_T^{t,x} + \lambda h \Delta_h^i X_T^{t,x}) \Delta_h^i X_T^{t,x} d\lambda$$

$$+ \int_s^T \int_0^1 \left[f'_x(\Xi_{r,\lambda}^{t,x,h}) \Delta_h^i X_r^{t,x} + f'_y(\Xi_{r,\lambda}^{t,x,h}) \Delta_h^i Y_r^{t,x} + f'_z(\Xi_{r,\lambda}^{t,x,h}) \Delta_h^i Z_r^{t,x} \right] d\lambda dr$$

$$- \int_s^T \Delta_h^i Z_r^{t,x} dW_r \ ,$$

where $\Xi_{r,\lambda}^{t,x,h} = (r, X_r^{t,x} + \lambda h \Delta_h^i X_r^{t,x},\ Y_r^{t,x} + \lambda h \Delta_h^i Y_h^{t,x},\ Z_r^{t,x} + \lambda h \Delta_h^i Z_r^{t,x})$.

Using again arguments similar to those in Lemma 2.1 and (16), we obtain that there exists c_p and q such that

$$E\left(\sup_{t \le s \le T} |\Delta_h^i Y_s^{t,x}|^p \right) \le c_p(1 + |x|^q + |h|^q) \ , \tag{22}$$

$$E\left[\left(\int_t^T ||\Delta_h^i Z_s^{t,x}||^2 \right)^{p/2} \right] \le c_p(1 + |x|^q + |h|^q) \tag{23}$$

Finally,

$$\Delta_h^i Y_s^{t,x} - \Delta_{h'}^i Y_s^{t',x'} = \left(\int_0^1 g'(X_T^{t,x} + \lambda h \Delta_h^i X_T^{t,x}) d\lambda \right) \Delta_h^i X_T^{t,x}$$

$$- \left(\int_0^1 g'(X_T^{t',x'} + \lambda h \Delta_{h'}^i X_T^{t',x'}) d\lambda \right) \Delta_{h'}^i X_T^{t',x'}$$

$$+ \int_s^T \int_0^1 [f'_x(\Xi_{r,\lambda}^{t,x,h}) \Delta_h^i X_r^{t,x} - f'_x(\Xi_{r,\lambda}^{t',x',h'}) \Delta_{h'}^i X_r^{t',x'}] d\lambda dr$$

$$+ \int_s^T \int_0^1 [f'_y(\Xi_{r,\lambda}^{t,x,h}) \Delta_h^i Y_r^{t,x} - f'_y(\Xi_{r,\lambda}^{t',x',h'}) \Delta_{h'}^i Y_r^{t',x'}] d\lambda dr$$

$$+ \int_s^T \int_0^1 [f'_z(\Xi_{r,\lambda}^{t,x,h}) \Delta_h^i Z_r^{t,x} - f'_z(\Xi_{r,\lambda}^{t',x',h'}) \Delta_{h'}^i Z_r^{t',x'}] d\lambda dr$$

$$- \int_s^T [\Delta_h^i Z_r^{t,x} - \Delta_{h'}^i Z_r^{t',x'}] dW_r$$

We claim that, again by the procedure of Lemma 2.1, using the properties of f and (17), (20), (21), we can deduce :

$$\begin{aligned} E\left(\sup_{t \wedge t' \le s \le T} |\Delta_h^i Y_s^{t,x} - \Delta_{h'}^i Y_s^{t',x'}|^p \right) &\le c_p(1 + |x|^q + |x'|^q + |h|^q + |h'|^q) \\ &\quad \times (|x - x'|^p + |h - h'|^p + |t - t'|^{p/2}) \end{aligned}$$

$$\begin{aligned} E\left[\left(\int_{t \wedge t'}^T ||\Delta_h^i Z_s^{t,x} - \Delta_{h'}^i Z_s^{t',x'}||^2 ds \right)^{p/2} \right] &\le c_p(1 + |x|^q + |x'|^q + |h|^q + |h'|^q) \\ &\quad \times (|x - x'|^p + |h - h'|^p + |t - t'|^{p/2}) \end{aligned}$$

Let us only indicate how we can treat the "hardest" term :

$$\Big|E\int_a^b \Big(\int_0^1 [f_z'(\Xi_{r,\lambda}^{t,x,h})\Delta_h^i Z_r^{t,x} - f_z'(\Xi_{r,\lambda}^{t',x',h'})\Delta_{h'}^i Z_r^{t',x'}]d\lambda\ ,$$

$$\Delta_h^i Y_r^{t,x} - \Delta_{h'}^i Y_r^{t',x'}\Big)\,|\Delta_h^i Y_r^{t,x} - \Delta_{h'}^i Y_{r'}^{t',x'}|^{p-2}dr\Big|$$

$$\leq cE\int_a^b ||\Delta_h^i Z_r^{t,x} - \Delta_{h'}^i Z_r^{t',x'}|| \times |\Delta_h^i Y_r^{t,x} - \Delta_{h'}^i Y_r^{t',x'}|^{p-1}dr$$

$$+cE\int_a^b ||\Delta_h^i Z_r^{t,x}||\Big(\int_0^1 |\Xi_{r,\lambda}^{t,x,h} - \Xi_{r,\lambda}^{t',x',h'}|d\lambda\Big)\,|\Delta_h^i Y_r^{t,x} - \Delta_{h'}^i Y_r^{t',x'}|^{p-1}dr$$

$$\leq \frac{1}{2}E\Big(\sup_{a\leq r\leq b} |\Delta_h^i Y_r^{t,x} - \Delta_{h'}^i Y_r^{t',x'}|^p\Big)$$

$$+\bar{c}(b-a)E\left[\Big(\int_a^b ||\Delta_h^i Z_r^{t,x} - \Delta_{h'}^i Z_r^{t',x'}||^2 dr\Big)^{p/2}\right]$$

$$+\bar{c}\sqrt{E\left[\Big(\int_a^b ||\Delta_h^i Z_r^{t,x}||^2 dr\Big)^{p/2}\right]}\sqrt{E\left[\Big(\int_a^b\int_0^1 |\Xi_{r,\lambda}^{t,x,h} - \Xi_{r,\lambda}^{t',x',h'}|^2 d\lambda dr\Big)^{p}\right]}$$

We note that the two first terms on the right are substracted from the left terms of the full inequality, with $(b-a)$ small enough, and the last term is estimated with the help of (23), (15), (20) and (21). Note also that we choose first $b=T$, $a=T-\alpha$, then $b=T-\alpha$, $a=T-2\alpha$, etc... □

As a by-product of the above proof, we obtain :

Corollary 2.11 *$\{(\nabla Y_s^{t,x},\ \nabla Z_s^{t,x}),\ t\leq s\leq T\}$, the unique solution of the BSDE (13), is the gradient of $\{(Y_s^{t,x},\ Z_s^{t,x}),\ t\leq s\leq T\}$ with respect to x.*

Proof : This follows easily from the fact that (17) holds true with $h,h'\in$ IR if we define $\Delta_0^i X_s^{t,x} = \frac{\partial X_h^{t,x}}{\partial x_i}$, and that by definition of the partial derivatives,

$$\frac{\partial Y_s^{t,x}}{\partial x_i} = \lim_{h\to 0}\Delta_h^i Y_s^{t,x},\ \frac{\partial Z_s^{t,x}}{\partial x_i} = \lim_{h\to 0}\Delta_h^i Z_s^{t,x}\ .$$

□

3 Backward SDEs and systems of quasilinear parabolic partial differential equations

We now relate our BSDE to the following system of quasilinear parabolic differential equations :

$$\begin{cases} \frac{\partial u}{\partial t}(t,x) + \mathcal{L}u(t,x) + f(t,x,u(t,x),(\nabla u\sigma)(t,x)) = 0 \\ u(T,x) = g(x) \end{cases} \qquad (24)$$

where $u\ :\ \mathbb{R}_+\times\mathbb{R}^d\to\mathbb{R}^k$, and

$$\mathcal{L}u = \begin{pmatrix} Lu_1 \\ \vdots \\ Lu_k \end{pmatrix},$$

$$L = \frac{1}{2}\sum_{i,j=1}^{d}(\sigma\sigma^*)_{ij}(t,x)\frac{\partial^2}{\partial x_i \partial x_j} + \sum_{i=1}^{d} b_i(t,x)\frac{\partial}{\partial x_i}.$$

Let us first recall a result from [4] :

Theorem 3.1 *If $u \in C^{1,2}([0,T]) \times \mathbb{R}^d)$ solves equation (24), then $u(t,x) = Y_t^{t,x}$, $t \geq 0$, $x \in \mathbb{R}^d$, where $\{(Y_s^{t,x},\ Z_s^{t,x});\ t \leq s \leq T\}_{t\geq 0,\, x\in\mathbb{R}^d}$ is the unique solution of the BSDE (2).*

Proof : Use Itô's formula applied to $u(s, X_s^{t,x})$ between $s = t$ and $s = T$, and note that $\{u(s, X_s^{t,x}), (\nabla u\sigma)(s, X_s^{t,x})\}$ solves the BSDE (2). □

We are now in a position to prove the converse to the above result :

Theorem 3.2 *Under the assumptions stated in section 1,*

$$u(t,x) \triangleq Y_t^{t,x};\ t \geq 0,\ x \in \mathbb{R}^d$$

is of class $C^{1,2}([0,T] \times \mathbb{R}^d)$, and solves equation (24).

Proof : From Theorem 2.9, $u \in C^{0,2}([0,T] \times \mathbb{R}^d)$ Let $h > 0$ be such that $t + h \leq T$. Clearly, $Y_{t+h}^{t,x} = Y_{t+h}^{t+h, X_{t+h}^{t,x}}$. Hence

$$\begin{aligned} u(t+h,x) - u(t,x) &= u(t+h,x) - u(t+h, X_{t+h}^{t,x}) + u(t+h, X_{t+h}^{t,x}) - u(t,x) \\ &= -\int_t^{t+h} Lu(t+h, X_r^{t,x})dr - \int_t^{t+h} (\nabla u\sigma)(t+h, X_r^{t,x})dW_r \\ &\quad - \int_t^{t+h} f(X_r^{t,x}, Y_r^{t,x}, Z_r^{t,x})dr + \int_t^{t+h} Z_r^{t,x} dW_r \end{aligned}$$

where we have used Itô's formula (and the fact that $u(t,\cdot) \in C^2(\mathbb{R}^d)$) and the BSDE. Let now $t = t_0 < t_1 < \ldots < t_n = T$. We have

$$\begin{aligned} g(x) - u(t,x) &= -\sum_{i=0}^{n-1}\int_{t_i}^{t_{i+1}} [Lu(t_{i+1}, X_r^{t_i,x}) + f(X_r^{t_i,x}, Y_r^{t_i,x}, Z_r^{t_i,x})]dr \\ &\quad + \sum_{i=0}^{n-1}\int_{t_i}^{t_{i+1}} [Z_r^{t_i,x} - (\nabla u\sigma)(t_{i+1}, X_r^{t_i,x})]dW_r \end{aligned}$$

It now follows from Theorem 2.9 and Lemma 2.5 that, if we take a sequence of meshes $t = t_0^n < t_1^n < \cdots < t_n^n = T$ such that $\lim_{n\to\infty} \sup_{i\leq n-1}(t_{i+1}^n - t_i^n) = 0$, we obtain in the limit :

$$u(t,x) = g(x) + \int_t^T [Lu(s,x) + f(s,x,u(s,x),\ (\nabla u\sigma)(s,x))]ds$$

hence $u \in C^{1,2}([0,T] \times \mathbb{R}^d)$ and satisfies the equation (24). □

Remark 3.3 In the classical case where $k = 1$ and

$$f(t, x, y, z) = c(t, x)y ,$$

our result reduces to the classical Feynman–Kac formula. Indeed, in that case the BSDE (2) has the explicit solution :

$$Y_s^{t,x} = e^{\int_s^T c(r, X_r^{t,x})dr} g(X_T) - \int_s^T e^{\int_s^r c(\alpha, X_\alpha^{t,x})d\alpha} Z_r^{t,x} dW_r$$

and

$$\begin{aligned} Y_t^{t,x} &= E(Y_t^{t,x}) \\ &= E\left[e^{\int_s^T c(r, X_r^{t,x})dr} g(X_T)\right] . \end{aligned}$$

□

4 Backward SDEs and viscosity solutions of quasilinear parabolic PDEs

We now restrict ourselves to the case $k = 1$, and we shall show that when the coefficients f and g are Lipschitz continuous, the BSDE provides the unique viscosity solution of a quasilinear parabolic PDE. The results of this section are particular cases of results in Peng [5]. However, we present then for the sake of completeness, and because the argument here is simpler than that in [5].

We first recall a technical Lemma. Let $f = \Omega \times [0, T] \times \mathbb{R}^d \times \mathbb{R}^{dd} \to \mathbb{R}^d$ be $\mathcal{P} \otimes \mathcal{B} \otimes \mathcal{B}_d / \mathcal{B}$ measurable, where $\mathcal{P}$ denotes the σ-algebra of $\{\mathcal{F}_t\}$-progressively measurable subsets of $\Omega \times [0, T]$. as usual, we shall write $f(t, y, z)$ instead of $f(\omega, t, y, z)$. We assume that

$$f(., 0, 0) \in M^2(0, T) \tag{25}$$

and that there exists $c > 0$ such that

$$|f(t, y, z) - f(t, y', z')| \le c(|y - y'| + |z - z'|) \tag{26}$$

Given $Q, \overline{Q} \in L^2(\Omega, \mathcal{F}_T, P)$ and $F \in M^2(0, T)$, let $\{(Y_t, Z_t), t \ge 0\}$ (resp.$\{(\overline{Y}_t, \overline{Z}_t), t \ge 0\}$) denote the unique solution of the BSDE

$$Y_t = Q + \int_t^T f(s, Y_s, Z_s)ds - \int_t^T Z_s dW_s \tag{27}$$

(resp. of the BSDE

$$\overline{Y}_t = \overline{Q} + \int_t^T [f(s, \overline{Y}_s, \overline{Z}_s) + F_s]ds - \int_t^T \overline{Z}_s dW_s). \tag{28}$$

We have the following comparison result :

Lemma 4.1 *Let $\overline{Q} \ge 0$ a.s., $F_t \ge 0$ a.s., t a.e. Then $\overline{Y}_t \ge Y_t$ a.s., t a.e.*

Proof : This result is proved in Proposition 2.4 of [5] under the additional assumption that f is C^1 in (y,z). The present result follows by a standard approximation argument. □

We now use again the notations from section 2, assuming only that b, σ, f, g are globally Lipschitz with respect to (x,y,z), uniformly in t -this last precision concerns only f. We define again, as in Theorem 3.1,

$$u(t,x) \triangleq Y_t^{t,x} \tag{29}$$

and note that the estimate (20) still applies, and hence u is locally Lipschitz in x and Hölder continuous in t, and therefore u is Hölder continuous in (t,x).

However, we do not expect that u is differentiable in (t,x). We are going to show that u is the viscosity solution of the backward parabolic PDE

$$\begin{cases} \frac{\partial v}{\partial t}(t,x) + Lv(t,x) + f(t,x,v(t,x),(\nabla v\sigma)(t,x)) = 0 \\ v(T,x) = g(x) \end{cases} \tag{30}$$

Definition 4.2 *Let $u \in C([0,T] \times \mathbb{R}^d)$ satisfy $u(T,x) = g(x), x \in \mathbb{R}^d$. u is said to be a viscosity sub-solution (resp. super-solution) of equation (4.6) if in addition for any $(t,x) \in (0,T) \times \mathbb{R}^d$ and $\varphi \in C^{1,2}((0,T) \times \mathbb{R}^d)$ such that $\varphi(t,x) = u(t,x)$ and (t,x) is a minimum (resp. maximum) of $\varphi - u$,*

$$\frac{\partial \varphi}{\partial t}(t,x) + L\varphi(t,x) + f(t,x,\varphi(t,x),(\nabla\varphi\sigma)(t,x)) \geq 0$$

$$(\text{resp.} \frac{\partial \varphi}{\partial t}(t,x) + L\varphi(t,x) + f(t,x,\varphi(t,x),(\nabla\varphi\sigma)(t,x)) \leq 0).$$

u is said to be a viscosity solution of (30) if it is both a viscosity sub-and super-solution of (30). □

We can now establish the main result of this section

Theorem 4.3 *The function u defined by (4.5) is the unique viscosity solution of the backward parabolic PDE (30).*

Proof : Uniqueness follows from Ishii-Lions [7]. We shall show that u is a viscosity sub-solution of (30). The property of being a super-solution could be proved analogously.

We first note that for $0 < t \leq t+h < T$ and $x \in \mathbb{R}^d$,

$$u(t,x) = u(t,h,X_{t+h}^{t,x}) + \int_t^{t+h} f(s,X_s^{t,x},Y_s^{t,x},Z_s^{t,x})ds - \int_t^{t+h} Z_s^{t,x}dW_s$$

Let now $\varphi \in C^{1,2}((0,T) \times \mathbb{R}^d)$ satisfy $\varphi(t,x) = u(t,x)$ and $\varphi \geq u$ on $(0,T) \times \mathbb{R}^d$.
We can without loss of generality assume that φ has bounded derivatives.
We then have

$$\begin{aligned} \varphi(t,h,X_{t+h}^{t,x}) - \varphi(t,x) \ &+ \int_t^{t+h} f(s,X_s^{t,x},Y_s^{t,x},Z_s^{t,x})ds \\ &- \int_t^{t+h} Z_s^{t,x}dW_s \geq 0 \end{aligned}$$

Let $\{(\overline{Y}_s,\overline{Z}_s), t \leq s \leq t+h\}$ denote the solution of the BSDE

$$\overline{Y}_s = \varphi(t+h,X_{t+h}^{t,x}) + \int_s^{t+h} f(r,X_r^{t,x},\overline{Y}_r,\overline{Z}_r)dr - \int_s^{t+h} \overline{Z}_r dW_r \tag{31}$$

From Lemma 4.1,

$$\varphi(t,x) \le \varphi(t+h, X_{t+h}^{t,x}) + \int_t^{t+h} f(s, X_s^{t,x}, \overline{Y}_s, \overline{Z}_s)ds - \int_t^{t+h} \overline{Z}_s dW_s$$

and from Itô's formula

$$\int_t^{t+h} [\frac{\partial \varphi}{\partial s}(s, X_s^{t,x}) \quad + \quad L\varphi(s, X_s^{t,x}) + f(s, X_s^{t,x}, \overline{Y}_s, \overline{Z}_s)]ds$$
$$+ \quad \int_t^{t+h} [\nabla\varphi\sigma(s, X_s^{t,x}) - \overline{Z}_s]dW_s \ge 0.$$

We note $\widehat{Y}_s = \overline{Y}_s - \varphi(s, X_s^{t,x}), \widehat{Z}_s = \overline{Z}_s - \nabla\varphi\sigma(s, X_s^{t,x})$.
The last inequality can be rewritten as :

$$\begin{aligned} \textstyle\int_t^{t+h}[(\frac{\partial\varphi}{\partial s} + L\varphi)(s, X_s^{t,x}) \quad &+ f(s, X_s^{t,x}, \varphi(s, X_s^{t,x}) + \hat{Y}_s, (\nabla\varphi\sigma)(s, X_s^{t,x}) + \hat{Z}_s)]ds \\ &- \textstyle\int_t^{t+h} \hat{Z}_s dW_s \ge 0 \end{aligned} \tag{32}$$

We deduce from (31) and Itô's formula that $\{(\hat{Y}_s, \hat{Z}_s), t \le s \le t+h\}$ is the unique solution of the BSDE

$$\hat{Y}_s = \int_s^{t+h} [(\frac{\partial\varphi}{\partial s} + L\varphi)(r, X_r^{t,x}) \quad + f(r, X_r^{t,x}, \varphi(r, X_r^{t,x}) + \hat{Y}_r, (\nabla\varphi\sigma)(r, X_r^{t,x}) + \hat{Z}_r)]dr$$
$$- \textstyle\int_s^{t+h} \hat{Z}_r dW_r.$$

We want to compare $\{(\hat{Y}_s, \hat{Z}_s)\}$ with the solution $\{(\widetilde{Y}_s, 0), (t \le s \le t+h\}$ of the BSDE

$$\widetilde{Y}_s = \int_s^{t+h} [(\frac{\partial\varphi}{\partial s} + L\varphi)(r,x) + f(r, x, \varphi(r,x) + \widetilde{Y}_r, (\nabla\varphi\sigma)(r,x))]dr \tag{33}$$

We note that since φ has bounded derivatives,

$$\sup_{t\le r\le t+h} E(|\varphi(r, X_r^{t,x}) - \varphi(r,x)|^2) \to 0 \text{ as } h \to 0,$$

$$\sup_{t\le r\le t+h} E(|(\sigma\nabla\varphi)(r, X_r^{t,x}) - (\sigma\nabla\varphi)(r,x)|^2) \to \text{ as } h \to 0.$$

It is then easy to deduce from the techniques of section 1 :

$$E\left(\sup_{t\le s\le t+h} |\hat{Y}_s - \widetilde{Y}_s|^2 + \int_t^{t+h} |\hat{Z}_s|^2 ds\right) = 0(h)$$

Now from (32) $\hat{Y}_s \ge 0, t \le s \le t+h$, hence

$$h^{-1}\widetilde{Y}_s \ge -\varepsilon(h)$$

where $\varepsilon(h) \to 0$ as $h \to 0$. Hence

$$\frac{1}{h}\int_t^{t+h} [(\frac{\partial\varphi}{\partial s} + L\varphi)(r,x) + f(r, x, \varphi(r,x) + \widetilde{Y}_r, (\nabla\varphi\sigma)(r,x))]dr \ge -\varepsilon(h)$$

Moreover it follows readily from (33) that there exists a constant $\overline{c}$ such that

$$|\widetilde{Y}_s| \le \overline{c}h, t \le s \le t+h.$$

We remark that $\widetilde{Y}$ clearly depends on h, also this was not made explicit. We finally conclude that

$$\frac{\partial\varphi}{\partial s}(t,x) + L\varphi(t,x) + f(t, x, \varphi(t,x), (\nabla\varphi\sigma)(t,x)) \ge 0$$

Remark 4.4 We note that, with the help of the techniques in Peng [6], it is possible to relax the assumption of uniform Lipschitz continuity of f with respect to y. □

References

[1] I. Karatzas, S. Shreve : *Brownian Motion and Stochastic Calculus*, Springer, 1988.

[2] D. Nualart, E. Pardoux : Stochastic Calculus with Anticipating integrands, *Prob. Theory and Rel. Fields* **78**, 535–581, 1988.

[3] E. Pardoux, S. Peng : Adapted Solutions of Backward Stochastic Equations, *Systems and Control Letters* **14**, 55–61, 1990.

[4] S. Peng : Probabilistic Interpretation for Systems of Quasilinear Parabolic Partial Differential Equation, *Stochastics*, **37**, 61–74, 1991.

[5] S. Peng : A Generalized Dynamic Programming Principle and Hamilton-Jacobi-Bellman Equation. *Stochastics*, to appear.

[6] S. Peng : Backward Stochastic Differential Equations and Applications to Optimal Control. *Applied Math. and Optimization*, to appear.

[7] H. Ishii, P.L. Lions : Viscosity Solutions of Fully Nonlinear Second Order Elliptic Partial Differential Equations, *J. Diff. Eqs.* **83**, 26–78, 1990.

[8] D.W. Stroock : *Topics in Stochastic Differential Equations*, Tata Institute of Fundamental Research Lecture Notes, Springer, 1982.

[9] A.S. Ustunel : Representation of the Distributions on Wiener space and Stochastic Calculus of Variation, *J. of Funct. Anal.* **70**, 126–139, 1987.

LYAPUNOV EXPONENT OF A STOCHASTIC WAVE EQUATION

Mark A. Pinsky
Mathematics Department, Northwestern University
Evanston, IL 60208

1. Statement of results. In this note we consider the solution of the problem

$$u_{tt} + 2\beta u_t + [\alpha + \epsilon N(t)]u = c^2 \Delta u$$

$$u = 0 \text{ on } \partial D$$

$$u(x;0) = f_1(x) \qquad u_t(x;0) = f_2(x)$$

where $\alpha > 0, \beta > 0, c > 0, \epsilon > 0$. $[N(t), t \geq 0]$ is a noise process with mean zero, which may be either the formal derivative of a Wiener process or a centered function of an ergodic Markov process. In the first case the equation is interpreted as an integral equation of the Itô type. D is a smoothly bounded domain in Euclidean space with Dirichlet eigenvalues $0 < \mu_1 < \ldots \leq \mu_n \leq \ldots$ and normalized eigenfunctions ϕ_n satisfying $\Delta\phi_n + \mu_n\phi_n = 0$ in D with $\phi_n = 0$ on the boundary ∂D and $\int_D \phi_n^2\,dx = 1$.

The energy of the solution u is defined by

$$E(t) = \frac{1}{2}\int_D [u_t^2 + c^2|\nabla u|^2 + \alpha u^2]\,dx$$

We have the following results.

Theorem 1. *Suppose that $\beta^2 - \alpha > \mu_1 c^2$. Then there exists the limit*

$$\lambda(\epsilon) = \lim_{t\uparrow\infty} t^{-1}\log E(t)$$

and when $\epsilon \downarrow 0$ we have

$$\lambda(\epsilon) = -2\beta + 2\sqrt{\beta^2 - \alpha - \mu_1 c^2} + O(\epsilon).$$

Theorem 2. *Suppose that $\beta^2 - \alpha \leq \mu_1 c^2$. Then there exists*

$$\bar{\lambda}(\epsilon) = \limsup_{t\uparrow\infty} t^{-1}\log E(t)$$

and we have $\limsup_{\epsilon\downarrow 0}\bar{\lambda}(\epsilon) \leq -2\beta$. (This includes the "underdamped case" $\beta^2 - \alpha \leq 0$).

We can also obtain results in the case of *continuous spectrum.* The initial-boundary-value problem for the wave equation (1.1) in the full Euclidean space R^d with the condition that the solutions have finite energy, defined by

$$E(t) = \frac{1}{2}\int_{R^d} [u_t^2 + c^2|\nabla u|^2 + \alpha u^2]\,dx.$$

The following theorems are obtained.

Theorem 3. *Suppose that* $\beta^2 - \alpha > 0$. *Then there exists the limit*

$$\lambda(\epsilon) = \lim_{t\uparrow\infty} t^{-1} \log E(t)$$

and when $\epsilon \downarrow 0$ *we have*

$$\lambda(\epsilon) = -2\beta + 2\sqrt{\beta^2 - \alpha} + O(\epsilon)$$

Theorem 4. *Suppose that* $\beta^2 - \alpha \leq 0$. *Then there exists*

$$\bar{\lambda}(\epsilon) = \limsup_{t\uparrow\infty} t^{-1} \log E(t).$$

and we have $\limsup_{\epsilon\downarrow 0} \bar{\lambda}(\epsilon) \leq -2\beta$.

2. Proof in white noise case

We introduce the Fourier components $u_n(t)$ through the eigenfunction expansion

$$u(x;t) = \sum_{n=1}^{\infty} u_n(t)\phi_n(x).$$

The stochastic wave equation translates into the system of ordinary differential equations

$$u_n'' + 2\beta u_n' + [\alpha + \mu_n c^2 + \epsilon N(t)]u_n = 0 \quad n = 1, 2, \ldots$$

and the energy is computed as

$$E(t) = \frac{1}{2}\sum_{n=1}^{\infty}[(u_n')^2 + (\alpha + \mu_n c^2)u_n^2].$$

Define $v_n(t) = u_n(t)e^{\beta t}$. The equation for u_n translates into

$$v_n'' + [\alpha - \beta^2 + \mu_n c^2 + \epsilon N(t)]v_n = 0$$

and the energy is

$$E(t) = e^{-2\beta t}\sum_{n=1}^{\infty}[(v_n' - \beta v_n)^2 + (\alpha + \mu_n c^2)u_n^2].$$

This is decomposed as $E(t) = E_1(t) + E_2(t)$ where $E_1(t)$ is the sum over those indices n for which $\mu_n c^2 \leq \sqrt{\beta^2 - \alpha}$ and $E_2(t)$ is the sum over the remaining indices. We will show that

Lemma 1. $\limsup_{t\uparrow\infty} t^{-1} \log E_2(t) \leq -2\beta^2$

For the proof introduce the functional

$$\hat{E}_2(t) := \frac{1}{2}\sum_{n=N_0}^{\infty}[(v_n')^2 + (\alpha - \beta^2 + \mu_n c^2)]$$

where N_0 is the first integer n for which $\mu_n c^2 > \beta^2 - \alpha$. By direct calculation we have

$$\hat{E}_2'(t) = \sum_{n=N_0}^{\infty} \epsilon N(t) v_n(t) v_n'(t).$$

Define a stochastic process

$$\Psi(t) := \sum_{n=N_0}^{\infty} \frac{v_n(t)v_n'(t)}{\hat{E}_2(t)}.$$

Clearly $|\sum_{n\geq N_0} v_n v_n'| \leq \frac{1}{2}\sum_{n\geq N_0}[(v_n)^2 + (v_n')^2] \leq M\hat{E}_2(t)$ for some constant M, so that $|\Psi(t)| \leq M$. In the Itô case we have $N(t) = w(t)$ a Wiener process and the SDE

$$d\hat{E}_2(t) = \epsilon\Psi(t)\hat{E}_2(t)\,dw.$$

The solution of this is given explicitly by

$$\hat{E}_2(t) = \exp[\epsilon\int_0^t \Psi(s)\,dw(s) - \frac{1}{2}\epsilon^2\int_0^t \Psi(s)^2\,ds]$$

from which we have

$$\limsup_{t\uparrow\infty} t^{-1}\log\hat{E}_2(t) = \limsup_{t\uparrow\infty} t^{-1}\int_0^t -\epsilon^2\Psi(s)^2\,ds \leq 0.$$

This proves that $\hat{E}_2(t) = e^{o(t)}, t\uparrow\infty$. The actual contribution to the energy is

$$\begin{aligned}
E_2(t) &= \frac{1}{2}e^{-2\beta t}\sum_{n\geq N_0}[(v_n' - \beta v_n)^2 + (\alpha - \beta^2 + \mu_n c^2)v_n^2] \\
&= \frac{1}{2}e^{-2\beta t}\sum_{n\geq N_0}[v_n'^2 - 2\beta v_n v_n' + (\alpha + \mu_n c^2)v_n^2] \\
&\leq \frac{1}{2}e^{-2\beta t}\sum_{n\geq N_0}[v_n'^2 + \beta(v_n^2 + v_n'^2) + (\alpha + \mu_n c^2)v_n^2] \\
&\leq const\,\hat{E}_2(t)
\end{aligned}$$

where the constant is the larger of $1+\beta$ and the $\sup_{n\geq N_0}\frac{\alpha+\beta+\mu_n c^2}{\alpha-\beta^2+\mu_n c^2}$. •

In order to analyze the contribution $E_1(t)$ we can appeal to the results of Pardoux and Wihstutz [PW; example 5.6, p.453] on the overdamped oscillator driven by white noise. According to these results there exists the limit

$$\lambda_n(\epsilon) = \lim_{t\uparrow\infty} t^{-1}[v_n^2(t) + v_n'(t)^2] \quad (n < N_0)$$

and it satisfies

$$\lambda_n(\epsilon) = \sqrt{\alpha - \beta^2 + \mu_n c^2} + \lambda_n^{(1)}\epsilon + O(\epsilon^2) \qquad (\epsilon\downarrow 0)$$

where the *first-order Lyapunov exponent correction* is given explicitly by

$$\lambda_n^{(1)} = \frac{\sigma^2/8}{\sqrt{\alpha + \mu_n c^2 - \beta^2}}.$$

The first Dirichlet eigenvalue is simple and the μ_n are non-decreasing. Since the sum defining $E_1(t)$ only contains a finite number of terms, we conclude that $\lim_{t\uparrow\infty} t^{-1}\log E_1(t) = \lambda_1(\epsilon)$ given above, from which the proof of theorem 1 follows.•

To prove theorem 2, we simply note that the sum defining $E_1(t)$ is empty and the above analysis can be directly applied to the entire sum defining the energy, from which the result follows.

3. Proof in the real noise case

In case $N(t) = F(\xi(t))$, a centered function of an ergodic Markov process, we may modify the above proof suitably to obtain the same results. For theorem 1 we define the terms $E_1(t), E_2(t)$ in exactly the same fashion. The replacement for lemma 1 is

Lemma 2. *In the real noise case the re-scaled energy* $\hat{E}_2(t)$ *satisfies* $\limsup_{t\uparrow\infty} t^{-1}\hat{E}_2(t) \leq \epsilon M$ *for a constant* M.

To see this we follow the steps of Lemma 1 above to obtain the equation $E_2'(t) = \epsilon\Psi(t)E_2(t)N(t) = \epsilon\Psi(t)F(\xi(t))$. The function F is bounded by a constant M_2. From Gromwall's inequality it is immediate that $\limsup_{t\uparrow\infty} t^{-1}\log E_2(t) \leq \epsilon M_1 M_2$, from which we conclude the result.•

REFERENCES

[PW] E. Pardoux and V. Wihstutz, Lyapunov exponent and small diffusion for linear stochastic differential equations, SIAM Journal of Applied Mathematics, 48(1988), 442-457.

ON STOCHASTIC ELLIPTIC BOUNDARY VALUE PROBLEMS ASSOCIATED WITH GAUSSIAN MARKOV RANDOM FIELDS

Loren D. Pitt and Daoyi Wang[1]
Department of Mathematics, University of Virginia
Charlottesville, VA 22903 USA

This paper resumes the study begun in [5] and initiates an investigation into the effective estimation of a Gaussian Markov random field on Euclidean space which has been observed on the complement of a domain. As described in [5], in certain circumstances, this problem reduces to solving an elliptic boundary value problem with stochastic boundary data. The boundary data is highly irregular and the required normal derivatives exist only as Schwartz distributions. In practice this boundary data must be estimated by making a finite set of observations of the random field. This raises both conceptual and sample design problems. How can the needed distributions, which are not functions, be estimates with a discrete set of observations, and which observations should be made? Once the observations are made, how can they be effectively used to estimate the values of the field?

Although these questions have interest in considerable generality, we only consider here what is probably the simplest example in dimensions greater than one. Namely, the real stationary mean zero Gaussian field $\Phi = \Phi(k,\beta) = \{\phi(x): x \in R^2\}$ introduced by Whittle [8] in 1954 which satisfies the stochastic Laplace equation:

$$(\Delta - k^2)\phi(x) = \beta\dot{W}(x). \tag{1}$$

Here β and k are real nonzero constants and $\{\dot{W}(x): x \in R^2\}$ is a real Gaussian white noise field on the plane, with

$$E[\int f(x)\dot{W}(x)dx]^2 = \int f(x)^2 dx. \tag{2}$$

This and a variety of related fields have been used to fit hydrological and geological data, see Jones [2] and Jones and Vecchia [3].

---

[1]Research sponsored in part by ONR Grant N0014-90-J-1639 and by NSF Grant DMS-9003073.

In the next section we present a brief introduction to this field, its sample functions, and its prediction theory, together with graphs of computer simulations. Section 2 discusses general questions of sample design for general Gaussian Markov fields. Section 3 contains the statements of our mains results, and Section 4 outlines the proofs. We mention that we have followed the reproducing kernel space formalism to the Markov property that was used in [5]. In our judgment this allows for the most direct use of the Sobolev space theory that is natural in this setting. Other approaches, such as those based on the theory presented in [6] are, of course, possible.

Section 1: Introduction to the field Φ

The spectral density of the field $\Phi(k,\beta) = \{\phi_{k,\beta}(x): x \in R^2\}$ satisfying equation (1) is given by

$$\rho(\lambda) = (\beta/2\pi)^2(k^2+|\lambda|^2)^{-2}. \tag{3}$$

The covariance function $R_{k,\beta}(x,y) = R(x,y) = E\phi(x)\phi(y)$ given by

$$R(x,y) = E\phi(x)\phi(y) = \int \exp[i(x-y)\cdot\lambda]\rho(\lambda)d\lambda,$$

and is a radial function $R(|x-y|)$. Scaling shows that $R_{k,\beta}(|x|) = (\beta/k)^2 R_{1,1}(|kx|)$, so the fields $\phi_{k,\beta}(x)$ and $(\beta/k)\phi_{1,1}(kx)$ are stochastically equivalent. The general case can thus be reduced to the special case of $\beta = \beta_0 = 2\pi^{1/2}$ and $k = k_0 = 1$. In what follows we shall only consider this case, for which we note that R is normalized with $R(0) = 1$. The covariance R is exactly calculated with

$$R(|x|) = |x|K_1(|x|),$$

where K_1 is the modified Bessel function of the second kind and first order. R satisfies the condition

$$R(x) \approx (1-|x|^2\log(1/|x|)), \quad \text{as } |x| \to 0.$$

Thus $1-R(x) \neq O(|x|^2)$, as $|x| \to 0$, and it follows that $\{\phi(x)\}$ is non-differentiable. A strong approximation theorem can be proved showing that for each fixed unit vector $x \in R^2$, as $t \to 0$ the difference quotient process $t \to [\phi(tx)-\phi(0)]/t$ approximates a one-dimensional Brownian motion $B(\log(1/t))$. In particular, it has the order of magnitude $(\log(1/t))^{1/2}$. The graph of $\phi(x)$ is thus a nondifferentiable surface which, however, satisfies a Hölder condition with exponent α for each $\alpha < 1$. We comment that the field belongs to the two-dimensional analogue of the quasi-smooth stochastic Λ^* processes introduced in [1]. This graph is much smoother than the usual examples of rough random curves and surfaces, and for a nondifferentiable surface is remarkably well approximated by piecewise linear surfaces. See Figures 1 and 2 below.

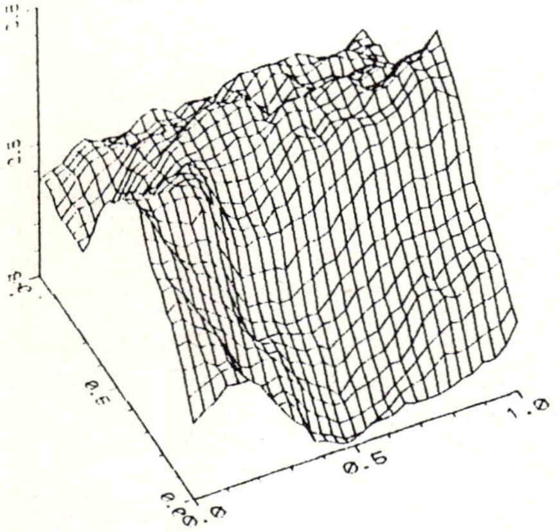

Figure 1

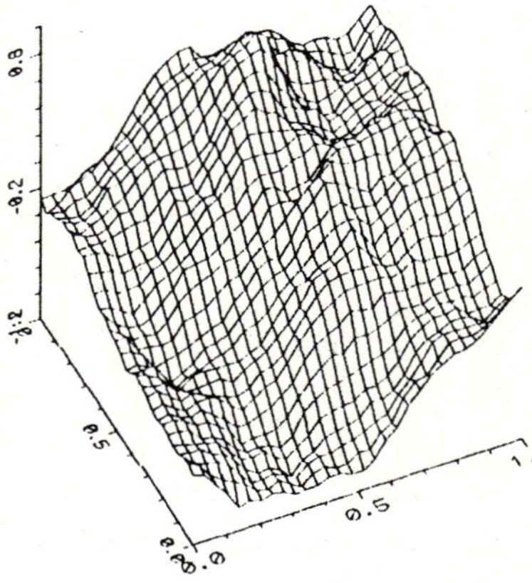

Figure 2

Simulations of $\Phi(1,2\pi^{1/2})$ on the unit square done on an IBM RS/6000S using ISML subroutines for random vector simulation and Surfer graphics software.

Following McKean's idea in [4] normal derivatives of Φ on smooth curves Γ can be interpreted as weak derivatives. Here and in what follows, "smooth" will always mean C^5, although considerably weaker conditions often suffice. For $x \in \Gamma$, we let $n = n(x)$ denote a smoothly varying unit normal vector to Γ at x (outward pointing if such exists) and let $d\sigma = d\sigma(x)$ denote arc length measure on Γ. For each smooth function of compact support $f(x)$ defined on Γ, the function,

$$F(t) \equiv \int_\Gamma f(x)\phi(x+tn(x))d\sigma(x),$$

defined for $t \in R^1$ is known to be once continuously differentiable for small t, see [5]. The normal derivatives $D_n\phi(x)$ of $\phi(x)$ on Γ are defined as the distributional derivative:

$$\int_\Gamma f(x)D_n\phi(x)d\sigma(x) \equiv F'(0).$$

The field Φ is known to be Markovian of order 2, see e.g., Proposition 10.2 in [5], in the following sense: For each bounded domain D with boundary Γ consisting of a finite number of smooth Jordan curves, the two boundary σ-fields, the germ-field

$$\Sigma_+(\Gamma) \equiv \cap\{\sigma\{\phi(x): d(x,\Gamma) < \delta\}: \delta > 0\}$$

and the differential field

$$\Sigma_2(\Gamma) \equiv \sigma\{F^{(k)}(0): \text{f smooth}, 0 \leq k < 2\}$$

are equal, and when conditioned on the boundary σ-field $\Sigma_+(\Gamma)$, the two σ-fields $\Sigma(D) = \sigma\{\phi(x): x \in D\}$ and $\Sigma(D_c) = \sigma\{\phi(x): x \notin D\}$ are conditionally independent.

A basic step in establishing the Markov property and the starting for this paper is the next theorem which provides a reformulation of the least squares prediction problem for Φ as a stochastic boundary value problem.

<u>Theorem 1</u> [5]. The conditional expectation field

$$\Phi_D: \phi_D(x) \equiv E[\phi(x) | \phi(y), y \in D]$$

is the unique solution U: $\{u(x): x \in D\}$ of the boundary value problem

$$(4) \qquad (\Delta-I)^2u(x) = 0, \quad x \in D,$$

$$u(x) = \phi(x), \quad x \in \Gamma,$$

$$(5) \qquad D_nu(x) = D_n\phi(x), \quad x \in \Gamma.$$

Here the normal derivative of u and the second equality in (5) are to be interpreted in the sense: the function $F(t) \equiv \int_\Gamma f(x)u(x+tn(x))d\sigma(x)$, is continuously differentiable on $t \leq 0$, and its left hand derivative at t = 0 agrees with the right hand derivatives of $F(t) \equiv \int_{\Gamma} f(x)\phi(x+tn(x))d\sigma(x)$ at $t = 0$. Our program here is to use Theorem 1 as an approach for calculating the field Φ_D.

Section 2: General Questions on Sample Design

To formulate these problems we let $D \subseteq R^2$ be a bounded domain, and suppose that the field $\Phi = \{\phi(x): x \in R^2\}$ is observable on the complement of D. It is desired to estimate Φ on D. Since Φ is Gaussian this can be done by computing the conditional expecation field

$$\Phi_D: \phi_D(x) \equiv E[\phi(x) \mid \phi(y), y \in D],$$

and the least squares prediction errors

$$\sigma_D^2(x) \equiv E[\phi(x)-\phi_D(x)]^2.$$

Now let a fixed finite set of observations $\{\phi(y_j): j=1,\cdots,N\}$, be made at observation sites $\{y_j: y_j \notin D, j = 1,\cdots,N\}$. We estimate $\phi_D(x)$ with

$$\phi_{D,N}(x) \equiv E[\phi(x) \mid \phi(y_j), j = 1,\cdots,N],$$

and denote the approximate prediction error with

$$\sigma_{D,N}^2(x) \equiv E[\phi(x)-\phi_{D,N}(x)]^2.$$

Taking the infinum over all N element subsets of $D_c = R^2-D$, the quantity

$$s_{D,N}^2(x) = \inf \sigma_{D,N}^2(x),$$

gives the theoretical (perhaps unattainable) prediction error bound for the best that can be achieved with N observations.

Since

$$\sigma_{D,N}^2(x) = E[\phi(x)-\phi_D(x)]^2+E[\phi_D(x)-\phi_{D,N}(x)]^2$$

$$= \sigma_D^2(x)+E[\phi_D(x)-\phi_{D,N}(x)]^2,$$

the two ratios

$$r_{D,N}^2(x) = [s_{D,N}^2(x)-\sigma_D^2(x)]/\sigma_D^2(x),$$

and

$$\rho^2_{D,N}(x) = [\sigma^2_{D,N}(x)-s^2_{D,N}(x)]/s^2_{D,N}(x)$$

measure the relative effectiveness of the observations. While $r^2_{D,N}(x)$ compares the theoretical accuracy achievable with infinitely many observations to that achievable with a particular set of N observations.

We list some basic questions of sample design which should guide investigations in this area. None of these questions are given complete answers here.

1. How large are the quantities $\sigma^2_{D,N}(x)$ and $r^2_{D,N}(x)$ which govern the accuracy that is achievable with N observations and which limit the additional accuracy that is achievable with increased numbers of observations?

2. How small is the quantity $\rho^2_{D,N}(x)$ and how does it depend on the particular set of observations? If it is small it indicates the observations are well chosen, and if it depends little on the particular set of observations, it indicates design stability.

3. How can the given set of observations $\{\phi(y_j): j = 1,\cdots,N\}$ be effectively used to compute $\phi_{D,N}(x)$, and if once $\phi_{D,N}(x)$ has been computed additional observations $\{\phi(y_j): j = N+1,\cdots,M\}$ are made, is it possible to effectively use the calculation of $\phi_{D,N}(x)$ as a step in the calculation of $\phi_{D,M}(x)$?

We note that these questions all have important variations. For example, the measures of error, the expected value of errors squared $\sigma^2_{D,N}(x)$ and $\sigma^2_{D,N}(x)$ may be replaced with other measures such as the expected square of the L^2-norm of the errors,

$$\sigma^2_{D,N}(x) \equiv E\int_D [\phi(x)-\phi_{D,N}(x)]^2dx = \int_D \sigma^2_{D,N}(x)dx,$$

and

$$\sigma^2_D \equiv E\int_D [\phi(x)-\phi_D(x)]^2dx = \int_D \sigma^2_D(x)dx.$$

In the next section we consider the particular field that we will study in this paper.

Section 3: The Results

(A) The Prediction Error. In order to evaluate the effectiveness of an approximation to $\phi_D(x)$, it is necessary to know, or at least estimate the prediction error $\sigma_D^2(x)$. Two answers to this problem are given in Theorem 2. We let $(\Delta-I)_0^2$ be the Friedrich's extension in $L^2(D)$ of the operator $(\Delta-I)^2$ restricted to the space $C_0^\infty(D)$ of smooth functions with compact support in D. Then $(\Delta-I)_0^2$ is invertible and the inverse is an integral operator R_D with kernel $R_D(x,y)$ that is called the Green's function of $(\Delta-I)^2$ on D. We write the error field as

$$e_D(x) = \phi(x)-\phi_D(x),$$

and observe that

$$\sigma_D^2(x) = E[e_D(x)]^2.$$

Moreover, the fields $\phi_D(x)$ and $e_D(x)$ are easily seen to be uncorrelated and hence are independent.

Theorem 2. Let $\Phi = \Phi(1, 2\pi^{1/2})$.

I. On the space $C_0^\infty(D)$ define the inner product

$$(f,g) = 1/(4\pi)\int_D (\Delta-I)f(x)(\Delta-I)g(x)dx. \tag{6}$$

Then

$$\sigma_D^2(x) = \sup\{f(x)^2\colon (f,f) \leq 1\}. \tag{7}$$

II. The Green's function $R_D(x,y)$ of $(\Delta-I)^2$ on D is the error covariance function:

$$R_D(x,y) = E[e_D(x)][e_D(y)]. \tag{8}$$

In particular, the prediction error is given by

$$\sigma_D^2(x) = R_D(x,x). \tag{9}$$

The problem of estimating the prediciton error is thus identified with two problems from the theory of partial differential equations and Sobolev inequalities. An example of results is obtainable by these methods is

<u>Corollary 3</u>. I. If D is smooth, there exists a $\delta = \delta(D) > 0$ such that

$$\sigma_D^2(x) \leq \text{minimum } \{1, 2d(x,\Gamma)^2\} \tag{10}$$

holds for all x in D with $d(x,\Gamma) \leq \delta$. If D is convex then (10) holds for all x in D.

II. For all $x \in D$,

$$\sigma_D^2(x) \geq 3d(x,\Gamma)^2/[16+6d(x,\Gamma)^2]. \tag{11}$$

Together, (10) and (11) show that as $d(x,\Gamma) \to 0$, $\sigma_D^2(x)$ is of the order $d(x,\Gamma)^2$. More refined results are possible. Below are graphs of $\sigma_D^2(x)$ on circular domains with radii 1 and 2.

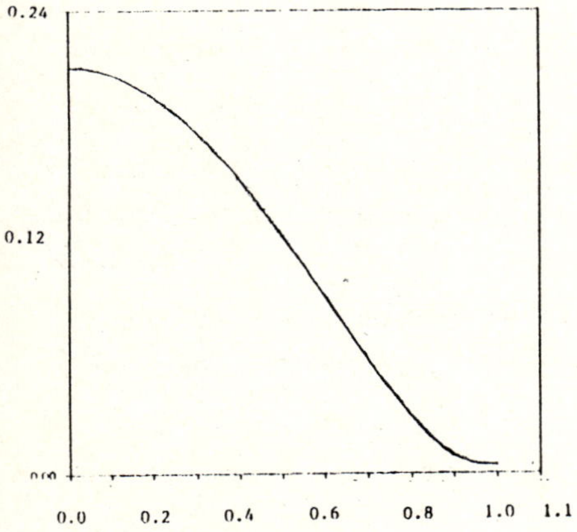

Figure 3
Calculation of $\sigma_D^2(x)$ for $\Phi(1, 2\pi^{1/2})$ on $\{|x| < 1\}$

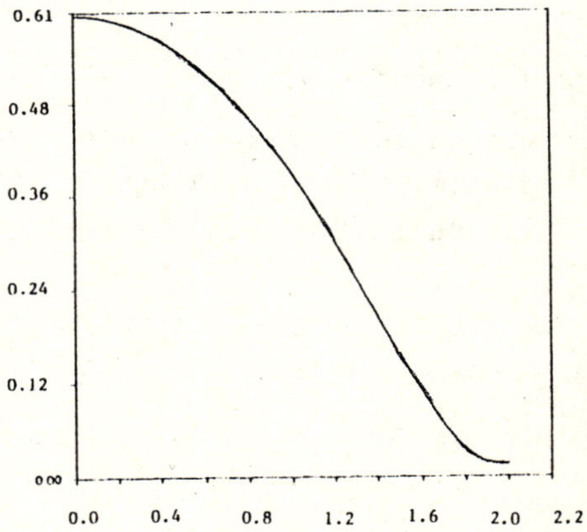

Figure 4
Calculation of $\sigma_D^2(x)$ for for $\Phi(1, 2\pi^{1/2})$ on $\{|x| < 2\}$

Done on IBM RS/6000S using IMSL subroutines and Surfer graphics software

An immediate and useful consequence of Theorem 2 is

<u>Corollary 4</u>. The expected square of the L^2-norm of the error is

$$\sigma_D^2 = \int_D \sigma_D^2(x)\,dx = \text{Trace}(R_D).$$

(B) The Prediction. If only N observations are made, where must the points $\{y_j\}$ be located if it is required that for each x in D the predictions $\phi_{D,N}(x)$ converge to $\phi_D(x)$ as $N \to \infty$? Our answer to this question which is based on the Poisson representation of $\phi_{D,N}(x)$ and $\phi_D(x)$ is practical but theoretically incomplete.

We consider a smooth domain D with boundary Γ of length ℓ. Recall, as was shown in [5], that there exist smooth functions $b_0(x,y)$ and $b_1(x,y)$ such that for x in D the prediction field $\phi_D(x)$ is given by the Poisson integral

(12) $$\phi_D(x) = \int_\Gamma [b_0(x,y)\phi(y)+b_1(x,y)D_n\phi(y)]d\sigma(y).$$

Since $\int_\Gamma b_1(x,y)D_n\phi(y)d\sigma(y)$ is defined as a limit integrals, (3) reduces the calculation of $\phi_D(x)$ to the estimation of integrals. We say that points $\{y_j: 1 \le j \le N\}$ are asymptotically uniformly distributed over Γ provided that with a uniform error of size $O(1/N^2)$ the points lie on Γ and are equally spaced. Similarly, if a sequence $\delta_N \to 0$ is given, and if with a uniform error of size $O(1/N^2)$ one half of the points lie on Γ and are equally spaced and one half of the points lie a distance δ_n outside D and are equally spaced, then we say that $\{y_j: 1 \le j \ge N\}$ is an asymptotically uniform double layer of thickness δ_N over Γ.

The basic technical results in this section are Lemmas 5 and 6.

<u>Lemma 5</u>. We consider a smooth domain D with boundary Γ of length ℓ. Let $\{y_j = y_{N,j}: 1 \le j \le N\}$ be asymptotically uniformly distributed over Γ, $\lambda_N = \ell/N$, and let f(y) be smooth. Then

(13) $$E[\lambda_N \Sigma f(y_j)\phi(y_j) - \int_\Gamma f(y)\phi(y)d\sigma(y)]^2 = O(1/N^3).$$

<u>Lemma 6</u>. With Γ and f as above, the function $F(t) \equiv \int_\Gamma f(x)\phi(x+tn(x))d\sigma(x)$ satisfies

(14) $$E[F'(t)-F'(s)]^2 = O(|t-s|).$$

From Lemmas 5 and 6 we deduce

<u>Theorem 7</u>. Let $\{y_j: 1 \leq j \leq N\}$ be an asymptotically uniform double layer over Γ of thickness δ_N over Γ. Then for all x,

$$\lim E[\phi_{D,N}(x)-\phi_D(x)]^2 = O((\delta_N+1/((\delta_N)^2N^3)). \tag{15}$$

In particular, if δ_N tends to 0 and $\lim N^{3/2}\delta_N = \infty$, then

$$\lim \phi_{D,N}(x) = \phi_D(x) \text{ in } L^2, \tag{16}$$

and if $\delta_N = 1/N$ over Γ, then

$$E[\phi_{D,N}(x)-\phi_D(x)]^2 = O(1/N). \tag{17}$$

<u>Remarks</u>. In Lemmas 5 and 6 the constants which are implicit in the "big oh" terms can be made explicit. They depend only on the size and smoothness of f, and on the curvature and length of Γ. Similar remarks hold for Theorem 7 with the modification that the constants also depend on x. We also remark that we believe that the results in Theorem 7 are sharp, or nearly so, but that no proof of this is known to us.

Section 4: Sketches of Proofs

<u>Proof of Theorem 2</u>. The reproducing kernel Hilbert space (RKHS) $\mathcal{H}$ with reproducing kernel R(x,y) may be identified with the function space of all functions f(x) that are Fourier transforms of functions $(F(\lambda)\rho(\lambda))$ where F is in $L^2(R^2,\rho)$. If g is the Fourier transform of another function $(G(\lambda),\rho(\lambda))$, then the inner product (f,g) in $\mathcal{H}$ is identified either as the inner product in $L^2(R^2,\rho)$ of F and G, or the inner product in $L^2(R^2,\rho^{-1})$ of $(F(\lambda)\rho(\lambda))$ and $(G(\lambda)\rho(\lambda))$. Standard Fourier arguments show that this inner product (F,g) agrees with the one in (6).

By the discussion in Section 3 of [5], the RKHS of the error field $\{e_D(x) = \phi(x)-\phi_D(x): x \in D\}$ is the closure in $\mathcal{H}$ of $C_0^\infty(D)$. Since $E[e_D(x)e_D(y)]$ is the reproducing kernel in this space, (7) follows form the identification, valid in any RKHS of the diagonal terms of the reproducing kernel with the square of the norm of point evaluation.

The proof of (8) follows from the fact that $(I-\Delta)^2$ is invertible in $L^2(R)$ and that $((I-\Delta)^2)^{-1}$ is the convolution operator with kernel $\beta_0 R(x-y)$, together with the observations that $\phi(x) = \phi_D(x)-e_D(x)$ while $(I-\Delta)^2\phi_D(x) = 0$ on D with $e_D(x)$ and its normal derivative vanishing on the boundary of D.

Proof of Corollary 3. We note that $\sigma_D^2(x) = E[\phi(x)-\phi_D(x)]^2 \le E[\phi(x)]^2 = 1$. Also, for any $x \in D$ and any vector y with both x+y and x+2y not in D we have

$$\sigma_D^2(x) \le E[\phi(x)+\phi(x+2y)-2\phi(x+y)]^2 = 1/\pi\int_{R^2}[1-\cos(y\cdot\lambda)]^2(1+|\lambda|^2)^{-2}d\lambda$$
$$= 2\int_{R^1}[1-\cos(|y|\xi)]^2(1+\xi^2)^{-3/2}d\xi .$$

Breaking this last integral into two parts with $|\xi| < 2|y|^2$. Now using the smoothness of G and choosing y as small as possible gives (10) if δ is sufficiently small.

Inequality (11) rests on the simple fact that $[f(x)]^2/(f,f) \le \sigma_D^2(x)$ holds for each f in the $\mathcal{H}$ closure of $C_0^\infty(D)$. The function $f(y) = [|x-y|^2-d(x,\Gamma)^2]^2$ is such a function. A simple calculation shows that for this f,

$$[f(x)]^2/(f,f) = 60d(x,\Gamma)^2/[320+40d(x,\Gamma)^2+d(x,\Gamma)^4]$$

from which (11) may be deduced.

The proof of Corollary 4 is clear.

Proof of Lemma 5. In the first instance we consider the special case where Γ is a segment of the real axis, say $\Gamma = [0,1]$, f(x) is an exponential, say $f(x) = f(u,v) = \exp(i\alpha u)$, and the points $\{y_j\}$ are $\{(j/N+1/2N,0): 0 \le j < N\}$. Expressing $E[1/N\Sigma f(y_j)\phi(y_j) - \int_\Gamma f(y)\phi_D(y)d\sigma(y)]^2$ in the spectral picture gives, after writing $\lambda = (\xi,\eta)$, that

$$E[1/N\Sigma f(y_j)\phi(y_j)-\int_\Gamma f(y)\phi_D(y)d\sigma(y)]^2$$

$$= \int_{R^2} |\exp(i\alpha/2N)[\Sigma\exp(i\alpha j/N)\exp(iy_j\cdot\lambda)]/N$$

$$- \int_0^1 \exp(i\alpha u)\exp(iu\xi)du|^2\rho(\lambda)d\lambda$$

$$= \int_{R^2} |\mathrm{Sin}(\alpha+\xi)/2)[[N\mathrm{Sin}((\alpha+\xi)/2N))]^{-1}-((\alpha+\xi)/2)^{-1}]|^2\rho(\lambda)d\lambda$$

$$= \frac{1}{2}\int_R |\mathrm{Sin}(\alpha+\xi)/2)[[N\mathrm{Sin}((\alpha+\xi)/2N))]^{-1}-((\alpha+\xi)/2)^{-1}]|^2(1+\xi^2)^{-3/2}d\xi .$$

We need to establish that this integral is $O(1/N^3)$. To do so we break the integral into pieces I_1 and I_2, where I_1 is the integral over $\{|\xi| \le N\}$. Applying Taylor series estimates to I_1 readily gives an estimate $I_1 = O(1/N^3)$. For I_2 we first observe that the integral $\int_N^\infty |(\alpha+\xi)^{-1}|^2(1+\xi^2)^{-3/2}d\xi = O(1/N^4)$, and hence those terms in I_2 involving $(\alpha+\xi)^{-1}$ may be ignored. It thus remains to show that

$$I_3 = \int_N^\infty |\mathrm{Sin}(\alpha+\xi)/2)[(N\mathrm{Sin}((\alpha+\xi)/2N))]^{-1}|^2\xi^{-3}d\xi = O(1/N^3).$$

Now write K_j for the interval $[N(1+4\pi j),N(1+4\pi(j+1))]$ and write I_3 as the sum

$$\sum_{j=0}^{\infty} \int_{K_j} |\mathrm{Sin}(\alpha+\xi)/2)[(N\mathrm{Sin}((\alpha+\xi)/2N))^{-1}|^2\xi^{-3}d\xi$$

$$\le \sum_{j=0}^{\infty} [N(1+4\pi j)]^{-3}\int_{K_j} |\mathrm{Sin}(\alpha+\xi)/2)[(N\mathrm{Sin}((\alpha+\xi)/2N))^{-1}|^2d\xi$$

$$= N^{-3}\sum_{j=0}^{\infty} (1+4\pi j)^{-3} \int_0^{4\pi N} |\mathrm{Sin}(\xi/2)[(N\mathrm{Sin}(\xi)/2N)]^{-1}|^2d\xi .$$

But

$$\int_0^{4\pi N}|\mathrm{Sin}(\xi/2)[(N\mathrm{Sin}(\xi)/2N)]^{-1}|^2 d\xi = 2/N\int_0^{2\pi}|\mathrm{Sin}(N\xi)[\mathrm{Sin}(\xi)]^{-1}|^2 d\xi,$$

and since $\mathrm{Sin}(N\xi)[\mathrm{Sin}(\xi)]^{-1}$ is a periodic function Plancherel's theorem shows that the last integral is just the sum of the squares of the Fourier coefficients, or $O(\Sigma_0^N 1) = O(N)$. Dividing by N we have $I_3 = N^{-3}\Sigma_0^\infty(1+4\pi j)^{-3}O(1) = O(1/N^3)$, thus completing the proof in this special case. We remark that this proof is sharp with the error bounded above and below by multiples of $1/N^3$.

For the next step we retain the condition that $\Gamma = [0,1]$, but we replace the hypothesis that f is an exponential, with the assumption that f(x) has a smooth periodic extension of period 2π. With this assumption it is elementary to show that the Fourier series for f converges with sufficient rapidity to allow the estimates obtained above for exponentials to be summed term by term with the result that (13) holds in this case as well.

We next use a change of variables argument, and the fact that if Θ is a smooth diffeomorphism of the plane onto itself and satisfies $\Theta(x) = x$ for all sufficiently large x, then the two reproducing kernel Hilbert spaces for the two fields $\Phi = \{\phi(x): x \in R^2\}$ and $\Psi = \{\psi(x) = \phi(\Theta(x)): x \in R^2\}$ contain the identical set of functions, and the spaces have equivalent norms. This fact follows from the observation that $f(x) \in \mathcal{H}$ iff f can be written in the form $f(x) = E\phi(x)X$ for some X in the L^2 closure $\mathcal{H}$ of $sp\{\phi(x): x \in R^2\}$, and $(f,f) = EX^2$. From this characterization it follows that g(x) is in the RKHS $\mathcal{K}$ of Ψ iff g has the form $g(x) = E\psi(x)X = E\phi(\Theta(x))X$ for some X in the L^2 closure of $sp\{\psi(x): x \in R^2\}$. But $sp\{\phi(x): x \in R^2\} = sp\{\psi(x): x \in R^2\}$, so $g \in \mathcal{K}$ iff $g(x) = f(\Theta(x))$ for some f in $\mathcal{H}$. But the space $\mathcal{H}$ is also identified with the Sobolev space $H_{2,2}$ and the form of the Sobolev norm in $H_{2,2}$ shows that $H_{2,2}$ is closed under composition with Θ. This shows that $\mathcal{H} \subseteq \mathcal{K}$, and the reverse inclusion is proved similarly. That the two norms are equivalent is a simple application of Garding's inequality.

To formalize the final stage in the argument we state another lemma, which surely is in the literature, but which we do not recall seeing earlier.

Lemma 8. Let $\Phi = \{\phi(x): x \in R^2\}$ and $\Psi = \{\psi(x): x \in R^2\}$ be two random fields with RKHS's $\mathcal{H}$ and $\mathcal{K}$ respectively. If the spaces $\mathcal{H}$ and

$\mathcal{K}$ coincide as function spaces and if the norms in H and K are equivalent, then there exist a positive constant C satisfying

$$C^{-1}E[\Sigma a_i\psi(x_i)]^2 \le E[\Sigma a_i\phi(x_i)]^2 \le CE[\Sigma a_i\psi(x_i)]^2 \tag{18}$$

for each finite set of constants $\{a_i\}$ and each set of points $\{x_i\}$.

<u>Proof</u>. As mentioned above, $f \in \mathcal{H}$ (resp. $g \in \mathcal{K}$), iff $f(x) = E\phi(x)X$, for some $X \in \mathcal{H}$ and all x in R^2 (resp. $g(x) = E\phi(x)Y$, for some $Y \in \mathcal{K}$ and all x in R^2). Since $\mathcal{H}$ and $\mathcal{K}$ are assumed equal, it follows that for each X in $\mathcal{H}$ there is a $Y = LX$ in $\mathcal{K}$ so that

$$E\phi(x)X = E\psi(x)\{LX\} \text{ holds for all x in } R^2.$$

Moreover, since the $\mathcal{H}$ norm and the $\mathcal{K}$ norm are assumed to be equivalent, and since the squared $\mathcal{H}$ norm of $f(x) = E\phi(x)X$ is EX^2 while the squared $\mathcal{K}$ norm of the same function $f(x) = E\psi(x)\{LX\}$ is $E[LX]^2$, there must exist a constant C such that

$$E[LX]^2 \le CEX^2 \text{ holds for all X in } \mathcal{H}.$$

Now, for any choice of $\{a_i\}$ and $\{x_i\}$,

$$\begin{aligned} \|\Sigma a_i\phi(x_i)\| &= \sup E\{\Sigma a_i\phi(x_i)X: X \in \mathcal{H}: \|X\| \le 1\} \\ &= \sup E\{\Sigma a_i\psi(x_i)LX: X \in \mathcal{H}: \|X\| \le 1\} \\ &\le \sup E\{\Sigma a_i\psi(x_i)Y: Y \in \mathcal{K}: \|Y\| \le C^{1/2}\} \\ &= C^{1/2}\|\Sigma a_i\psi(x_i)\|, \end{aligned}$$

thereby proving one half of (18). The opposite inequality follows similarly and the proof is complete.

If we now apply Lemma 8 to any field $\Psi = \{\psi(x) = \phi(\Theta(x)): x \in R^2\}$, where $\Theta(x)$ is a smooth diffeomorphism satisfying but the condition $\Theta(x) = x$ for all sufficiently large x and the condition $|\partial_u\Theta(u,0)| = 1$ for $0 \le u \le 1$, we may argue from the above result that for $\{y_j = (j/n + 1/2N, 0): 0 \le j < N\}$, $\{x_j = \Theta(y_j)\}$ and $\Gamma = \Theta[0,1]$ that

$$E[\lambda_N \Sigma f(x_j)\phi(x_j) - \int_\Gamma f(x)\phi(x)d\sigma(x)]^2$$

$$= E[1/N\Sigma f((\Theta(y_j))\phi(\Theta(y_j)) - \int_0^1 f((\Theta(y))\Psi(y)dy]^2 = O(1/N^3).$$

In general, since each smooth function on the smooth boundary Γ may be an approximate identity argument be written as a finite sum involving smooth functions $\{f_i\}$ with compact supports $\{\Gamma_i\}$ that may be mapped onto segments of the real line with diffeomorphisms $\{\Theta_j\}$ that satisfy the above conditions, and since the L^2 modulus of continuity for Φ is $|x|(\log(1/|x|))^{1/2}$ is small enough to allow for perturbations of size $O(1/N^2)$ without affecting the vlaidity of Lemma 5, the result follows.

<u>Proof of Lemma 6</u>. Lemma 8 and change of variables arguments again reduce this to the special case that Γ is the unit interval. Now $F(v) \equiv \int_\Gamma f(x)\phi(x+vn(x))d\sigma(x) = \int_0^1 f(u,0)\phi(u,v)du$, and after setting $g(\xi) = \int_0^1 f(u,0)\exp(iu\xi)du$ and passing to the spectral picture we have

$$E[F'(v)-F'(w)]^2 = \int_{R^2} |g(\xi)\exp(i(v-w)\eta)-g(\xi)|^2\eta^2\rho(\lambda)d\lambda$$

$$= \int_{R^2} 2|g(\xi)\eta|^2[1-\mathrm{Cos}((v-w)\eta)]\rho(\lambda)d\lambda$$

$$= \text{constant}\int_{R^1}[1-\mathrm{Cos}((v-w)\eta)]\eta^2/(1+\eta^2)^2 d\eta = O(|v-w|).$$

<u>Proof of Theorem 7</u>. Our proof is based on the Poisson integral representaiton of $\phi_D(x)$ in (12), which expresses $\phi_D(x)$ as two integrals over Γ, $\int_\Gamma b_0(x,y)\phi(y)d\sigma(y)$ and $\int_\Gamma b_1(x,y)D_n\phi(y)d\sigma(y)$. By Lemma 5, because $\{y_j: 1 \le j \le N\}$ is an asymptotically uniform double layer over Γ, we may approximate the first integral to within an error of L^2 norm $O(1/N^{3/2})$. Using Lemma 6 to estimate the second integral, we find that

$$E\Big[\int_\Gamma b_1(x,y)D_n\phi(y)d\sigma(y)-(1/\delta_N)\int_\Gamma b_1(x,y)[\phi(y+\delta_N n(y))-\phi(y)]d\sigma(y)\Big]^2$$

$$= O(\delta_N).$$

Using again the fact that $\{y_j: 1 \le j \le N\}$ is an asymptotically uniform double layer over Γ, the two integrals here each with an error of L^2 norm $O(1/N^{3/2})$. Thus, using the observations $\{y_j: 1 \le j \le N\}$ we may approximate $\phi_d(x)$ with a combined error of L^2 not greater than $O((\delta_N)^{1/2}+1/\delta_N N^{3/2})$, thus establishing (15).

Remark. Since all strongly elliptic positive fourth-order operators have associated quadratic forms that are equivalent to the form for $(\Delta-I)^2$, it follows that the results presented here are valid for all elliptic Gaussian Markov fields of order 2. These results, additional details and extensions will appear in [6]. Further computational studies and more refined theoretical results are the subject of ongoing work.

References

[1] Anderson, J. M., Horowitz, J., and Pitt, L. D. "On the existence of local times: A geometric study," J. Theoretical Prob. 4, 563-603 (1991).

[2] Jones, R. H. "Fitting a stochastic partial differential equation to aquifer data," Stochastic Hydrology and Hydraulics 3, 85-96 (1989).

[3] Jones, R. H. and Vecchia, A. V. "Fitting continuous ARMA models to unequally spaced spatial data," preprint 1991, presented to ONR workshop, Santa Barbara, California, February 1991.

[4] McKean, H. P. "Brownian motion with a several dimensional time," Theory Prob. Appl. 8, 335-354 (1963).

[5] Pitt, L. D. "A Markov property for Gaussian processes with a multidimensional parameter," Arch. Ration. Mech. Anal. 43 367-391 (1971).

[6] Rozanov, Yu. A., *Markov Random Fields*, Springer-Verlag: New York, 1982.

[7] Wang, Daoyi, "On the prediction theory of Gaussian Markov random fields," University of Virginia dissertation, 1991.

[8] Whittle, P. "On stationary processes in the plane," Biometrika 41, 434-449 (1954).

White Noise Methods for Stochastic Partial Differential Equations

Jürgen Potthoff†

Department of Mathematics,
Louisiana State University,
Baton Rouge, LA 70803, USA

1. Introduction.

This paper presents some methods and tools from white noise analysis which appear to be useful for the treatment of stochastic partial differential equations (SPDE's).

The basic idea of the present paper has already been applied in [KP 90] to the solution of stochastic differential equations with anticipating coefficients. It consists of two main ingredients: the first is the application of a linear transformation (the "$\mathcal{S}$–transform") which diagonalizes the Hitsuda–Skorokhod integral (or divergence operator). The result will often be an equation which can be treated with standard analytical methods. The second ingredient is the reversal of the $\mathcal{S}$–transform by help of a theorem in [PS 89] which characterizes the space of Hida distributions (a space of generalized Brownian functionals) in terms of its $\mathcal{S}$–transform.

It is shown in two examples of SPDE's how these tools apply. Within variants of the white noise framework presented here, these SPDE's have already been studied with different methods in [Ch 89] and [LØU 90], respectively.

In Section 2, we review some basic notions and results from white noise analysis which are necessary for the remainder of the paper. In Section 3 we present an approach to stochastic integration which generalizes the Skorokhod integral to processes with values in the space of Hida distributions. Such integrals will be called Hitsuda–Skorokhod integrals. In Sections 4 and 5 the above mentioned examples will be studied.

---

† Supported by the National Science Foundation under grant DMS–9001859, and by the Louisiana Education Quality Support Fund under grant (91–93) RD–A–08.

2. Some Tools from White Noise Analysis.

In this section we present some pertinent facts from white noise analysis. For more background, results and details we refer to [Hi 80, HKPS, HP 90, Kuo 90, PS 89, PY 89] and the literature quoted there.

Consider the space $(L^2) := L^2(\mathcal{S}'(\mathbb{R}), \mathcal{B}, \mu)$, where $(\mathcal{S}'(\mathbb{R}), \mathcal{B}, \mu)$ is the standard white noise probability space. On (L^2) we introduce the following transformations. For $\varphi \in (L^2), \xi \in \mathcal{S}(\mathbb{R})$, we set

$$\mathcal{S}\varphi(\xi) := \int \varphi(x+\xi)d\mu(x), \tag{2.1}$$

$$\mathcal{T}\varphi(\xi) := \int e^{i<x,\xi>}\varphi(x)d\mu(x), \tag{2.2}$$

where $< \cdot, \cdot >$ denotes dual pairing. We have

$$\mathcal{S}\varphi(\xi) = \int :e^{<x,\xi>}: \varphi(x)d\mu(x), \tag{2.3a}$$

with the notation

$$:e^{<\cdot,\xi>} := e^{<\cdot,\xi>-\frac{1}{2}|\xi|_2^2}. \tag{2.3b}$$

Here we denote the norm of $L^2(\mathbb{R}, du)$ (du denoting Lebesgue measure) by $|\cdot|_2$.

Let $\varphi \in (L^2)$ have chaos decomposition given as follows:

$$\varphi = \sum_{n=0}^{\infty} I_n(f^{(n)}),$$

$$\|\varphi\|_{(L^2)}^2 \equiv \|\varphi\|_2^2 = \sum_{n=0}^{\infty} n!|f^{(n)}|_{L^2(\mathbb{R}^n)}^2,$$

where $I_n(f^{(n)})$ is the n-fold Wiener integral of $f^{(n)} \in \widehat{L^2(\mathbb{R}^n)}$ ($\widehat{\cdot}$ denotes symmetrization). Then

$$\mathcal{S}\varphi(\xi) = \sum_{n=0}^{\infty} \int_{\mathbb{R}^n} f^{(n)}(u)\xi^{\otimes n}(u)d^n u. \tag{2.4}$$

Let A be the self-adjoint extension on $L^2(\mathbb{R})$ of the differential operator $A\xi(u) = -\xi''(u) + (1+u^2)\xi(u)$, $u \in \mathbb{R}$. We shall use the following notation: for $n \in \mathbb{N}$, $f \in \mathcal{S}(\mathbb{R}^n)$, $p \in \mathbb{R}$,

$$|f|_{2,p} := |(A^{\otimes n})^p f|_{L^2(\mathbb{R}^n)}.$$

Moreover, $\mathcal{S}_p(\mathbb{R}^n)$ denotes the the completion of $\mathcal{S}(\mathbb{R}^n)$ with respect to the Hilbertian norm $|\cdot|_{2,p}$. It is well-known (e.g., [Si 71]) that $\mathcal{S}(\mathbb{R}^n)$ is the projective limit of the family $\{\mathcal{S}_p(\mathbb{R}^n,\ n \in \mathbb{N}\}$, while $\mathcal{S}'(\mathbb{R}^n)$ equipped with the strong topology is equal to the inductive limit of $\{\mathcal{S}_{-p}(\mathbb{R}^n), p \in \mathbb{N}\}$.

The second quantization $\Gamma(A)$ of A on (L^2) is by definition the (unique) self-adjoint extension of the operator determined by

$$\Gamma(A) : e^{<\cdot,\xi>} :=: e^{<\cdot,A\xi>} :, \quad \xi \in \mathcal{S}(\mathbb{R}).$$

Mimicking the above indicated construction of $\mathcal{S}(\mathbb{R}^n)$, $\mathcal{S}'(\mathbb{R}^n)$, we set for $p \in \mathbb{R}$,

$$\|\varphi\|_{2,p} := \|\Gamma(A)^p \varphi\|_2,$$

for example for φ belonging to the algebra $\mathcal{A}$ generated by exponential functions $e^{<\cdot,\xi>}$, $\xi \in \mathcal{S}(\mathbb{R})$. Let $(\mathcal{S})_p$ denote the completion of $\mathcal{A}$ with respect to $\|\cdot\|_{2,p}$, $p \in \mathbb{R}$, and set

$$(\mathcal{S}) := \operatorname*{proj-lim}_{p \in \mathbb{N}} (\mathcal{S})_p.$$

The space $(\mathcal{S})$ of white noise test functionals has been studied in a number of papers (e.g., [HP 90, Ko 80, KPS 91, KY 89, PY 89] and literature quoted there). It is a nuclear Fréchet algebra. Its elements have a unique version which is pointwise defined and strongly continuous on $\mathcal{S}'(\mathbb{R})$, and we will usually not distinguish between a class and this representative.

$(\mathcal{S})^*$ is the dual of $(\mathcal{S})$, and its elements are called *Hida distributions*. It is not hard to see that

$$(\mathcal{S})^* = \bigcup_{p \in \mathbb{N}} (\mathcal{S})_{-p} = \bigcup_{p \in \mathbb{N}} (\mathcal{S})_p^*,$$

and that $(\mathcal{S})^*$, equipped with the strong topology, is equal to the inductive limit of the family $\{(\mathcal{S})_{-p}, n \in \mathbb{N}\}$.

The dual pairing between $(\mathcal{S})$ and $(\mathcal{S})^*$ is also denoted by $< \Phi, \varphi >$, $\Phi \in (\mathcal{S})^*$, $\varphi \in (\mathcal{S})$. (We make the convention that this dual pairing is antilinear in the first factor. Therefore it is an extension of the inner product of (L^2) under the natural imbedding of (L^2) into $(\mathcal{S})^*$: $< \Phi, \varphi >= (\Phi, \varphi)_{(L^2)}$, whenever $\Phi \in (L^2)$.) The chaos decomposition of $(\mathcal{S})$ as a subspace of (L^2) induces a chaos decomposition of $(\mathcal{S})^*$: every $\Phi \in (\mathcal{S})^*$ is of the form

$$\Phi = \sum_{n=0}^{\infty} I_n(F^{(n)}), \quad F^{(n)} \in \widehat{\mathcal{S}'(\mathbb{R}^n)},$$

where the sum converges weakly in $(\mathcal{S})^*$, and $I_n(F^{(n)})$ acts on $\varphi = \sum_n I_n(f^{(n)})$, $f^{(n)} \in \widehat{\mathcal{S}(\mathbb{R}^n)}$, as

$$< I_n(F^{(n)}), \varphi >= n! < F^{(n)}, f^{(n)} >,$$

the last pairing being the one between $\mathcal{S}'(\mathbb{R}^n)$ and $\mathcal{S}(\mathbb{R}^n)$. Moreover, $\Phi \in (\mathcal{S})^*$ if and only if there exists a $p \in \mathbb{N}$ so that

$$\|\Phi\|_{2,-p}^2 = \sum_{n=0}^{\infty} n! |F^{(n)}|_{2,-p}^2 < +\infty.$$

We extend $\mathcal{S}-$ and $\mathcal{T}-$transform to $(\mathcal{S})^*$ via:

$$\mathcal{T}\Phi(\xi) = <\overline{\Phi}, e^{i<\cdot,\xi>}>, \tag{2.5}$$

$$\mathcal{S}\Phi(\xi) = <\overline{\Phi}, :e^{<\cdot,\xi>}:>, \tag{2.6}$$

where $\xi \in \mathcal{S}(\mathbb{R})$. (2.5) and (2.6) make sense, because for all $\lambda \in \mathbb{C}$, $\xi \in \mathcal{S}(\mathbb{R})$, we have $\exp(\lambda < \cdot, \xi >) \in (\mathcal{S})$. Note that for all $\Phi \in (\mathcal{S})^*$, $\mathcal{T}\Phi$ and $\mathcal{S}\Phi$ are everywhere defined functions on $\mathcal{S}(\mathbb{R})$. Actually, $\mathcal{T}\Phi$ and $\mathcal{S}\Phi$ are analytic functions of at most second order exponential growth, and $(\mathcal{S})^*$ is characterized by these properties. This is the content of a result in [PS 89] which we are going to describe now.

Consider a complex-valued function F on $\mathcal{S}(\mathbb{R})$. F might have the following two properties:

(A) "Analyticity": F has a ray entire analytic extension on $\mathcal{S}(\mathbb{R})$, i.e., for all $\eta, \xi \in \mathcal{S}(\mathbb{R})$, the mapping $\lambda \longrightarrow F(\eta + \lambda\xi)$ has an entire analytic extension from $\mathbb{R}$ to $\mathbb{C}$.

(B) "Bound": The entire extension of F is at most of order two at zero, i.e., there exist $p \geq 0$, and constants $K_1, K_2 > 0$ so that for all $z \in \mathbb{C}$, $\xi \in \mathcal{S}(\mathbb{R})$,

$$|F(z\xi)| \leq K_1 \exp(K_2 |z|^2 |\xi|^2_{2,p}). \tag{2.7}$$

Definition 2.1. Let $F : \mathcal{S}(\mathbb{R}) \longrightarrow \mathbb{C}$ be a mapping satisfying (A) and (B). Then F is called a *U-functional.*

Theorem 2.2.

(i) Let $\Phi \in (\mathcal{S})^*$. Then $\mathcal{S}\Phi$ and $\mathcal{T}\Phi$ are U-functionals.

(ii) Assume that F is a U-functional. Then there exist unique Φ_S, Φ_T in $(\mathcal{S})^*$ so that $\mathcal{S}\Phi_S = \mathcal{T}\Phi_T = F$.

Remarks.

a) Theorem 2.2 has been localized, generalized and extended in several papers, see [MY 90, SW 91, Ya 90];

b) $(\mathcal{S})$ has been characterized by analytic properties of its $\mathcal{S}$-transform, too. We refer to [KPS 91, Ko 80, Ya 90].

c) Theorem 2.2 has been applied to quantum field theory [PS 90a], Feynman integrals [FPS 91], Cameron–Martin theory [PS 90b], intersectional local times of Brownian motion [SW 91, Wa 91], and anticipating stochastic differential equations [KP 90].

d) Since the Meyer–Watanabe space $\mathcal{D}^*$ (e.g., [Po 88, Wa 83]) of generalized Brownian functionals is imbedded into $(\mathcal{S})^*$ ([PY 89]), its $\mathcal{S}-$ and $\mathcal{T}-$transforms are U-functionals, too. However, an equivalent characterization as in Theorem 2.2 is not known.

One can find a result in [PS 89] concerning the problem whether the convergence of a sequence of U-functionals implies the convergence of the associated elements in $(\mathcal{S})^*$. Using the arguments given in [PS 89], and an application of Vitali's theorem one can prove the following stronger result.

Theorem 2.3. Assume that $\{F_n, n \in \mathbb{N}\}$ is a sequence of U–functionals. Let $\Phi_{S,n}, \Phi_{T,n}$, denote the respective Hida distributions. If for every $\xi \in \mathcal{S}(\mathbb{R})$, $\{F_n(\xi), n \in \mathbb{N}\}$ is Cauchy, and if for every $n \in \mathbb{N}$ the bound (2.7) holds for F_n uniformly in $n \in \mathbb{N}$, then there exist unique elements $\Phi_S, \Phi_T \in (\mathcal{S})^*$ so that $\Phi_{S,n}$ converges strongly in $(\mathcal{S})^*$ to Φ_S, and $\Phi_{T,n}$ converges strongly to Φ_T.

Let us consider two examples. (For other examples we refer to [PS 89, PS 90, KPS 91].)

Example 2.4. Let $n \in \mathbb{N}$, $t_1, \ldots, t_n \in \mathbb{R}$, and let $\delta_{t,m}$, $m \in \mathbb{N}$, be a regularization (say, in $L^2(\mathbb{R})$) of the Dirac distribution δ_t at $t \in \mathbb{R}$, which converges strongly to δ_t as m tends to infinity. Consider the (L^2)–random variable given by $I_n(\delta_{t_1,m}\widehat{\otimes}\cdots\widehat{\otimes}\delta_{t_n,m})$. Its $\mathcal{S}$–transform at ξ is given by $\prod_k (\delta_{t_k,m}, \xi)_{L^2(\mathbb{R})}$, cf. (2.4). Clearly, as m tends to infinity this expression converges for every $\xi \in \mathcal{S}(\mathbb{R})$ to $\prod_k \xi(t_k)$. It is trivial to check the conditions of Theorem 2.3 to conclude that $I_n(\delta_{t_1,m}\widehat{\otimes}\cdots\widehat{\otimes}\delta_{t_n,m})$ converges strongly in $(\mathcal{S})^*$ to some element with $\mathcal{S}$–transform given by $\prod_k \xi(t_k)$. We denote this generalized random variable by $:\dot{B}(t_1)\cdots\dot{B}(t_n):$.

If $n = 1$, $t_1 = t \in \mathbb{R}$, we also write $:\dot{B}(t): \equiv \dot{B}(t)$. It is clear that $\dot{B}(t)$ has to be interpreted as white noise at time t, which has a rigorous meaning as an element in $(\mathcal{S})^*$. Also, if $t_1 = \ldots = t_n = t$, then $:\dot{B}(t)^n:$ has an interpretation as the n–th "renormalized" power of white noise at time t.

Example 2.5. Let $y \in \mathcal{S}'(\mathbb{R})$ and suppose that $\{y_m, m \in \mathbb{N}\}$ is a sequence in $L^2(\mathbb{R})$, which converges strongly to y in $\mathcal{S}'(\mathbb{R})$ as m tends to infinity. We have

$$\mathcal{S} :e^{\lambda<\cdot,y_m>}:(\xi) = e^{\lambda(y_m,\xi)_{L^2(\mathbb{R})}}, \quad \xi \in \mathcal{S}(\mathbb{R}),\ m \in \mathbb{N},$$

as an elementary computation shows. The right hand side of the last equation converges in the limit $m \to \infty$ for every $\xi \in \mathcal{S}(\mathbb{R})$ to $\exp(\lambda < y, \xi >)$. Again, it is straightforward to check the conditions of Theorem 2.3, with the result that $:\exp(\lambda < \cdot, y_m >):$ converges strongly in $(\mathcal{S})^*$ to an element which is denoted by $:\exp(\lambda < \cdot, y >):$. In the special case where $y = \delta_t$, $t \in \mathbb{R}$, we obtain a "renormalized" exponential $:\exp(\lambda \dot{B}(t)):$ of white noise at time $t \in \mathbb{R}$ as a generalized random variable in $(\mathcal{S})^*$.

We remark in passing that $:\exp(< \cdot, y >):$ can be interpreted as the generalized Radon–Nikodym derivative (in $(\mathcal{S})^*$) of the white noise measure translated by y with repect to μ (cf. also [PY 89, PS 90b]).

Let us discuss another consequence of Theorem 2.2. Let F, G be two U–functionals, corresponding (via the $\mathcal{S}$–transform) to the Hida distributions Φ, Ψ, respectively. Obviously, also $F \cdot G$ is a U–functional, and hence there exists a Hida distribution denoted by $\Phi \diamond \Psi$ in $(\mathcal{S})^*$ with $\mathcal{S}(\Phi \diamond \Psi) = F \cdot G$. The product $\Phi \diamond \Psi$ is called the *Wick product* of Φ and Ψ. In particular, $(\mathcal{S})^*$ is an algebra with respect to the Wick product (a result which had already been proved in [MY 89]). In order to get an interpretation of the Wick product, we consider the chaos decompositions of Φ and Ψ:

$$\Phi = \sum_{n=0}^{\infty} I_n(F^{(n)}), \quad F^{(n)} S'(\widehat{\mathbb{R}^n}),$$

$$\Psi = \sum_{n=0}^{\infty} I_n(G^{(n)}), \quad G^{(n)} S'(\widehat{\mathbb{R}^n}).$$

Then the Wick product $\Phi \diamond \Psi$ has chaos decomposition given by

$$\Phi \diamond \Psi = \sum_{n=0}^{\infty} I_n(H^{(n)}),$$

$$H^{(n)} = \sum_{k=0}^{n} F^{(n-k)} \widehat{\otimes} G^{(k)}.$$

Let us now turn to the differential calculus of functions of white noise.

Denote by D_y the Gâteaux derivative in direction $y \in \mathcal{S}'(\mathbb{R})$, i.e. for suitable $\varphi : \mathcal{S}'(\mathbb{R}) \longrightarrow \mathbb{C}$,

$$(D_y \varphi)(x) = \frac{d}{ds} \varphi(x + sy) \Big|_{s=0}.$$

The following result was proved in [PY 89].

Theorem 2.6. For every $y \in \mathcal{S}'(\mathbb{R})$, D_y acts continuously on $(\mathcal{S})$. In particular, $(\mathcal{S})$ is C^∞ with respect to Gâteaux differentiation in every direction of $\mathcal{S}'(\mathbb{R})$.

Let δ_t, $t \in \mathbb{R}$, denote the Dirac distribution concentrated at t. Denote $\partial_t := D_{\delta_t}$. The differential operator ∂_t on $(\mathcal{S})$ is called *Hida derivative.*

Let $\varphi \in (\mathcal{S})$, $x \in \mathcal{S}'(\mathbb{R})$, $t \in \mathbb{R}$, and put

$$(\nabla \varphi)(t, x) := (\partial_t \varphi)(x).$$

Then $\nabla \varphi \in \mathcal{S}(\mathbb{R}) \otimes (\mathcal{S})$, and one can show the following result (e.g., [KPY 90]).

Theorem 2.7. Every $\varphi \in (\mathcal{S})$ is infinitely often Fréchet differentiable, and its Fréchet derivative is given by $\nabla \varphi$. In particular, for all $y \in \mathcal{S}'(\mathbb{R})$,

$$D_y \varphi(x) = < y, \nabla \varphi(x) >, \quad x \in \mathcal{S}'(\mathbb{R}).$$

Since for every $t \in \mathbb{R}$, ∂_t acts continuously on $(\mathcal{S})$, its adjoint ∂_t^* acts (weakly – and in fact strongly – continuously) on $(\mathcal{S})^*$. ∂_t^* is closely related to the divergence operator (cf., e.g., [Kr 79, Kr 83, NP 88, NZ 86, Oc 84] and the next section). It is useful to observe the next relation which follows from a straightforward computation:

$$(\mathcal{S} \partial_t^* \Phi)(\xi) = \xi(t) \mathcal{S} \Phi(\xi), \tag{2.8}$$

holding for all $\Phi \in (\mathcal{S})^*$, $\xi \in \mathcal{S}(I\!R)$, $t \in I\!R$. Since ∂_t^* appears quite frequently in stochastic calculus (for example, cf. the following sections), relation (2.8) makes the $\mathcal{S}$–transform quite useful in this domain.

Also, it is worthwhile to mention the *CCR* (canonical commutation relations), which are stated (slightly informally) as follows. For $s, t \in I\!R$,

$$[\partial_t, \partial_s] = [\partial_t^*, \partial_s^*] = 0, \tag{2.9a}$$

$$[\partial_t, \partial_s^*] = \delta(t-s). \tag{2.9b}$$

3. Hitsuda–Skorokhod Integrals.

In this and the next section, we will only consider real (generalized) random variables. However, we continue to use the notation of Section 2.

Consider the adjoint ∂_t^* of the Hida derivative ∂_t, $t \in I\!R$. Let X be a process indexed by an interval T, on the white noise probability space $(\mathcal{S}'(I\!R), \mathcal{B}, \mu)$, which belongs to $L^2(T, \mathcal{D}^{2,1})$. Here $\mathcal{D}^{2,1}$ denotes the (L^2)–domain of $N^{1/2}$, N being the number (or Ornstein–Uhlenbeck operator). Then an explicit calculation using the chaos decomposition of X shows that $\int_T \partial_t^* X(t)dt$ is in (L^2), and that this expression is the white noise version of the Skorokhod integral or divergence of X (cf. [GT 82, Hi 72, Ki 75, Kr 79, Kr 83, KT 81, KR 88, NP 88, NZ 86, Sk 75] and literature quoted there). Moreover, it is well-known (*loc. cit.*) that in the case that X is adapted to the filtration of Brownian motion, $\int_T \partial_t^* X(t)dt$ is equal to the Itô integral of X. In this section we shall allow for generalized processes X indexed by T, so that for every $t \in T$, $X(t) \in (\mathcal{S})^*$.

Let $(T, \mathcal{F}, dm)$ be a measure space, and assume that $X : T \longrightarrow (\mathcal{S})^*$.

Theorem 3.1. Suppose that X is weakly in $L^1(T, dm)$, i.e., for every $\varphi \in (\mathcal{S})$, the mapping $t \longmapsto < X(t), \varphi >$, $t \in T$, belongs to $L^1(T, dm)$. Then the Pettis integral $\int X(t)dm(t)$ belongs to $(\mathcal{S})^*$, and for all $\varphi \in (\mathcal{S})$,

$$< \int X(t)dm(t), \varphi > = \int < X(t), \varphi > dm(t). \tag{3.1}$$

In particular, $\mathcal{S}$–transform and Pettis integral commute:

$$\mathcal{S}\left(\int X(t)dm(t)\right) = \int \mathcal{S}(X(t))dm(t). \tag{3.2}$$

Sketch of the Proof. $(\mathcal{S})$ is a Fréchet space, and consequently the closed graph theorem is valid on $(\mathcal{S})$. Now it suffices to mimick the arguments in [HP 57, III.3.7] to conclude the proof. □

From now on we choose T as some closed, finite interval, and dm as Lebesgue measure dt on T. Since for every $t \in T$, $\partial_t^* X(t) \in (\mathcal{S})^*$, we can apply Theorem 3.1 and formula (2.8) to obtain the following result.

Theorem 3.2. Assume that $X : T \longrightarrow (\mathcal{S})^*$ is such that $t \longmapsto \partial_t^* X(t)$ is weakly in $L^1(T, dt)$. Then $\int_T \partial_t^* X(t)dt$ belongs to $(\mathcal{S})^*$. Moreover, for every $\xi \in \mathcal{S}(I\!R)$,

$$\mathcal{S}\left(\int_T \partial_t^* X(t)dt\right)(\xi) = \int_T \xi(t)\mathcal{S}(X(t))(\xi)dt. \tag{3.3}$$

The weak stochastic integral $\int_T \partial_t^* X(t)dt$ will henceforth be called *Hitsuda–Skorokhod integral of X over T*. Here is a simple sufficient condition for X to have a Hitsuda–Skorokhod integral in $(\mathcal{S})^*$, whose proof we leave as an exercise to the interested reader.

Lemma 3.3. Assume that $X : T \longrightarrow (\mathcal{S})^*$ is piecewise strongly continuous. Then it admits a Hitsuda–Skorokhod integral over T in $(\mathcal{S})^*$.

Equation (3.3) can be used to give a very simple proof of Itô's formula. For convenience of notation, we shall state it here only in one dimension. The general case follows by obvious modifications. Let $F \in \mathcal{S}'(I\!R)$, and let $B(t)$, $t \geq 0$, denote Brownian motion. It is well-known (cf., e.g., [Wa 83]) that for $t > 0$, $F \circ B(t)$ makes sense in $\mathcal{D}^* \subset (\mathcal{S})^*$. Moreover, it follows either by direct computation or from the proof in [Wa 83] (and a simple hypercontractivity argument, e.g. [PY 89]) that for all $s, t \in I\!R$ with $0 < s \leq t$, $F \circ B(\cdot)$ satisfies the condition of Lemma 3.3 with $T = [s, t]$, and therefore admits a Hitsuda–Skorokhod integral in $(\mathcal{S})^*$. Itô's formula for the composition of a tempered distribtuion F with Brownian motion has been proved first in [Ku 83]. We give a different proof here.

Theorem 3.4. For all $s, t \in I\!R$ with $0 < s \leq t$, the following formula holds:

$$F \circ B(t) - F \circ B(s) = \int_s^t \partial_u^* F' \circ B(u)du + \frac{1}{2}\int_s^t F'' \circ B(u)du, \tag{3.4}$$

where F', F'' denote the first and second order distributional derivatives of F, respectively.

Proof. Consider first the function $F(y) = e^{iky}$, $y, k \in I\!R$. Then

$$\int_s^t \partial_u^* F' \circ B(u)du = ik \int_s^t \partial_u^* : e^{ikB(u)} : e^{-\frac{k^2}{2}u} du.$$

By (3.3) and integration by parts we find

$$\begin{aligned}\mathcal{S}\left(\int_s^t \partial_u^* F' \circ B(u)du\right)(\xi) &= \exp\left(ik\int_0^t \xi(v)dv - \frac{k^2}{2}t\right) - \exp\left(ik\int_0^s \xi(v)dv - \frac{k^2}{2}s\right)\\ &\quad + \frac{1}{2}k^2 \int_s^t \exp\left(ik\int_0^u \xi(v)dv - \frac{k^2}{2}u\right)du.\end{aligned}$$

But the last term is just the $\mathcal{S}$–transform of

$$e^{ikB(t)} - e^{ikB(s)} + \frac{1}{2}k^2 \int_s^t e^{ikB(u)}du$$

at $\xi \in \mathcal{S}(I\!R)$. Thus we have shown (3.4) for this special choice of F.

Now let $F \in \mathcal{S}(I\!R)$, and write $F(y) = (2\pi)^{-1/2} \int e^{iky} \widetilde{F}(k) dk$, where $\widetilde{F} \in \mathcal{S}(I\!R)$ is the Fourier transform of F. Then one can interchange of the Hitsuda–Skorokhod integral of $F \circ B(u)$ with the Fourier integral, as can be justified for example by using the $\mathcal{S}$–transform and Fubini's theorem. This way, one obtains equation (3.4) for $F \in \mathcal{S}(I\!R)$. Finally, one proves (3.4) for $F \in \mathcal{S}'(I\!R)$ by a standard limiting argument. □

It is worthwhile to make the following observation. Let X be as above. Then we have for $\xi \in \mathcal{S}(I\!R)$,

$$\begin{aligned} \mathcal{S}\Big(\int_T \dot{B}(t) \diamond X(t) dt\Big)(\xi) &= \int_T \Big(\mathcal{S}\dot{B}(t)\Big)(\xi)\,(\mathcal{S}X(t))\,(\xi) dt \\ &= \int_T \xi(t)\,(\mathcal{S}X(t))\,(\xi) dt \\ &= \mathcal{S}\Big(\int_T \partial_t^* X(t) dt\Big)(\xi). \end{aligned}$$

Thus we may also write the Hitsuda–Skorokhod integral as $\int_T \dot{B}(t) \diamond X(t) dt$, giving it an intuitive meaning (cf. also [LØU 91]).

4. The LØU Equation.

In this section we employ the tools of Sections 2 and 3 to treat a stochastic partial differential equation, which has been investigated recently by Lindstrøm, Øksendal and Ubøe [LØU 90]. In the context of problems from hydrodynamics, they considered an equation of the following form for a function $p : I\!R_+ \times I\!R^d \longrightarrow I\!R$,

$$\begin{aligned} \operatorname{div}\Big(k(x) \operatorname{grad} p(x,t)\Big) &= f(x,t), \quad x \in D_t, \\ p(x,t) &= 0, \quad x \in \partial D_t, \end{aligned} \tag{4.1}$$

where $p(x,t)$ is the pressure of a fluid in a porous medium at the point x at time t. In (4.1), divergence and gradient are taken with respect to x, f is a given function, D_t is the region in $I\!R^d$ which is saturated by the liquid at time t. Finally, k is the porosity of the medium. In cases of interest, k varies rapidly from point to point in an irregular way, and is usually not accessible to measurements. Hence, the porosity is a source of randomness and should be considered as a noise indexed by x: thus p becomes a function of chance, too.

The first mathematical problem is to how model a noise k which is *positive*; – porosity being a positive quantity. Let us (over)simplify (4.1) by considering the following toy model in one dimension, without time variable. In order to use the notation of the preceding sections, let us also rename the space variable x into $t \in I\!R$, i.e. let us consider an equation of the form

$$\frac{d}{dt}\Big(k(t)\frac{d}{dt}p(t)\Big) = f(t), \tag{4.2}$$

where f is a given deterministic function. In [LØU 90] it is proposed to choose as a model for the positive noise k the renormalized exponential of white noise (cf. Example 2.5): $k(t) =: \exp(\lambda \dot{B}(t)) :, t \in \mathbb{R}$. Then $k(t)$ is positiv in the sense that for every $\varphi \in (\mathcal{S})$ with $\varphi \geq 0$, we have $< k(t), \varphi > \geq 0$ for all $t \in \mathbb{R}$. Moreover, since the product of the generalized random variable $k(t) \in (\mathcal{S})^*$ with other random variables is in general not well-defined, it is proposed in this paper to replace the product in (4.2) by the Wick product. (Also, this is consistent with the remark at the end of Section 3.) Thus we are led to

$$\frac{d}{dt}\Big(k(t) \diamond \frac{d}{dt}p(t)\Big) = f(t), \quad k(t) =: \exp(\lambda \dot{B}(t)) :, \tag{4.3}$$

which we call the LØU equation. This equation is studied in [LØU 90] within a variant of the white noise framework described in Section 2. In particular, the authors use there the Hermite transform to solve (4.3). Here we shall use the $\mathcal{S}$–transform combined with Theorem 2.2 and Theorem 3.1 instead.

We shall assume throughout, that $f \in L^1_{loc}(\mathbb{R}, dt)$. Perform a first integration in (4.3), and take the $\mathcal{S}$–transform, informally interchanging it with the second t–derivative. The result is

$$\frac{d}{dt}(\mathcal{S}p(t))(\xi) = \exp(-\lambda\xi(t))\Big[\int_0^t f(s)ds + \mathcal{S}p_1(\xi)\Big], \quad \xi \in \mathcal{S}(\mathbb{R}),$$

where p_1 is a constant of integration which might be a (generalized) random variable. Another integration gives

$$(\mathcal{S}p(t))(\xi) = \int_0^t \exp(-\lambda\xi(s_1)) \int_0^{s_1} f(s_2)ds_2\, ds_1 + \mathcal{S}p_1(\xi) \int_0^t \exp(-\lambda\xi(s))ds + \mathcal{S}p_2(\xi),$$

where $\mathcal{S}p_2(\xi)$ is another constant of integration. It is not hard to see that if $p_1 \in (\mathcal{S})^*$, and if $\mathcal{S}p_2(\xi)$ is a U–functional, then we can apply Theorem 2.2 to ensure the existence of an element $p(t)$ in $(\mathcal{S})^*$ whose $\mathcal{S}$–transform is given by the right hand side of the last equation for every $t \geq 0$. Moreover, informally this element has the form

$$p(t) = \int_0^t : \exp(-\lambda\dot{B}(s_1) : \Big(\int_0^{s_1} f(s_2)ds_2\Big)ds_1 + p_1 \diamond \int_0^t : \exp(-\lambda \dot{B}(s)) : ds + p_2. \tag{4.4}$$

Now we are going to give a rigorous meaning to the last equation. First we need the following result.

Lemma 4.1. Let $p > 5/12$. Then $t \mapsto \delta_t$ is uniformly continuous from $\mathbb{R}$ into $\mathcal{S}_{-p}(\mathbb{R})$. In particular, $t \mapsto \delta_t$ is uniformly strongly continuous from $\mathbb{R}$ into $\mathcal{S}'(\mathbb{R})$.

Sketch of the proof. Expand $\delta_t = \sum_n e_n(t)e_n$, where $\{e_n, n \in \mathbb{N}_0\}$ is the system of Hermite functions (the sum being strongly convergent in $\mathcal{S}'(\mathbb{R})$). Now the result follows from a straightforward computation and estimate 21.3.3 in [HP 57]. □

Lemma 4.2. For every $p > 5/12$ there exists a constant $C_p > 0$, so that for all $\lambda, t, s \in \mathbb{R}$,

$$\| : e^{\lambda\dot{B}(t)} : - : e^{\lambda\dot{B}(s)} : \|_{2,-p} \leq \lambda|\delta_t - \delta_s|_{2,-p}e^{C_p\lambda^2}.$$

In particular, $t \mapsto : \exp(\lambda \dot{B}(t)) :$ is strongly uniformly continuous from $\mathbb{R}$ into $(\mathcal{S})^*$.

Proof. Note that $: \exp(\lambda \dot{B}(t)) :=: \exp(\lambda < \cdot, \delta_t >:$ has chaos decomposition given by the sequence $\{(n!)^{-1}\lambda^n \delta_t^{\otimes n};\, n \in \mathbb{N}_0\}$. Thus we get from Section 2 that

$$\| : e^{\lambda \dot{B}(t)} : - : e^{\lambda \dot{B}(s)} : \|_{2,-p} = \sum_{n=0}^{\infty} \frac{\lambda^{2n}}{n!} |\delta_t^{\otimes n} - \delta_s^{\otimes n}|_{2,-p}^2$$

$$\leq |\delta_t - \delta_s|_{2,-p}^2 \sum_{n=1}^{\infty} \frac{\lambda^{2n}}{n!} n (\sup_{t \in \mathbb{R}} |\delta_t|_{2,-p})^{2(n-1)}.$$

It is easy to show as indicated in the sketch of the proof of Lemma 4.1 that for $p > 5/12$, $\sup_t |\delta_t|_{2,-p} = 2C_p^{1/2} < +\infty$, for some constant $C_p > 0$. □

It is clear that we may apply now the results of Section 3, in particular Theorem 3.1, to conclude that $p(t)$ defined as in (4.4) is indeed in $(\mathcal{S})^*$, the integral being interpreted as a Pettis integral (and in this case it is actually a Bochner integral). Moreover, using the commutativity of the Pettis integral with the $\mathcal{S}$–transform one sees that $p(t)$ solves the LØU equation in weak sense. Namely, $p(t)$ solves equation (4.3) when paired with $: \exp(< \cdot, \xi >) :$, $\xi \in \mathcal{S}(\mathbb{R})$, and these functions form a total set in $(\mathcal{S})$. Therefore we have proved the following result.

Theorem 4.3. Among all $(\mathcal{S})^*$–valued processes which are twice weakly differentiable, $p(t)$ as given in (4.4) is the unique weak solution of the LØU equation (4.3) with initial conditions $p(0) = p_1 \in (\mathcal{S})^*$, $\frac{d}{dt}p(0) = p_2 \in (\mathcal{S})^*$.

5. An Equation from Turbulent Transport.

In this section we are going to recover some results (in slightly weaker form) which have been proved by Chow in [Ch 89]. For a related work we refer also to [NZ 89]. We shall use again the method of employing the $\mathcal{S}$-transform and Theorems 2.2 and 2.3 to "undo" it, and obtain a rather simple proof.

Chow considers a stochastic partial differential equation for a function $u : \mathbb{R}_+ \times \mathbb{R}^d \times \Omega \longrightarrow \mathbb{R}$, which reads informally as follows:

$$\frac{\partial}{\partial t} u(t,x) + \dot{\eta}(t,x) \circ \nabla u(t,x) = \frac{1}{2} \nu(t) \Delta u(t,x), \tag{5.1}$$

where $(t, x, \omega) \in \mathbb{R}_+ \times \mathbb{R}^d \times \Omega$, and t stands for time, x for position, and ω is a random parameter (which is, as usually, suppressed). Moreover, $\dot{\eta}$ is the noise given by

$$\eta(t,x) = \int_0^t \sigma(s,x) dB(s),$$

where σ is a deterministic function of space and time, and $\circ$ in (5.1) indicates Stratonovic multiplication. This equation arises in modelling the transport of a substance in a turbulent medium. In this context, $\nu \geq 0$ has an interpretation as the molecular viscosity of the medium. Following Chow, we shall consider (5.1) with the initial condition of a unit point source at the origin: $u(0,\cdot) = \delta_0$. (Other initial conditions are easily implementable, too.)

Of particular interest is the singular limit (cf. [Ch 89]) as the molecular viscosity ν tends to zero.

We will only discuss here the homogeneous case: namely, we assume that the noise coefficient σ does not depend on the space variable x. The non–homogeneous case will be treated with the present methods in a forthcoming paper [CCP 91]. Moreover, for notational simplicity we restrict ourselves to the situation $d = 1$. We suppose that $\nu \in L^1_{loc}(I\!R, dt)$, $\sigma \in L^2_{loc}(I\!R, dt)$, $\sigma \not\equiv 0$.

As in [Ch 89], we first solve (5.1) with the mentioned initial condition by taking the Fourier transform with respect to the space variable x. Then equation (5.1) becomes ($\widetilde{\ }$ denoting Fourier transform):

$$\frac{\partial}{\partial t}\widetilde{u}(t,p) + ip\dot{\eta} \circ \widetilde{u}(t,p) = -\frac{1}{2}p^2\nu(t)\widetilde{u}(t,p), \quad p \in I\!R,$$

while the initial condition becomes $\widetilde{u}(0,p) = (2\pi)^{-1/2}$. We can easily rewrite this equation into a stochastic differential equation for every $p \in I\!R$ and solve it. The result is

$$\widetilde{u}(t,p) = \frac{1}{\sqrt{(2\pi)}}\exp\Big(-\int_0^t [ip\sigma(s)dB(s) + \frac{1}{2}p^2\nu(s)ds]\Big). \tag{5.2}$$

It is not hard to invert the Fourier transform, and to show that for all $x \in I\!R$, $t > 0$ the solution $u(t,x)$ belongs to (L^2). The S–transform of (5.2) is

$$S\widetilde{u}(t,p)(\xi) = \frac{1}{\sqrt{2\pi}}\exp\Big(-i\int_0^t p\sigma(s)\xi(s)ds - \frac{1}{2}p^2\tau_\nu(t)\Big),$$

where we have set $\tau_\nu(t) = \int_0^t [\sigma(s)^2 + \nu(s)]ds$. Inversion of the Fourier transform (using Fubini's theorem) yields:

$$Su_\nu(t,x)(\xi) = \frac{1}{\sqrt{2\pi\tau_\nu(t)}}\exp\Big(-\frac{1}{2\tau_\nu(t)}\Big[x - \int_0^t \sigma(s)\xi(s)ds\Big]^2\Big). \tag{5.3}$$

In the last equation we have given u the index ν in order to emphasize its ν–dependence. Now let ν tend in $L^1_{loc}(I\!R)$ to zero (through a sequence). Then $\tau_\nu(t)$ tends to $\tau_0(t) = \int_0^t \sigma(s)^2 ds > 0$, and the right hand side of (5.3) converges for every $\xi \in S(I\!R)$ to the U–functional:

$$\frac{1}{\sqrt{2\pi\tau_0(t)}}\exp\Big(-\frac{1}{2\tau_0(t)}\Big[x - \int_0^t \sigma(s)\xi(s)ds\Big]^2\Big). \tag{5.4}$$

Moreover, it is obvious that we may apply Theorem 2.3. Hence, $u_\nu(t,x)$ converges strongly in $(S)^*$ to a generalized functional $u_0(t,x)$, whose S–transform is given by (5.4). One can show that actually $u_0(t,x)$ is equal to $\delta_x \circ \eta(t)$ (for example by using again Theorems 2.2 and 2.3). So we have proved the following result for the homogeneous case.

Theorem 5.1. Assume that $x \in \mathbb{R}^d$, $t > 0$, $\sigma \in L^2_{loc}(\mathbb{R}, dt), \sigma \not\equiv 0$, and that $\nu \in L^1_{loc}(\mathbb{R}, dt)$. Suppose that ν tends through a sequence to zero in $L^1_{loc}(\mathbb{R}, dt)$. Then the solution of (5.1) converges strongly in $(\mathcal{S})^*$ to $\delta_x \circ \eta(t)$.

We mention without explicit proof, that one can also use the expression (5.4) for the $\mathcal{S}$–transform of $u_0(t, x) = \delta_x \circ \eta(t)$ to show that $u_0(t, x)$ satisfies (5.1) for $\nu = 0$ in weak sense.

References.

[AP 91] Asch, J. and Potthoff, J.: Itô's lemma without non-anticipatory conditions; *Probab. Th. Rel. Fields* **88** (1991) 17–46

[Ch 89] Chow, P.L.: Generalized solution of some parabolic equations with a random drift; *J. Appl. Math. Optimization* **20** (1989) 81–96

[CCP 91] Chow, P.L., Cochran, G. and Potthoff, J.: On parabolic stochastic partial differential equations; in preparation

[FPS 91] de Faria, M., Potthoff, J. and Streit, L.: The Feynman integrand as a Hida distribution; *J. Math. Phys.* **32** (1991) 2123–2127

[GT 82] Gaveau, B. and Trauber, P.: L'intégrale stochastique comme opérateur de divergence dans l'espace fonctionnel; *J. Funct. Anal.* **46** (1982) 230–238

[Hi 80] Hida, T.: *Brownian Motion.* Berlin, Heidelberg, New York: Springer (1980)

[HKPS] Hida, T., Kuo, H.-H., Potthoff, J. and Streit, L.: *White Noise: An Infinite Dimensional Calculus.* Monograph in preparation

[HP 90] Hida, T. and Potthoff, J.: White noise analysis – an overview; in *White Noise Analysis – Mathematics and Applications*, T. Hida, H.-H. Kuo, J. Potthoff and L. Streit (eds.). Singapore: World Scientific (1990)

[HP 57] Hille, E. and Phillips, R.S.: *Functional Analysis and Semigroups.* Providence: American Mathematical Society Colloq. Publ. (1957)

[Hi 72] Hitsuda, M.: Formula for Brownian partial derivatives; in *Proc. 2nd Japan-USSR Symp. Probab. Th.* **2**, Academy of Science USSR (1972) 111–114

[Ki 75] King, R.: Stochastic integrals and metadistributions. Applications to stochastic partial differential equations and quantum field theory. Ph. D. dissertation (1975), Cornell Univ.

[Ko 80] Kondrat'ev, Ju.G.: Nuclear spaces of entire functions in problems of infinite–dimensional analysis; *Soviet Math. Dokl.* **22** (1980) 588–592

[Kr 79] Krée, P.: Calcul d' intégrales et de dérivées en dimension infinie; *J. Funct. Anal.* **31** (1979) 150–186

[Kr 83] Krée, M. and Krée, P.: Continuité de la divergence dans les espaces de Sobolev relatifs à l'espace de Wiener; *C. R. Acad. Sci. Paris* **296** (1983) 833–836

[Ku 83] Kubo, I.: Itô formula for generalized Brownian functionals; in *Theory and Application of Random Fields*; G. Kallianpur (ed.). Berlin, Heidelberg, New York: Springer (1983)

[KT 81] Kubo, I. and Takenaka, S.: Calculus on Gaussian white noise III; *Proc. Japan Acad.* **57A** (1981) 433–437

[KY 89] Kubo, I. and Yokoi, Y.: A remark on the space of testing random variables in the white noise calculus; *Nagoya Math. J.* **115** (1989) 139–149

[Kuo 90] Kuo, H.-H.: Lectures on white noise analysis; *Preprint* (1990)

[KP 90] Kuo, H.-H. and Potthoff, J.: Anticipating stochastic integrals and stochastic differential equations; in *White Noise Analysis – Mathematics and Applications*, T. Hida, H.-H. Kuo, J. Potthoff and L. Streit (eds.). Singapore: World Scientific (1990)

[KPS 91] Kuo, H.-H., Potthoff, J. and Streit, L.: A characterization of white noise test functionals; *Nagoya Math. J.* **121** (1991) 185–194

[KPY 90] Kuo, H.-H., Potthoff, J. and Yan, J.-A.: Continuity of affine transformations of white noise test functionals and applications; *Preprint* (1990), to appear in *Stoch. Proc. Appl.*

[KR 88] Kuo, H.-H. and Russek, A.: White noise approach to stochastic integration; *J. Multivariate Analysis* **24** (1988) 218–236

[LØU 90] Lindstrøm, T., Øksendal, B. and Ubøe, J.: Stochastic differential equations involving positive noise; *Preprint* (1990), to appear in *Stochastic Analysis*, M. Barlow and N. Bingham (eds.)

[LØU 91] Lindstrøm, T., Øksendal, B. and Ubøe, J.: Wick multiplication and Itô–Skorokhod stochastic differential equations; *Preprint* (1991), to appear in *Ideas and Methods in Mathematical Physics*, S. Albeverio et al. (eds.)

[MY 89] Meyer, P.A. and Yan, J.–A.: Distributions sur l'espace de Wiener (suite); *Séminaire de Probabilités* **XXIII**; J. Azéma, P.A. Meyer and M. Yor, eds. Berlin, Heidelberg, New York: Springer (1989)

[MY 90] Meyer, P.A. and Yan, J.–A.: Les "fonctions caractéristiques" des distributions sur l'espace de Wiener; *Preprint* (1990)

[NP 88] Nualart, D. and Pardoux, E.: Stochastic calculus with anticipating integrands; *Probab. Th. Rel. Fields* **78** (1988) 535–581

[NZ 86] Nualart, D. and Zakai, M.: Generalized stochastic integrals and the Malliavin calculus; *Probab. Th. Rel. Fields* **73** (1986) 255–280

[NZ 89] Nualart, D. and Zakai, M.: Generalized Brownian functionals and the solution to a stochastic partial differential equation; *J. Funct. Anal.* **84** (1989) 279–296

[Oc 84] Ocone, D.: Malliavin's calculus and stochastic integral representation of functionals of diffusion processes; *Stochastics* **12** (1984) 161–185

[Po 88b] Potthoff, J.: On Meyer's equivalence; *Nagoya Math. J.* **111** (1988) 99–109

[PS 89] Potthoff, J. and Streit, L.: A characterization of Hida distributions; *Preprint* (1989), to appear in *J. Funct. Anal.*

[PS 90a] Potthoff, J. and Streit, L.: Invariant states on random and quantum fields: ϕ–bounds and white noise analysis; *BiBoS Preprint* (1990)

[PS 90b] Potthoff, J. and Streit, L.: Generalized Radon – Nikodym derivatives and Cameron – Martin theory; *Preprint* (1990), to appear in *Proc. Int. Conf. Gaussian Random Fields*, Nagoya, 1990

[PY 89] Potthoff, J. and Yan J.A.: Some results about test and generalized functionals of white noise; to appear in *Proc. Singapore Probab. Conf.* (1989), L.Y. Chen (ed.)

[Si 71] Simon, B.: Distributions and their Hermite expansions; *J. Math. Physics* **12** (1971) 140–148

[Sk 75] Skorokhod, A. V.: On a generalization of a stochastic integral; *Theory Probab. Appl.* **20** (1975) 219–233

[SW 91] Streit, L. and Westerkamp, W.: A generalization of the characterization theorem for generalized functionals of white noise; *Preprint* (1991)

[Wa 91] Watanabe, H.: The local time of self–intersection of Brownian motions as "generalized Brownian functionals"; *Preprint* (1991)

[Wa 83] Watanabe, S.: Malliavin's calculus in terms of generalized Wiener functionals; in *Theory and Application of Random Fields*; G. Kallianpur (ed.). Berlin, Heidelberg, New York: Springer (1983)

[Ya 90] Yan, J.–A.: A characterization of white noise functionals; *Preprint* (1990)

Lecture Notes in Control and Information Sciences

Edited by M. Thoma and A. Wyner

Vol. 137: S. L. Shah, G. Dumont (Eds.)
Adaptive Control Strategies for Industrial Use
Proceedings of a Workshop
Kananaskis, Canada, 1988
VI, 360 pages. 1989

Vol. 138: D. C. McFarlane, K. Glover
Robust Controller Design Using Normalized Coprime Factor Plant Descriptions
X, 206 pages. 1990

Vol. 139: V. Hayward, O. Khatib (Eds.)
Experimental Robotics I
The First International Symposium
Montreal, June 19-21, 1989
XVII, 613 pages. 1990

Vol. 140: Z. Gajic, D. Petkovski, X. Shen
Singularly Perturbed and Weakly Coupled Linear Control Systems
A Recursive Approach
VII, 202 pages. 1990

Vol. 141: S. Gutman
Root Clustering in Parameter Space
VIII, 153 pages. 1990

Vol. 142: A. N. Gündeş, C. A. Desoer
Algebraic Theory of Linear Feedback Systems with Full and Decentralized Compensators
V, 176 pages. 1990

Vol. 143: H.-J. Sebastian, K. Tammer (Eds.)
System Modelling and Optimization
Proceedings of the 14th IFIP Conference
Leipzig, GDR, July 3-7, 1989
X, 960 pages. 1990

Vol. 144: A. Bensoussan, J. L. Lions (Eds.)
Analysis and Optimization of Systems
Proceedings of the 9th International Conference, Antibes, June 12-15, 1990
XII, 992 pages. 1990

Vol. 145: M. B. Subrahmanyam
Optimal Control with a Worst-Case Performance Criterion and Applications
X, 133 pages. 1990

Vol. 146: D. Mustafa, K. Glover
Minimum Entropy H_∞ Control
IX, 144 pages. 1990

Vol. 147: J. P. Zolésio (Ed.)
Stabilization of Flexible Structures
Third Working Conference,
Montpellier, France, January 1989
V, 327 pages, 1990

Vol. 148: In preparation

Vol. 149: K. H. Hoffmann, W. Krabs (Eds.)
Optimal Control of Partial Differential Equations
Proceedings of the IFIP WG 7.2 International Conference, Irsee, April 9-12, 1990
VI, 245 pages. 1991

Vol. 150: L. C. G. J. M. Habets
Robust Stabilization in the Gap-topology
IX, 126 pages. 1991

Vol. 151: J. M. Skowronski, H. Flashner, R. S. Guttalu (Eds.)
Mechanics and Control
Proceedings of the 3rd Workshop on Control Mechanics
in Honor of the 65th Birthday of George Leitmann
January 22-24, 1990, University of Southern California
IV, 497 pages. 1991

Vol. 152: J. D. Aplevich
Implicit Linear Systems
XI, 176 pages. 1991

Vol. 153: O. Hájek
Control Theory in the Plane
X, 272 pages. 1991

Vol. 154: A. Kurzhanski, I. Lasiecka (Eds.)
Modelling and Inverse Problems of Control for Distributed Parameter Systems
Proceedings of IFIP (W.G. 7.2)-IIASA Conference
Laxenburg, Austria, July 24-28, 1989
VII, 179 pages. 1991

Vol. 155: M. Bouvet, G. Bienvenu (Eds.)
High-Resolution Methods in Underwater Acoustics
V, 249 pages. 1991

Lecture Notes in Control and Information Sciences

Edited by M. Thoma and A. Wyner